TRANSMITTER HUNTING
RADIO DIRECTION
FINDING SIMPLIFIED

JOSEPH D. MOELL, K0OV, AND THOMAS N. CURLEE, WB6UZZ

TAB Books
Division of McGraw-Hill, Inc.
New York San Francisco Washington, D.C. Auckland Bogotá
Caracas Lisbon London Madrid Mexico City Milan
Montreal New Delhi San Juan Singapore
Sydney Tokyo Toronto

pbk 8 9 10 11 12 13 14 15 16 FGR/FGR 9 0 0 9 8 7 6 5

Library of Congress Cataloging-in-Publication Data

Moell, Joseph D.
    Transmitter hunting.

    Bibliography: p. 315
    Includes index.
    1. Radio direction finders. I. Curlee, Thomas N.
II. Title.
TK6565.D5M64 1987   796.2   86-30140
ISBN 0-8306-2701-4 (pbk.)

Contents

Preface   vi

Introduction   viii

1 RDF Is Born   1

The Very Beginning—World War II—Hams Aid the War Effort—Military RDF Today—Sport Hunting Takes Off—FM Takes Over

2 Getting Started   7

Directive Gain Antennas—Homing DFs—Dopplers—Making the Choice for VHF—Marine DF Units— Getting Good Maps—The Compleat Hunter

3 VHF Mobile Hunting Techniques   13

The Importance of High Ground—The Initial Bearing—Magnetic Declination—Getting Bearings Without a Compass—Ready to Roll—Closing In—Navigating on the Run—Guessing the Distance—The End Game—Hunting Like a Pro—Hunting as a Team—Coordinating a Cooperative Hunt

4 VHF Hunting with Directional Antennas   26

Simple VHF Antennas—Two Meter Loops—Other Instant Hunting Ideas—Aluminum Yagis—Quad Antennas—Measuring Beam and Quad Performance—Phased Arrays—The ZL Special

5 All About S-Meters   44

VHF-FM S-Meter Circuits—External S-Meters—The Amplified External Meter—Linearity Evaluation— LED Meters—An Audible Signal Strength Indicator—S-Meters for HF AM/SSB Receivers—The Bridge Circuit—FM Gain Block ICs

6 Knocking Down the Signal   **55**

External Attenuators—The Magic Antenna Switcher—Internal Attenuators—The Automatic Attenuation Control

7 Equipping Your Vehicle   **71**

Through-the-Window Mounts—Mirror Mounts—Door Mounts—Sun Roofs and Convertibles—The Hole in the Roof—Window Brackets—Window Coverings—Window Inserts—Direction Indicators

8 Homing DF Units   **99**

The Double-Ducky Direction Finder—Switched Cardioid Pattern Homing Units— The Little L-Per Direction Finder—The Happy Flyers DF—Determining Bearing Inaccuracies—Side-Step and Baseline Averaging Techniques—The K6BMG SuperDF—Testing Homing DFs

9 Doppler DF Units   **120**

How Doppler DFs Work—Doppler DFs for Radio Amateurs—The Roanoke Doppler for VHF—Modifications and Improvements—Hunting with Doppler DFs—Commercial Doppler DFs—Other Doppler DF Applications

10 Search and Rescue Hunting   **142**

ELTs and EPIRBs—The Civil Air Patrol—The US Coast Guard Auxiliary—The Happy Flyers—Amateur Detection of ELT Alarms—Airborne Hunting—Advanced Interferometer Techniques

11 Weak Signal Hunting   **151**

Why Bother with Weak Signals?—Grabbing the Signal—A Build-It-Yourself Preamp—Using FM Quieting—A Noise Meter for FM Receivers—Sideband Detectors for AM and FM Hunts—An Add-On Sideband Detector—Signature Analysis

12 Sniffing Out the Bunny   **165**

Homing DF Units—The Body Fade—Sealing Up the Receiver—Primitive Sniffers—The Sniff-Amp—Antennas for Sniffing—Systematic Sniffing—Deluxing Your Sniffer—Listening to the Signal—Other Sniffer Uses

13 Planning For Hunts in Your Community   **182**

When and Where—Writing the Rules—Novelty Hunts—Finding Your Batting Average—Getting Greater Attendance

14 You're the Fox   **190**

Finding the Perfect Spot—Preparing the T—The Tape Recorder—A Simple Transmitter Cycler—Tone Boxes—Antennas for Hiding—Other Antenna Possibilities—Powering the Transmitter—Remote Control—Some Final Thoughts on Hiding

15 The Bunny Box: A Cigar Box Sized Rig for Hiding   **219**

Physical Layout—Audio and Timing—The RF Synthesizer—Triplers and Final Amplifier—Power and Keying—Crystal Control—Other Bands and Variations

16 Hunting Without a Vehicle   **234**

Real Radiosport—Championship Rules—Asia, Too—Amateur Radio's Sporting Goods—Radiosport for Americans

17 Hunting Below 50 MHz   **239**

Locating DX Signals—Loops for 15 to 50 MHz—Preamps—Unidirectional Loop Systems—Loops for 2 to 15 MHz—Setting Up for Loop Hunting—Adcocks for Base Stations—An Inexpensive Oscilloscope Display

18 Direction Finding from Fixed Sites   **256**

VHF Vagaries—HF Headaches—Setting Up a Fixed Site DF Station—Calibrating the Station

19 Commercial and Military Direction Finding Systems   **260**

Rotating Antennas—Lens Antennas—State of the Art Adcock Systems—Wide Aperture Wullen-weber Systems—New Tactical Mobile DFs—Triple Channel Interferometers—Time-Difference-of-Arrival DFs—Countermeasures Against RDF

20 A Mobile Computerized Triangulation System   **269**

The Coordinate System—Hunting with a Computer—Explanation of the Programs—The Mobile Set-Up—Correcting Errors—Improving the System—Other Uses

21 Dealing with Mischief and Malice   **281**

Organizing the Hunting Team—Is It Really Jamming?—Bait-ers, Bait-ees, and Reverse Jamming—Shut Off the Repeater?—Psychology May Help—Resisting the Urge—Keeping Him On the Air—Prosecuting the Offender—We're Fourth on the List—Volunteers to the Rescue—Submitting Your Own Reports—Generating the Hue and Cry—Technical Tricks—Are You Able to Turn It Off?

22 Other Uses for Your RDF Skills   **294**

Stalking the Wild Fox—Navigating by Wire—Hunting Cable Television Leakage—Hunting Power Line Noise

23 Looking Ahead

T-Hunting from Orbit—The Psychology of Sport Hunting

Appendix A Manufacturers and Organizations   **313**

Appendix B References   **315**

Index   **319**

Preface

Ever since radio direction finding became an important technology in World War I, its practitioners have mostly been engineers and scientists who spent much time learning the esoteric concepts of antenna theory, wave propagation, and signal processing. They have written prolifically of their discoveries, but little of it has been understandable to the average electronics hobbyist without a math or engineering background.

Amateur transmitter hunting is now going on in various forms all over the country—indeed, all over the world. Surprisingly there is little standardization of equipment or technique. In one locale, everyone uses Dopplers; in another, quads predominate. Switched antenna units are standard in a few spots. Hams, who are known for being able to communicate, don't seem to be talking outside their own towns about T-hunting. Try to get a discussion going in a QSO or on a computer bulletin board about it—it's not easy! More interchange is needed to help hunters improve their odds in the malicious interference and search and rescue battles.

Articles on transmitter hunting now appear more and more often in the Amateur Radio press. Very good booklets have been prepared for specialized aspects of DFing by the Happy Flyers, the National Association of Search and Rescue, the American Radio Relay League, and others. But somehow no one has put all together in one place everything the beginning hobbyist needs to "bootstrap" himself into successful competitive or public service hunting. That is the first objective of this book.

The second objective is a book that's easy to understand and use. You don't need to know anything about the esoteric concepts of antenna theory, wave propagation, and signal processing to start out. But when you've finished the last chapter, you'll be surprised at what you've learned about these and other topics.

Although there are many circuits and techniques here that are new and never before published, much of this volume is possible because of the individual efforts of many experimenters, both amateur and professional, over the years. We don't claim to cover every single DF technique or concept applicable to hobbyists, although we have scoured the libraries and beaten the bushes looking. Surely the next few years will bring a plethora of new ones. We want to be a part of that and hope you do too.

A couple of points about the style of this book. When you see "we," it's not an editorial idiom. There really are two authors, who have hunted with and against each other enough to know they'll never agree 100% about the best way to do it. And although we frequently refer to hunters in this book as "he" or "him," please don't consider us sexist. We fully realize that many YLs are active and proficient hunters, and some of the best hunting teams are OM/YL combinations.

In the same vein, we want to publicly acknowledge our families, April Moell (WA6OPS), Karleen Curlee, and son David Curlee. It is in large measure to their credit that this book has come about. They have not only actively participated in T-hunting; they have patiently endured the writing of this book.

According to a saying in the electronics industry, "There comes a time in the life of every program when you have to shoot the engineer and begin production." Engineers are always trying to make one last improvement in their designs, to the dismay of manufacturing managers. No doubt publishers feel that way about authors, too. And when the authors are engineers, it's a wonder that books like this ever get finished.

There are many new discoveries of interest to hunters being made right now, particularly in the areas of computer-aided triangulation and mobile navigation. There's a great temptation to pursue these advances rather than sit at the word processor. Well, the advances will come faster with more hobbyists working on them. Perhaps you, by being part of the T-hunting scene, will make new contributions for the second edition of this book.

T-hunters may be competitive, but they certainly aren't glory seekers. Those that have brought DF to its present state, whether for business, pleasure, or public service, have done so with little fanfare. They all deserve recognition, but we can only thank those who helped us directly with this particular book. They include, in alphabetical order: Stas Andrzejewski W6UCM, Russ Andrews K6BMG, J. Scott Bovitz N6MI, Frank Crowe WB6UNH, Dave Cunningham W7BEP, Ken Diekman WA6JQN, Jorge DiMartino KI6MD, W.B. Skip Freely K6HMS, Gary Frey W6XJ, John Gallegos W6EQ, Richard Gehle WD6Y, Rick Goodman W5ALR, Jim Grove N6AXN, Larry Guy K6EZM, Albert Hamilton AG1F, Clarke Harris WB6ADC, Lloyd Harwood WB6ULU, Bob Hastings K6PHE, Dale Heatherington WA4DSY, Paul Hower WA6GDC, Carl O. Jelinek, John Klein WA6TQT, Ted Kramer NB6N, Paul Lambert N8ABS, Chuck Lobb KN6H, Walt Le Blanc WB6RQT, Sanford Mills K6PPO, John Moore NJ7E, Dick Reimer W6ET, Austin Rudnicki K6IA, Vince Stagnaro WA6DLQ, Chuck Tavaris N4FQ, Cleyon Yowell AD6P, Paul Wirt W6AOP, and the transmitter hunters of southern California.

We are also grateful to Diehl and Monica Martin (N5AQ and WD5JCW) for the use of their excellent darkroom facilities for the photographs in this book.

Many thanks to each of you. Surely there are some who have been overlooked; we hope you accept our thanks as well.

Introduction

Amateur radio is not just a hobby, it is a collection of many hobbies. Hams communicate, experiment, build, test, and provide public service. They work DX, vhf, SSTV, ATV, AM, SSB, FM, CW, and RTTY. They handle traffic. They also hunt transmitters. This aspect of Amateur Radio is not new, but has largely been ignored by ham radio publications. Only recently has interest in amateur DFing grown, spurred perhaps by the problems of jamming and malicious interference in certain areas.

Jammer hunting is an important use of RDF techniques, but there's more to T-hunting than that. The skills of hams and non-hams are needed by search and rescue groups to help save lives. The Civil Air Patrol, US Coast Guard Auxiliary, and similar groups welcome volunteer support in their valuable work. And sometimes DF capabilities are needed by Citizens Band users.

Transmitter hunting goes by a number of names around the world. In our locale, it's almost always "T-hunting." In Europe, the term "fox-hunting" seems to predominate. The origin of "bunny hunting" and its derivatives is unknown. Searching for illegal operators is commonly called "jammer hunting" but this is not entirely correct, as the search is as often for unidentified or stolen rigs as it is for malicious interferers. The terms "turkey hunt" or "maverick hunt" used by ham radio magazines seem more appropriate.

If you like competition, the DXing and contesting aspects of amateur radio have probably caught your interest. At the risk of offending the DXers and contesters in our hobby, may we humbly suggest that T-hunting now offers an even greater challenge. To the ham with a good location, a first class rig, and a vast antenna farm, getting good DX signal reports is almost a sure thing. Once on the DXCC Honor Roll, there's not much new DX to compete for. Each T-hunt, on the other hand, is a fresh start.

No single hunting setup is superior to all others in all situations. The skill of the participants, not their gear, is the major factor in determining the winner. If a hunter gains an edge by having superior gear, chances are he has built it himself, adding an extra measure to his satisfaction. There are even scoring systems to determine who the best hunters in town are.

Just because this book is large, with lots of gadgets and "secret weapons," don't get the idea that T-hunting is complicated or only for engineers. It can be as simple or complex as you and your friends want. Simple equipment gives very good results. Many hams have put off getting started, thinking that a lot of work lies ahead. Actually, you can be ready for this evening's hunt with a couple of hours of work this afternoon. Most hunters get started this way, using mostly gear they already have, adding to it as interest grows and needs become clear.

If you have wanted to build a ham project, but a piece of transmitting or receiving equipment is too formidable a first project, there are many small items used in DFing that are easy to build and will give you experience and satisfaction. Expensive test equipment is not required. It's fun and educational. If you'd like to interest a youngster in electronics, T-hunting is a great way to do it.

The emphasis in this book is on circuits that are easy to build, and parts gathering information is included whenever possible. Most parts can be found locally. Radio Shack (RS) part numbers are given for many critical parts. A list of manufacturers and a bibliography is in the back of the book. A few concepts are presented without detailed instructions, to inspire more experienced experimenters to improve the state of the ham art in DF; we are very interested in any such advances you may make.

In the first chapters, we'll review how RDF has played an important role in peace and war, and provided fun for experimenting hams for many years. You'll then be guided through the basics of setting up to hunt. Should you build your own setup or buy a commercial RDF unit?

After you've gotten started, and the bug has bitten hard, you'll want to become a serious competitor. There are lots of new tricks and secrets in the following chapters to help you deal with concealed transmitters, very weak and very strong signals, and other stunts of perverse hiders. When it's your turn to hide, the chapters on hiding will help you to put out a real challenge. There are ideas for getting more hams in your area into hunting and how to work up rules and scoring schemes. There's also some help in dealing with jammers that you may round up with your new-found skills.

Finally, you'll learn how RDF will benefit from new technology, and how you can use RDF to solve some common RFI problems.

There's a lot more to T-hunting than just driving around with a funny antenna. It's a public service, a technical challenge, a club builder, and a whole lot of fun. So let's get started!

Chapter 1

RDF Is Born

It is May 30, 1916, the eve of the great naval battle of World War I, Jutland. A chain of radio direction finding (RDF) stations along the east coast of England detects unusual movements of units of the German High Seas Fleet. Admiral Sir Henry Jackson, First Sea Lord, acting on this information, commits the British fleet to action. RDF has come of age.

Ask five people the uses of RDF and you'll probably get as many different answers. The average person most likely thinks about locating spies with clandestine transmitters. Hams and CBers may think of the FCC van loaded with radio gear, looking for illegal or bootleg operators. The pilot or boater knows that he can be found in time of trouble with RDF.

In this book we try to touch on all facets of RDF, but the main emphasis is on the practical end of things—how to do it, what equipment is available, how to build gear. The thrust is on ham radio RDF, commonly known as T-hunting, fox hunting, or bunny hunting, both for sport and serious use.

THE VERY BEGINNING

Directional antennas date back to the earliest days

of radio. Hertz and Marconi used them before 1900 to concentrate their transmitted energy at about 200 MHz. Some of the earliest recorded work on the use of antennas for direction finding was done by J. Stone in 1904 and improved by Bellini and Tosi. The classic Bellini-Tosi direction finder, shown in Fig. 1-1, is the predecessor of the Adcock of today. Marconi acquired the patents in 1912 and began installing RDF equipment on commercial ships. These vessels could then take bearings on known shore-based wireless stations and plot their location. F. Adcock patented his system in 1919.

During World War I, the first direction finding loop was developed by Dr. F. A. Kolster of the National Bureau of Standards. This loop, with minor modifications, was the standard for years and is still being used in some applications. Figure 1-2 shows another early use for loops.

RDF technology was improved by Captain H. J. Rounds of the Royal Navy. He installed a series of RDF stations along the east coast of Great Britain for Room 40, the code-breaking intelligence service of the British Admiralty. These stations were active in tracking the German fleet during World War I.

For instance, just before the battle of Jutland, Captain Rounds ordered a traffic watch on the 28,000 ton German battleship Bayern, one of the principal wireless units

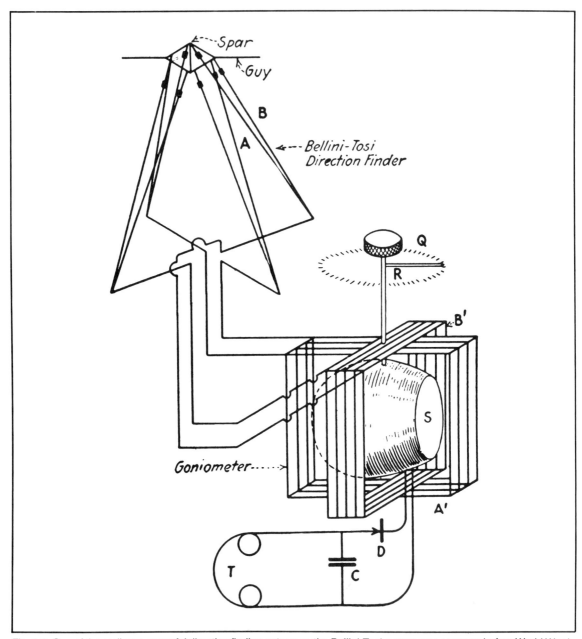

Fig. 1-1. One of the earliest successful direction finding antennas, the Bellini-Tosi system was common before World War I.

of the German High Seas Fleet. RDF established that the Bayern and other ships in her communication net had moved dramatically northward overnight. Admiral Sir Henry Jackson, First Sea Lord, committed the British fleet to battle at Jutland based on this information. Success at Jutland meant that the British would not be challenged at sea again during that war.

The vessels that did not have DF equipment on board (because of inaccuracy, complexity, expense, patent rights, and military security problems) relied on shore-based DF stations. Established in the USA by the Navy in 1921, these stations were designed to give bearings

Fig. 1-2. Another use for the loop antenna—this 1924 mobile radiotelephone (photo courtesy Paul Lambert, N8ABS).

to any ship that called in.

However, by the mid 1920s, an increasing number of ships had their own RDF installations.

WORLD WAR II

During World War II, RDF was used by the intelligence services on both sides—and those in the middle. Counter-intelligence agencies used these devices to locate clandestine transmitters. In Germany, what the Abwehr (the secret service) lacked in sophistication, they made up for in brute force. They would DF a signal down to a certain area of town. Then the transmitter would be located by turning off the power, block by block, until it went off the air.

Switzerland, the only non-Axis country in central Europe, was a hotbed of intelligence activity for the Allied powers. The neutral Swiss counter-intelligence force found itself DFing transmitters from a number of countries, Axis and Allied alike.

The Japanese fear of DF caused strict radio silence to be observed on the way to Pearl Harbor. CW operators from some of the fleet's larger ships were left in Japan to operate shore stations simulating ship stations. These phantom ships were duly DFed in their own home waters by the U.S. Navy.

British Naval Intelligence DF efforts, along with cryptography and traffic analysis, helped defeat the German U-boat campaign in the Atlantic. The German wolfpack tactics spread U-boats across hundreds of miles of ocean, looking for convoys. When a convoy was spotted, the U-boat radioed the convoy's position and course to other boats in the pack. Naval Intelligence was able to DF these transmissions, giving a fifty percent probability that the U-boat was within a 100 mile diameter circle. Later, DF equipment was installed on convoy ships, giving even better DF fixes. General intelligence information often gave the times the U-boat would surface. A radar-equipped airplane would often be waiting.

HAMS AID THE WAR EFFORT

Though they could not operate their stations during the war, RDF techniques were used by hams at the Radio Intelligence Division (RID) of the FCC. Banks of receivers and special antennas there were used to ferret out clandestine radio operations and assist pilots who were lost, disabled, or forced down. About three quarters of the employees there at the time were amateur radio licensees.

Many antenna systems were used, including rhombics and folded dipoles. The heart of most installations was one or more Adcock direction finders. The "balanced H" type, shown in Fig. 1-3, was the most successful. While capable of high accuracy, these DFs were extremely sensitive to their surroundings and critical in adjustment. Some of these problems were pointed out by W9ETI in a 1944 QST article:

☐ A spider web across the transmission line could cause an error of one or two degrees.

☐ Calibration could be upset by a change of one sixteenth of an inch in the line spacing at the junction.

☐ Filters were needed on the power lines to the unit. It was necessary to bury the ac supply line 15 feet down and enclose it in lead.

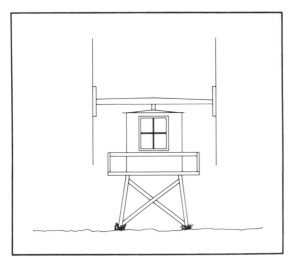

Fig. 1-3. Balanced H Adcock installations similar to this drawing were widely used by the Radio Intelligence Division (RID) of the FCC during World War II.

☐ To completely eliminate ac line effects, the engineers had to resort to battery operation—not an easy task in the days of tubes and dynamotors.

A dedicated network of teletype machines tied all twelve RID sites together for instant triangulation of bearings. Mobile units and portable field strength meters called "snifters" were used to close in.

The system was very successful. It resulted in about 400 unlicensed stations being put off the air during the war, and quickly stopped all attempts at subversive radio activities within our borders. In addition, the system was called upon several times a day for fixes on lost aircraft, saving many lives.

It was during World War II that a new Adcock DF system came out of the lab. It did not require a rotating antenna and thus could see very short and intermittent transmissions such as radar. It was immediately pressed into service on U.S. Navy patrol bombers, which were searching for German submarines. The radar-equipped planes were already successfully keeping the subs down, but the Navy hoped to use the new DF to pinpoint the exact location of the subs by tracking their air defense radar units.

Intelligence reports had indicated that the radar in the U-boats was capable of very high performance, and was to be used to alert the subs of the presence of the planes in sufficient time to allow the sub to submerge. But no signals from this new super-radar were ever heard, though the DF set was functioning perfectly.

It was not until after the war that interrogations of the U-boat crews determined that the Germans were afraid to use their new super-radar. They were aware of the potential for Allied countermeasures and chose not to risk detection by turning it on. RDF thus played a significant role in the Allied victory at sea.

MILITARY RDF TODAY

In the decades since World War II, with the onset of the Cold War, intelligence services on both sides have installed large strategic DF systems. Some employ German technology, such as the Wullenweber system. A number are operated around the world by the National Security Agency (NSA).

A typical Wullenweber system may have 96 broadband hf verticals set in a ring one half to one kilometer in diameter. Some sources state bearing accuracy on hf signals is on the order of one half degree, but 3 to 5 degrees is more likely on most signals. Obviously, most detailed information about this type of equipment is classified.

The system shown in Fig. 1-4 is an impressive example of strategic DF gear. The AN/FLR-9 Countermeasures Receiving System was built by Fischbach and Moore and installed in southeast Asia in 1970. One of a number of such systems installed throughout the world since the early 1960's, it had coverage from 0.5 to 32 MHz.

There were 96 verticals in the outside ring, each in excess of 100 feet tall. The ring was hundreds of feet across, circling an inner shielding screen 120 feet high. Those outside antennas were used for work up to 6 MHz. There were 48 more verticals for 6 to 12 MHz. For 12 to 32 MHz, an "X" array was placed in the center of the ring shield. A grounding screen extended hundreds of feet out from the antennas in all directions.

Antenna control was done by a pair of Lockheed MAX-16 16-bit minicomputers, allowing the system to be used by up to 800 operators simultaneously. Each operator had at his command a highly directional receiving antenna system with direction finding capability. This particular installation was dismantled in the mid 1970s.

Other state of the art DF equipment uses various phase measurement, Doppler, and interferometric techniques. One interesting system uses the time difference of arrival (TDOA), where the exact time of arrival of the signal is measured at three or more widely separated sites and the location determined by computer processing.

Modern tactical DF systems (short range, usually at vhf and uhf) often use Doppler or Adcock/Watson-Watt techniques. Most of these systems are mounted in jeeps, trucks, tracked vehicles, and aircraft, although some units can be carried and set up by personnel in the field.

The common denominator of all the latest state of the art communication intelligence (COMINT) equipment is computer control. Many systems have digitally tuned receivers capable of scanning well over 500 channels per

Fig. 1-4. The AN/FLR-9 Countermeasures Receiving System, an excellent example of a Wullenweber DF array. The array, commonly called an "Elephant Cage," is 900 feet in diameter. The 120 foot high girder towers hold a shield screen. The large white capped cylinders are 105 foot high monopole antennas. The smaller cylinders are antennas for the higher frequency band.

second. The number of channels is limited only by the computer software and the number of times per second each channel is to be scanned. These systems can be remotely operated, with the digitized COMINT data from widely located sites sent to a central processing location.

The received data is processed in near real time and the location of the emitter determined. These systems can use computer graphics to display a map of the region of interest and pinpoint the transmitter or transmitters. Under ideal conditions some DF systems can locate a transmitter within a square mile area from 25 miles away. Imagine what a chain of these units would do for jamming problems!

SPORT HUNTING TAKES OFF

DFing as a sport began to gain popularity in the early fifties, as hams discovered the fun of having a ham station in the family car. "Mobiling" became a new aspect of the hobby, and magazine articles abounded for small AM rigs for various bands.

While a few adventurous hams tried 6 or 2 meters, and others retuned their car radios for 160 meters, the majority of the action was on 75 and 10. Rather than build an entire receiver, most mobileers built a simple converter to feed the car radio. Those who wanted to hear CW, or those new-fangled single sideband signals, added a beat frequency oscillator (BFO) in the car radio i-f stage.

CQ Magazine's "Mobile Corner" told of dynamotors, loading coils, rallies, and hamfests. It also told of the transmitter hunts which were becoming popular in several urban areas. Most hunters used a loop of some sort, but many other ingenious devices were tried. Advanced hunters used rf amplifiers at the loop to overcome the low antenna sensitivity.

Directional loops have broad peaks and sharp nulls. Hunters found that much more bearing accuracy was obtained by hunting with the null instead of the peak. As the hidden T was approached, however, the null became shallow due to overloading.

This was cleverly solved by adding a 200-ohm rheostat in series with the filament of the preamplifier tube as a sensitivity control. It not only cut the gain but helped conserve the "A" battery. Hunters know they were within a half mile or so of the bunny, because they could turn off the filament entirely and continue to hunt.

Because the loop is a bidirectional antenna, it has a 180 degree ambiguity in both peaks and nulls. The resourceful hunters of the early fifties used their standard 8-foot mobile whips to decide which was the correct null. Most mobileers had these whips mounted on the rear of the car, and knew that the car body interaction caused signals received on the whip from the front to read two or three S units higher than signals from the rear. Part of their starting bearing ritual was to drive slowly in a circle and observe the broad peak on the whip, and then correlate it with one of the sharp nulls on the loop. As you can imagine, the hunts always started in large empty parking lots!

Transmitter hunting in the 1950s wasn't limited to ten meters. Southern California hams liked to use 75 meters as well. During the daytime, the band is fairly inactive. There also are fewer signal reflections. On the other hand, QRM and re-radiation from power lines may be more of a problem on 75.

One southern California club held its AM hunts simultaneously on 75 and 160 meters. Anyone, even a non-ham, could readily hunt the 160 meter signal by using a retuned AM broadcast receiver. It was a great way to get started in DFing (and still is, as we describe in a later chapter).

FM TAKES OVER

Today, DFing is a more popular sport than ever with hams, but it's done almost entirely on the 2 meter band. This can be far more challenging than hf hunting because vhf reflections from hills and buildings are much more pronounced. Vhf hunting isn't brand new—it was providing fun for southern California hams when inexpensive commercial rigs such as the famous Gonset "Gooney Bird" were becoming popular. As converted business band radios began to take over the top end of 2 meters in the late 1960s, hiders began to use FM instead of AM signals.

From the outset, the quad or beam was the ham's antenna of choice for vhf DFing. At first, its high gain was needed to overcome the poor noise figure of tube-type receiver front ends. As receivers improved, hiders used less power, and gain antennas remained the mainstay for serious sport hunters.

Hams certainly are not using the latest hush-hush military technology in their DFing, but they now have some good commercial DF gear available at reasonable prices. While the technology for Doppler DF units, for example, has been around for some time, it took integrated circuit technology to make it inexpensive and easy for the average ham to build.

In many ways, transmitter hunting hasn't changed a lot over the years. It's still educational, potentially lifesaving, and a lot of fun.

Chapter 2

Getting Started

Every weekend all over the country small groups of radio enthusiasts gather for friendly competition. None of them knows for sure where the contest will take him, how long it will take, or sometimes even what physical device he is looking for. It's a test of both man and machine, and one is never sure what to expect.

There are few prerequisites for being a good hunter. All it takes is interest and imagination, plus a little skill. There are young and old hunters, Novice and Amateur Extra class hunters. Disability needn't be a bar to T-hunting. One of the top-notch Santa Barbara (CA) hunters, Dennis Schwendtner, WB6OBB, is sightless. He uses the audible S-meter, the receiver noise, and his knowledge of the territory, and is an excellent navigator.

Good-natured treachery and perfidy are all part of the hunting/hiding game. "Never trust another sport hunter," is a good rule to follow. Don Root, WB6UCK, tells how he and a friend had some fun at the expense of another hider. Don volunteered to hide the rig for the unwitting fellow (whose name is withheld to spare him further embarrassment). That way he could confuse the other hunters by pretending to participate in his own hunt.

"No problem," said Don, "We'll just stick your rig beneath an underpass on the 605 freeway. You can drive right to it." But instead, Don and his buddy hung it from the Harbor Freeway, 20 miles away from the spot on the 605, in downtown Los Angeles. Imagine the fun they had watching from the top of a nearby parking garage as the hapless victim was forced to scour the county for his own transmitter!

The uncertainty about the choice of gear can be the biggest stumbling block to a new hunter. In many areas all hunters seem to use the same thing, but in others there's a big variety. There's no perfect setup for everyone. This chapter describes some of the choices for the vhf bands, to familiarize you with what's available. In later chapters each type is covered in detail, along with equipment for the lower frequencies and for other DF uses.

Don't run out and buy the first commercial DF you see in a magazine ad. Study the various types and consider your particular needs before spending any money. Give serious thought to building some or all of your own gear. You're more likely to get just what you want. Building your own can be a lot of fun, and a real source of pride.

Talk to other DFers to find out what works best for their needs, and how they like or dislike the gear they use. See if you can ride along or help out on a hunt or two. But don't just go out and duplicate the local champion hunter's setup. His choice of gear may not be best for you.

Don't be surprised if you end up wanting an assortment of gear. It's not unusual for us to set out with a quad on the mast, a Doppler on the roof, and a switched antenna DF and an amplitude sniffer in the trunk. In addition, we bring along numerous gadgets and "secret weapons" that are described later. Part of this is for research and comparison, but mostly it's just a matter of being prepared, like a good Scout.

There is no single DF setup or method that's ideal for every hunt situation. Much of the remainder of this book is spent explaining the pros and cons of various equipment and techniques. An even greater percentage of your knowledge will be gained by "trial and error." No, let's be positive, and say "trial and success."

DIRECTIVE GAIN ANTENNAS

A directional antenna favors signal from one (and sometimes more) directions over others. Quads and beams predominate. They are mounted or held so that they rotate in a full circle, and the S-meter or other indicator is used to determine the direction or directions that correspond to some feature of the antenna, such as a forward lobe or null.

Gain antennas are popular because they are simple, easy to understand, and readily available. If you have a 3 or 4 element 2 meter beam, and a receiver with an S-meter, you're almost ready to hunt. Gain antennas can be used on any mode that the receiver can receive, as long as the transmissions are long enough for the observer to turn the antenna and get a bearing.

To increase gain (the ability to hear very weak signals), antennas must get larger. Just like the perpetual motion machine, someone is always trying to invent a tiny antenna with more efficiency than a full sized one for a given band, but success has been elusive. Fortunately, a relatively small antenna will hunt foxes very well at vhf.

Gain antennas do have disadvantages. They are relatively slow to turn, do not give instantaneous bearing readout, and are not by themselves capable of fractional-degree reading accuracy. Null reading antennas may have more reading accuracy, but have lower sensitivity. Also, at 2 meters, a 4 element quad is a bit hard to carry around on foot when closing in.

HOMING DFs

Homing DFs are characterized by a pair of antennas and some specialized electronics that switch them to give a very distinct indication when aimed right at the signal source. It can be either a meter that reads exactly zero or a pair of lights that indicate right/left. The indicators tell which way to turn to be pointing right at the source.

Homing DFs can be very small and light. For airborne use, they are the DF of choice. There are models for both AM and FM receivers and models with receivers built in.

The primary advantages are relatively small size (compared to gain antennas), and the ability to give a very sharp and accurate (within a degree or so) bearing when the signal is in the clear and free of reflections. They are a good home construction project, or they can be purchased from several suppliers.

DOPPLERS

Unlike the two types just discussed, a Doppler DF has a direct readout of bearing to the apparent signal source, relative to the observer. The readout can be a ring of lights, a digital display in degrees, or even a synthesized voice or computer-compatible data stream. No manual antenna turning is required. Doppler DFs are less sensitive than gain antenna methods, and give only one direction at any instant, making it difficult in some cases to get good bearings when reflections are very strong.

Once the unit is mounted in a vehicle, it can be left in place permanently, to be ready to go when needed. Their ease of use makes them a good choice for locating reasonably strong signals in urban areas, particularly when hunting alone. The convenience of a Doppler DF has a relatively high cost when bought commercially, at least compared to the other two types. But Doppler DFs for the ham market are some of the cheapest commercial direct-bearing-reading RDF equipment available, when compared to such gear for the government/commercial market. It's also possible to build your own at a considerable saving.

MAKING THE CHOICE FOR VHF

The decision of gain antenna versus Doppler versus homing system is not always easy. You may wish we had put together a table with "scores" for each type, with the clear winner emerging with the most points. But that wouldn't take into account your particular circumstances, such as the type of terrain, what the competition is using, your vehicle type, and so forth. Instead, we'll cover the things to consider, both in terms of the signal environment and user convenience. (These considerations aren't necessarily in order of importance.)

You can make up your own table. Assign point scores to each system for each of the following criteria, based on your expected needs. A clear choice may then emerge.

□ Performance in multipath. When reflections environment that can cause wrong bearings for a moment, consider the ability of the DF to read out to the nearest degree or so. This is most important for triangulation or very long distance work, but may be relatively unimportant if you're just going to follow the signal in a car, boat, or airplane. A combination of high gain and readout accuracy is possible with gain antenna interferometer techniques (described in Chapter 10).

□ Performance in Multipath. When reflections abound, a gain antenna is better at showing you each signal source, direct or reflected, and its relative intensity. All you have to do is figure out which is direct and which are reflected. The other types are unaffected when reflections are very weak, but when they are of moderate strength, false indications occur. In general, the stronger the reflections, the more error results.

□ Weak signal performance. A gain antenna (peak, not null) is best for very weak signals, say from ELTs or the All Day Hunt. Some Doppler DFs have considerable loss due to switching noise. Weak signal performance is very important, unless all you want to do is hunt very strong jammers.

□ Strong signal performance. A good attenuator is required for use with a gain antenna, and there is risk that the signal can come through the coax. This is also true with an AM detection-based homing DF. FM-based homing DFs and Dopplers are less affected by very strong signals.

□ Polarization sensitivity. Dopplers and most switched antenna units have real problems when the signal is predominately horizontally polarized, unless there are no reflections around. Properly configured gain antennas can hunt either or both polarizations.

□ Compactness. Will you want to carry your set around? Will you be moving it from vehicle to vehicle?

□ Use on foot. Homing DFs can be configured to transition rapidly from vehicle to foot use; you may want separate systems for foot and vehicle application.

□ Modes. Get a DF set that will cover the mode or modes you'll be using—AM, SSB, FM, or something else. Sets with dedicated receivers can sometimes hunt modes other than those they can demodulate, but the signal may be unintelligible during the DFing.

□ Use in motion. It's harder to take bearings on the fly with a gain antenna and S-meter due to fluctuations. Homing and Doppler DFs actually work best in motion.

□ Use when stopped. See above.

□ Short transmissions. The Doppler is best for catching short transmissions. The switched pattern DF can zero in on a series of short transmissions faster than can a quad, which takes several seconds to get a good accurate fix.

□ Changing amplitudes. If the hider is changing power rapidly, he can make it really tough for the quad/beam hunters to get a good bearing using their S-meters. Null antennas fare better, and homing and Doppler units are virtually unaffected, as long as they can detect the signal.

□ Price and availability. The best high-tech DF is of no use if it is unavailable or too expensive. There are plenty of very inexpensive methods that work very well. Of the units discussed in this chapter, the Doppler is usually by far the most expensive.

Made a choice yet? Here's another idea to mull over.

MARINE DF UNITS

A marine DF unit is a quick and potentially inexpensive way to get into RDF. The prices of these units when new are the same or higher than Doppler DF units for the amateur radio market, but you may find a used bargain if you nose around in an area where boaters abound. They are designed for rugged use, but are easy to hook up and may have some extra features not on ham units, such as bearing storage.

Unless you want it for boating, look for a unit intended to be used with an external vhf/FM radio. With appropriate plugs it can be hooked to a base, mobile, or even handheld 2 meter transceiver. Most units draw less than half an ampere at 12 volts, so a do-all system for all-terrain-vehicle hunting is not out of the question. Normally there's no problem operating a marine DF in the amateur 2 meter band or the high band business frequencies, and no retuning is required.

The standard marine antenna unit supplied with these DFs is a set of four vertical dipoles (Fig. 2-1). It can be used as is on a mast out the car window. Care must be taken to make sure that the antenna unit is always "pointed" correctly. Some have filters to eliminate the Doppler tone, others don't.

Companies making these units include Apelco (Raytheon), Intech, Regency, Sitex, and Taiyo. They're listed in marine catalogs. As an example, the Regency Electronics NC6000 (Fig. 2-2) is an add-on for any FM transceiver in its frequency range.

Only two connections are required to the receiver, at the antenna and external speaker jacks. The unit contains automatic switching for a separate transmitting an-

Fig. 2-1. A typical marine DF antenna unit mounted on the roof of a 16 story building for use by the U.S. Coast Guard Auxiliary. Santa Catalina Island can be seen in the distance.

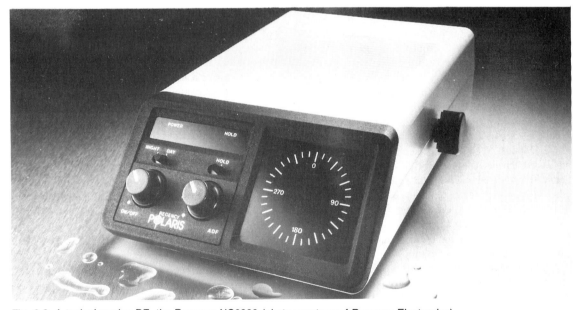

Fig. 2-2. A typical marine DF, the Regency NC6000 (photo courtesy of Regency Electronics).

tenna. This separate antenna is also used for better receiver performance when the DF unit is turned off.

GETTING GOOD MAPS

Maps are a necessary part of the hunt. Finding a good map, or maps, to cover the area of the hunt with sufficient detail can be difficult, but the effort is worth it. Some hunters buy different maps that cover only a section of the T-hunting area and select the proper map once the initial bearing is taken at the start of the hunt. (An example of this is shown in the chapter on computerized triangulation, where four maps cover an approximately square boundary area with the starting point in a corner of each.) Other hunters use a general map covering the entire area for the initial bearing, and then switch to detailed maps as they close in on the transmitter.

The maps you decide to use depend on the actual placement of the boundaries of your local T-hunt. For example, the area inside the bounds of one southern California hunt is mostly in Orange county, but the northern 20% or so (heavily used by hiders) is in Los Angeles county. The eastern boundary goes through both Riverside and San Bernardino counties. A single detailed street map covering this entire area is impossible to find.

By checking different sources, you may be able to find maps that cover a specific area better than others. You may find that the auto club maps are generally outstanding over most of the hunting area, but that one of the service station maps covers one segment with much more detail. Few things are more frustrating than closing in on the hidden T on a mileage hunt and discovering that the proper roads to it are just off the edge of your map.

The United States Geological Survey (USGS) topographical (topo) maps are probably the most detailed that you can get. They show not only the roads but everything else: buildings, lakes, streams, power lines, pipe lines, towers, and on and on. They also have contour lines that allow you to visualize the actual terrain, complete with heights of hills, mountains, and valleys. These maps are great when trying to figure out how a signal is bouncing around or knife edge refracting over a hill. They also cover all roads—everything from the smallest trails to freeways—in exquisite detail. They're vital for ELT searching in the wilderness.

As good as these maps are, they have two shortcomings. They are only updated every 10 years or so. This is not a problem if you are hunting out in the desert or in a national forest, but can be disastrous in an urban or suburban environment. Second, these maps are designed

around a strict grid system. Don't expect your town to be right in the center of the map. Chances are several maps will be needed. One map of the California coast has a thin sliver of the coastline in one corner and the rest of the map is only blue ocean. Except in rural situations, the USGS map will in most cases be used to complement other maps.

The local auto clubs affiliated with the American Automobile Association are excellent map sources. They generally have a broad selection and update the maps regularly, especially in areas with growth. There are local maps and maps of a county or portions of a county. There may also be a local region map covering portions of more than one county. Unfortunately, different county maps usually have differing scales, making it impossible to paste them together to cover unusual boundary areas.

If you are just interested in your city limits, you may find detailed local maps available through your city or county engineering departments. We found one municipal map that's 25 by 36 inches—big enough that the streets are about a sixteenth of an inch wide on it. With this kind of detail you can pinpoint each house. The local Chamber of Commerce is another good source. These maps usually do not cover any rural areas around your town.

Many areas of the country have independent map companies that publish local and regional maps. The accuracy of these maps can range from outstanding to poor. In California and a few other west coast areas, Thomas Brothers maps are the standard of excellence for T-hunting. Their map books are issued in 10 × 8.5 inch spiral bindings and are updated yearly. They have up to 450 pages in combined county books, with each urban area page covering 15 square miles. These or similar maps may be available in your local area. Be prepared to spend ten to twenty dollars or more, depending on the area covered, to get all the maps you need.

As good as these premium maps are, they aren't perfect. Hiders are always searching for the little parks, dirt roads, or new housing areas that aren't on them. On the other hand, the maps frequently show roads in new housing tracts that don't exist or aren't open yet. You may follow your expensive map only to find yourself facing a locked gate or dead end where the map says the road ought to be.

So here's some more advice: Don't hesitate to mark up your maps. As you discover discrepancies between the maps and the actual terrain, either on a hunt or while searching for a hiding spot, make the corrections right away. By doing so, you may save yourself some time or mileage later.

For a first class setup, mount your maps on corrugated cardboard or thin particle board and cover them with clear plastic. Large sheets with adhesive can be bought in stores selling artists' and drafting supplies. Several finishes are available—get one which allows ordinary felt tip pen ink to wash right off. Before putting the plastic on, draw the important lines that you use regularly, such as lines to reference repeaters. Mark the boundaries and the location of any important starting or bearing taking points so that they can be located rapidly.

Large map boards can be unwieldy in a small car. Larry Starkweather, WD6EJN, mounted his to the ceiling with plastic fasteners that allow it to be rapidly removed. By leaning back and looking up, both passenger and driver can read the map and plot bearings. Don't try this if you have weak neck muscles!

THE COMPLEAT HUNTER

As in any other sport, proper equipment distinguishes the rookie from the veteran. Besides radio gear and maps, here's what the well-outfitted mobile hunter will want to have:

☐ Compass. A cheapie car compass will help but will have an error (due to the influence of the car body) that cannot be fully compensated for. Find a good hiker's compass and get out of the car to use it when bearing accuracy is important. How about a surplus aircraft gyrocompass?

☐ Flashlight. An obvious requirement for night hunting. Also consider a high intensity "portable sun" that plugs into the car's 12 volt system.

☐ Map lights. It's hard to drive at night with the dome light on. A small high intensity lamp for just the maps and direction pointer will make it easier to see outside.

☐ Shoes. You may end up tramping through the brush or splashing through a creek. Don't wear your best wing-tips. And don't forget to carry a jacket, sunscreen, and foul weather gear.

☐ Spare equipment. If your radio fails or your antenna gets mashed, are you out of the hunt? If you have a second hunting system, bring it. Don't forget your handie-talkie.

☐ Snacks. On a lengthy hunt for a jammer or a bunny, you'll be in the car for a long time and perhaps over varied terrain. You may want to pack a lunch, a thermos, or an ice chest.

☐ Glove. Some antenna systems are a bit hard to turn, and could cause sore muscles or even raise a blister during an all day or all night hunt. A garden or handball glove with a good gripping surface might be just what is needed to avoid "T-hunt wrist syndrome."

☐ Vehicle. When shopping for the next family car, give a thought to its suitability for hunting. This may be the excuse to get that four-wheel drive off-road buggy you've thought about. But there may be some disadvantages. Rich Krier, N6MJ, loves to hunt in an open jeep, but you won't see him at the restaurant afterwards because of the potential problem of theft of his gear.

Chapter 3

VHF Mobile Hunting Techniques

Before you set out on your first hunt with all that shiny new gear, there's one principle you must understand thoroughly: Your DF set *does not* tell you the direction to the transmitter. It doesn't know that direction. If it did, T-hunting would be a lead pipe cinch.

No matter what kind of set-up you have, all it can be expected to indicate in azimuth is the direction (or directions) from which it is sensing the signal. You can hope that the signal is arriving from the same direction as the direction to the T, but you can't count on it. Hunting with that in mind is what separates the art from the science of DFing.

Technique is without question the most important component of successful DFing. A good hunter can DF with some degree of success with the poorest equipment. A poor hunter, on the other hand, is not the one you'd want coming to rescue you even with the best equipment. The tips in this chapter will guide you through your first hunt experiences and help you make the most of whatever gear you use.

First we'll look at some of the things that make hunting a challenge, both from a psychological and a technical standpoint. Then we'll go step by step through the fine points of good navigation, including figuring out which bearings and locations are most trustworthy, get-

ting the best starting bearing, distance quessing, and closing in on the fox. After that there are tips on how to team up, both within the car and with other teams when appropriate.

The techniques in this chapter are geared primarily to hunting vhf signals with gain antennas. Hunting techniques with other types of gear may be a bit different, but the same general concerns about bearing accuracy and multipath apply. There is more detail about the special techniques associated with switched antennas, Dopplers, interferometers and the like in the chapters to follow, but read this chapter first before jumping ahead.

Some people are what might be called seat-of-the-pants hunters. Their tires screech as they roar off the hill after deciding only that the signal was sort of north, south, east, or west. They only follow the antenna, driving continuously in the apparent direction of the transmitter until the roads run out. Others, usually the more successful hunters, take the time to plot the signals on a good map before starting, and continue plotting regularly throughout the hunt, even on a timed hunt.

In sport hunting there is very little to take for granted. Hiders are treacherous. It's all part of the game. When you hide, you try to be just as sneaky, right? You'll hear it many times: "Never trust anything the hider says." Another safe assumption is to never trust anything another hunter says during a competitive hunt. Why should he help you? Put your trust only in your gear and your team members. Don't try to guess where the bunny is, based on the type of spots where that hider has put it before. The odds are against you.

THE IMPORTANCE OF HIGH GROUND

Anything that stands between you and the transmitter, or is close enough to the signal path to cause reflection or distortion, can make the DF unit read a direction of signal that is not in the direction of the transmitter. Oftentimes these terrain features are close to you, and you can take them into account. For example:

☐ You are on a road right next to a high embankment or hill on your right. The signal can't come through the hill, so if the T is to your right, you will either lose the signal or it must come from another direction by reflection.

☐ You are on a road under long power lines or a high metal fence. The metal picks up and re-radiates the signal, and your DF indication continues to point to the lines. While this is most common on the hf bands, it can also occur close-in on vhf, particularly with fences.

☐ You are driving down a long road through a canyon or cut in the hills, which are high on either side. The signal from one side can't easily get to you, so it may enter the canyon at one end and ping-pong its way along between the hills. Your gear's direction indication might be ahead of you or behind you, no matter what the actual direction to the source is.

☐ You are on top of a high hill—perhaps the highest around. But there are other high hills in view. Your set can't "see" down into every pocket and crevice in the terrain between you and the horizon. If the source is very low and obscured, the best signal path from it may be via a bounce from one of the other high hills. Your beam could sense the greatest amount of signal from a hillside that is not in the correct direction to the transmitter.

DFing is beginning to look a little more difficult, isn't it? How do you know when the direction of signal isn't the direction to the bunny? Experience will help a lot, but while you're learning, watch for situations such as those above. Generally the higher you are, the more trustworthy is the bearing, though there are exceptions as just noted. But you can't stay on the hill forever—you have to go where you think it is and flush out the T.

THE INITIAL BEARING

There are three important points to good triangulation: good maps, an accurate direction indicator on the antenna, and an antenna that gives consistently accurate bearings (i.e., doesn't have a skewed main lobe). Deficiencies in any of these areas degrade bearing accuracy.

Though other conventions are used, stating bearings on a 360 degree basis minimizes the chance of ambiguity, and is used here. The bearing from any point on the earth to another is somewhere between 0 and 359.99 degrees. Zero degrees is due north, 90 degrees east, 180 degrees south, and 270 degrees west. Most maps are drawn with north at the top.

The direction indicators described in this book all give the direction relative to the front of the vehicle, where 0 degrees is straight ahead, 180 degrees is to the rear, 90 degrees to the right, and 270 degrees to the left. Either the antenna direction must be read independently (with a compass) or the direction indication relative to the car must be converted to map bearings.

Getting a beam/quad bearing with a compass is relatively easy. Turn the antenna for the greatest signal reading, making sure that it doesn't move after you let go of the mast. Stand exactly behind the antenna and sight down the boom with the compass. A fancy method of using a compass is shown in Fig. 3-1. Note this bearing on a piece of paper.

MAGNETIC DECLINATION

Before you can use this bearing, magnetic declination must be taken into account. True north from any point on the earth can be thought of as a line drawn from that point to the north pole. However, your compass points to the north magnetic pole which is located in the general area of Hudson Bay. Depending on where you are located, there may be a deviation between the direction to the north pole (true north) and the magnetic pole (magnetic north). This is known as declination or variation.

Determine the declination in your local area. The map shown in Fig. 3-2 gives approximate declinations for the USA. More accurate declinations can be determined from a USGS topographical map, but trust only a current map. The declination changes slowly over the years, drifting about 1 degree every 6 years. Once you have the magnetic bearing and the declination for your area, the rest is easy.

Suppose that your compass bearing is 127 degrees

Fig. 3-1. Jim Hostetler, K6SXD, getting the initial bearing for an All Day Hunt.

and your local declination is 15 degrees east. Add the 15 degrees declination to your compass bearing getting 142 degrees true. If your local declination is 10 degrees west, then subtract the 10 degrees from your compass bearing, getting 117 degrees true. Just remember: ADD east declination, and SUBTRACT west declination. If your magnetic bearing is near 360 degrees, say 355, then just carry over past 0 degrees by subtracting 360 from an answer over 360, and adding 360 to an answer less than 0. With 15 degrees east declination the true bearing would be 10 degrees.

Once you have the true bearing, plot it on your map using a 360 degree protractor. Place the center of the protractor directly over your current location on the map. Rotate the protractor until 0 degrees is pointing to north on the map (usually toward the top). Mark the true bear-

ing on the edge of the protractor and draw a line from your location to the edge of the T-hunt boundary with a straight edge.

GETTING BEARINGS WITHOUT A COMPASS

The best method of determining your initial bearing without a compass is to reference everything to a known signal source. First take a bearing on a repeater or fixed station that is at a known location in the clear about 5 to 10 miles away from you. The signal should be line of sight and strong enough to get a good bearing on. Read the direction indicator mounted on the antenna mast with everything referenced to the front of the vehicle. Write down this bearing. Draw a line on the map between your present location and the repeater.

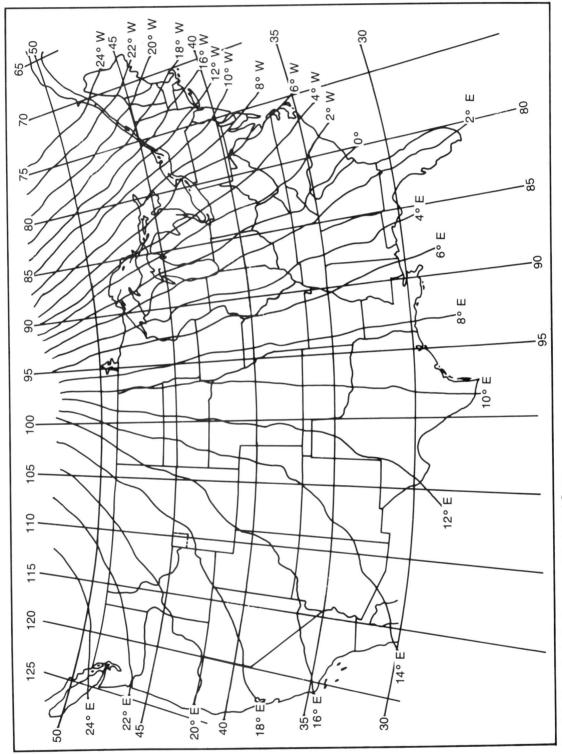

Fig. 3-2. Magnetic declination map of the United States.

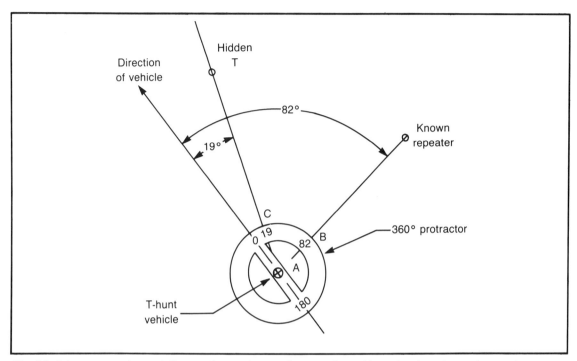

Fig. 3-3. A known repeater or station can be used as a reference when taking a bearing of an unknown signal.

Now take a bearing on the hidden T and write it down. Place a 360 degree protractor on the map with the center over your present location. Rotate the protractor until the reference bearing number is over the line to the reference station. Without moving the protractor, locate the bearing to the hidden T and mark it on the map. Draw a line from your current location through this mark to the T-hunt boundary. This line is the bearing to the hidden T.

In the example shown in Fig. 3-3, the bearing to the known repeater, relative to the vehicle, is 82 degrees and the bearing to the bunny is 19 degrees. The protractor is placed directly over the location of the observer and rotated until 82 degrees is over the line to the reference (point B). Then the bearing to the hidden T is marked (point C) and a line is drawn from point A through point C out to the hunt boundary. Assuming that the signal is directly propagated (unlikely with many hiders) and the bearings are 100% accurate (unlikely with most hunters), the object of the hunt is somewhere exactly along this line.

All this is easier to do than to explain. You will notice that when the protractor is aligned, the north (0 degree) point is pointing the same direction as the vehicle. By taking a bearing off of a known station you have calibrated the direction of the vehicle. The big advantage of this reference system is that any errors in the main lobe of your antenna (such as a skewed lobe) are automatically cancelled out.

READY TO ROLL

The start bearing is the primary reference for the entire hunt. If you lose the signal as you leave the starting point, you can follow the original bearing line until the signal is heard again. This is particularly important on long distance weak signal hunts where it's common to lose the signal for 15 to 30 minutes after leaving the start. On the All Day Hunt you might lose the signal for hours!

Now check the hidden signal on horizontal and vertical polarization if your equipment is capable of both. If one polarization gives a substantially higher S-meter reading, use it. If both are about equal:

☐ The bearing could be to a reflection from another hill or mountain.

☐ The transmitter could be behind a hill or otherwise obstructed from your line of sight. Propagation might be by "knife-edge" refraction over the hill or around it.

☐ The hider could be using a very creative antenna system.

As you swing the beam or quad across his bearing,

is the peak sharp, just like it was when you measured its pattern? (You did do that, didn't you?) Or is the indication fairly broad? A diffuse response can indicate either a reflection or an obstruction, as above.

Now we've learned all about the signal, but there's one more thing to do before starting. Check the maps for best routes in the direction of the signal. Choosing potential paths and identifying critical intersections in advance is one mark of an above-average hunting team.

As you drive off the hill, keep paying close attention to the signal.

☐ Does it drop off rapidly or stay at relatively the same strength? Depending on the terrain in the direction of the signal, this may give you a clue as to whether the transmitter is in a high or low location.

☐ Does the true bearing stay the same? If it changes rapidly you may be the potential victim of a close-in hider—BEWARE!

CLOSING IN

Amplitude variations often make it difficult to get good bearings with a quad or beam while moving. Stopping occasionally to take a bearing may be helpful, and indeed is often necessary for decision-making at a crossroads. If you must stop in an area where there are lots of nearby buildings or power lines, take two or more bearings from about 20 feet apart, to judge the effect of the reflections.

Conversely, units such as Dopplers, L-Pers, and BMGs usually perform best if bearings are taken while moving. Motion tends to average out the nearby reflections and it's easier to see the trends develop. With these, it may be best to always take bearings "on the fly."

Add attenuation to the system as necessary to keep the S-meter in the mid-scale region. This is generally the most sensitive area of the S-meter scale. Some meters are very non-linear toward the full scale area, and a few never reach full scale no matter how much signal is present. Bearings taken with such meters near top scale will appear very broad. Reflections and sidelobes will appear just as strong as the main signal. When this occurs, add more attenuation.

NAVIGATING ON THE RUN

Using a bearing system that sets the front of the vehicle at 0 degrees makes taking running bearings easy. As you come to an intersection, take a bearing. Find your location on the map and place your protractor (360 de-gree) over the intersection with the 0 degree point facing the direction the vehicle is heading. Mark the bearing you took and draw the line. Note that the streets do not need to be exactly north-south or east-west. As long as the streets are straight enough to line up the protractor, these bearings will be almost as accurate as those taken with a compass, but much easier and faster to do.

Unless the hidden T is directly out one street or highway (a bad idea for the hider), the initial bearing on the map will cross most of the streets at an angle, forcing you to zig-zag to the transmitter. It's a good idea to take new bearings as you cross the original bearing line. You may find that your first bearing was off for one reason or another. After plotting the bearings from a few points along the way you may find that one of these new lines seems to be more accurate than the original.

As an example, assume that the starting bearing was just a few degrees south of due west. Decide what is the most direct east-west road and start driving. The signal is mostly to the west but a small amount south. As you drive along notice the signal swing to the south. This starts gradually at first and increases as you come abreast of the signal. Try to look ahead and determine what streets you can turn south on before it is necessary.

As the signal starts to swing, keep taking and plotting new bearings. In a perfect world all these bearings would cross at exactly one point—the transmitter. That just won't happen, unfortunately. But as you take more and more bearings most will start to cross in the same general area. The closer you get, the smaller the error circle the crossing points form, assuming there are no major local site errors. For example, at 25 miles, a 5 degree bearing error will cause the bearing line to miss the transmitter by 2.2 miles. At one mile the five degree error will cause a miss by 462 feet.

GUESSING THE DISTANCE

When the bunny is almost straight ahead, and there's no cross bearing to help, how do you tell how far away he is? In a hilly area, or an area of tall buildings or other good reflectors, there's probably no telling. On the straightaway, or in a residential neighborhood, your S-meter can be of some help if the signal is steady.

The intensity of a wave in space at various points from the transmitter is predicted by the inverse square law. Simply put, every time the distance from the transmitter doubles, the signal power drops to one fourth, all other factors being equal. This makes the voltage at the receiver input terminals drop to one half. If the receiver S-meter could have a perfectly linear scale, or read directly in microvolts, we could use it to "guesstimate" our progress to the signal source.

The receiver input voltage increases as a function of relative distance from the starting point. When the strength is twice that of the starting point reading (6 dB greater), we have gone half way, all other factors being equal. When it is twice the halfway reading, we have gone halfway from the halfway point, or three quarters of the distance.

Receivers with calibrated S-meter scales, called Field Strength Meters, do exist. Ham receivers, however, don't have them. You can make a calibration chart for your S-meter using a signal source with known output level. Figure 3-4 shows typical S-meter characteristics of several commercial vhf FM rigs. Of course your radio may vary widely from these numbers, but they give an idea of how linear, or non-linear, typical FM rig's meters are. By making a similar chart for your hunt radio, you can get an idea of how signal strength increases correspond to distance from the bunny.

As the figure shows, the range from no reading to full scale reading on most S-meters is only about 30 dB. This limits the usefulness of this distance measuring technique. To augment it, or to get distance estimates without S-meter calibration, use your attenuator. For example, as you move closer, and you find that the meter reading now with 12 dB attenuation is the same as it read earlier with 6 dB attenuation, you know that you have approximately halved the distance since the previous reading, as the signal intensity has gone up by 6 dB.

This technique is accurate only for direct line of sight signals of the same polarization. When the signal is cross polarized, reflected, or obscured, your distance predictions won't be very close. Results are thus best with powerful jammers and worst with tricky foxes.

Sometimes the signal goes down as you move along, while moving toward a hill. This usually indicates that the hidden T is behind or over the hill. If the signal jumps up suddenly, it could be that the hider has changed power. Or perhaps you've come from behind a shielding terrain feature, or come over a hill.

Keep the rules of the hunt in mind at all times. On a mileage-only sport hunt, it pays to take it very slowly, stopping at any high or clear point for a fresh bearing with an accurate check of the vehicle heading. Watch the map for any dead end roads that could add unwanted mileage. Use all the distance-measuring tricks described earlier to avoid accidentally going too far.

Sometimes time is the major factor, as on a timed hunt, when hunting an ELT, or when a jammer may not stay on the air. Less time should be spent in map reading under these circumstances. Keep moving, get bearings on the fly, and take more risks on checking possible hiding spots on non-through streets. But try to remain as methodical as possible. Don't abandon the maps and triangulation and succumb to hunting only by instinct.

On rectangular streets when time is of essence, the stairstep method is a favorite of some hunters. Say that the bearing is 20 degrees relative to the vehicle as you're driving down a city street. Continue forward until the bearing is exactly 90 degrees, then turn right. Continue until the bearing is either 90 degrees again or 270 degrees, and turn right or left as appropriate. Keep doing this and you'll be there after only a few turns. Though this method may not result in minimum mileage, it is excellent for speed.

The "45/90" technique is an extension of the stairstep method. It uses the geometric principles of right triangles to help hunters estimate how close they are getting to the hidden T. As we set out on our course, shown in Fig. 3-5, at point A the bearing is nearly in front of us. The indicator shows 15 degrees to the right. We continue straight ahead, watching carefully for the point where the bearing is exactly 45 degrees right. When we reach that point (B), we note the odometer reading and continue without turning.

As the bearing approaches 90 degrees right, we find an appropriate through street on which to turn right. If possible, we turn right when the bearing is exactly 90 degrees, and take an odometer reading. If we must overshoot the 90 degree crossing slightly to find an appropriate road, we note the odometer reading at the 90-degree bearing point.

In the illustration, points B, C, and D form a right isosceles triangle, since angle CBD is 45 degrees and BCD is 90 degrees. Therefore, side BC is equal in length to CD. Since we measured the mileage from B to C, we can say that it is also the mileage from C to D, and that the hidden T must be very close to D.

Sure enough, as we approach D, the signal gets very loud, and the bearing swings right. Once again we note the 45 degree bearing point (E) and measure the mileage to the 90 degree point (F). We proceed east an amount equal to EF, and there's the bunny!

If you're familiar with seagoing navigation, you may have heard the 45/90 method being called the "bow and beam" bearing technique, which is a special case of a geometric solution called "doubling the angle on the bow." It works for other angles besides 45/90, though these larger angles produce best accuracy. The general case is used when the roads aren't all at right angles or when there aren't a lot of cross streets.

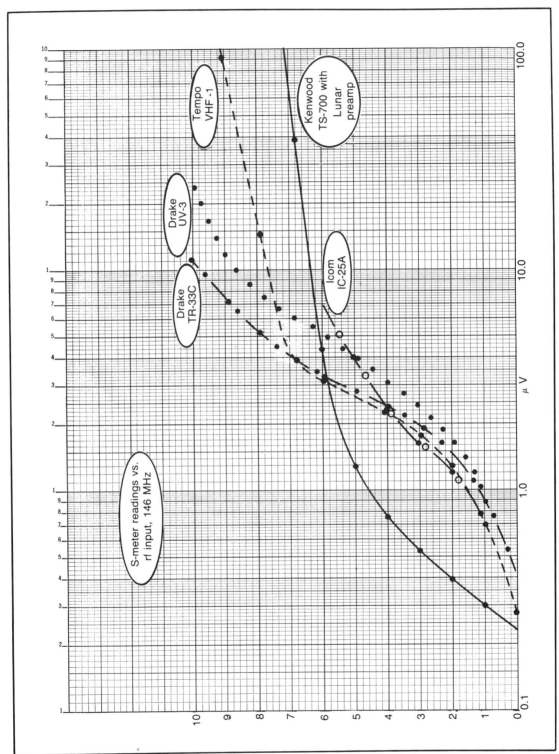

Fig. 3-4. Typical S-meter performance curves of some 2 meter receivers.

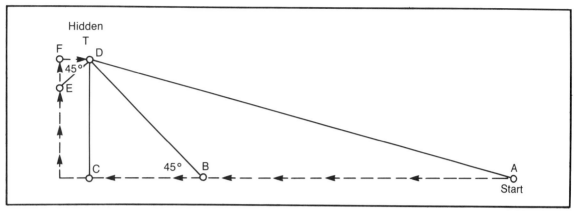

Fig. 3-5. Estimating the distance to a hidden transmitter using the "45-90" method.

In Fig. 3-6, the bearing angle at point Y is twice the angle at point X. Thus distance XY equals distance YZ, and the airline distance from Y to the transmitter at Z equals the distance traveled from X to Y. Unless your mast direction indicator has excellent accuracy, this generalized method is good primarily for rough estimates as you travel along.

THE END GAME

Continuing to take bearings in the last half mile or less is very important. Human nature urges us to rush on and find the hidden transmitter, even if the hunt is mileage only. But take these last bearings carefully, plotting them on the map that has the most detail. Try to plot bearings from as many different locations as possible without time or mileage loss. When the bearing to the bunny is changing rapidly, it may be desirable to take a bearing as often as every 50 feet.

Why this extra care? Without it you may overshoot the right cross street, forcing the choice between extra blocks on the odometer, an illegal U-turn (tsk, tsk), or backing up while hoping no one's watching (TSK! TSK!).

The hypothetical example of Fig. 3-7 shows the im-

portance of careful bearings in the end game. It is a time only hunt, and N6JSX has hidden his transmitter south of the large park with a high gain beam pointing to the embankment, which scatters his signal. Following the initial bearing you come north on 5th Street, and swing onto State Street and the T intersection, noting that the signal source has suddenly swung from the northwest to the southwest, as the direct signal becomes predominant over the embankment reflection.

Stopping at the entrance to the large park (point F), you see that the bearing is exactly 90 degrees left (south). Aha! He's in the park! Grabbing the portable DF, you saunter into the park. After 10 minutes and a half mile of walking around the lake you find yourself at the 8 foot chain-link razor-wire fence on the other side of the park. But Kuby is outside the park and inaccessible. Sorry, but the rules say you must touch the transmitter for your time to count. Now you have to walk all the way back to the car and drive around to his hiding spot.

At point A the predominant signal was from the embankment, but you should have become suspicious at point B. There the double indication of the direct and reflected signals becomes apparent. Had you then carefully taken and triangulated bearings at points B, D, F,

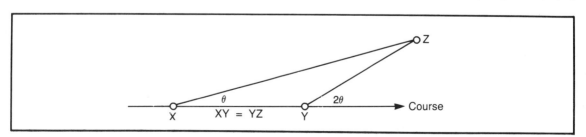

Fig. 3-6. "Doubling the angle on the bow" method of estimating distance.

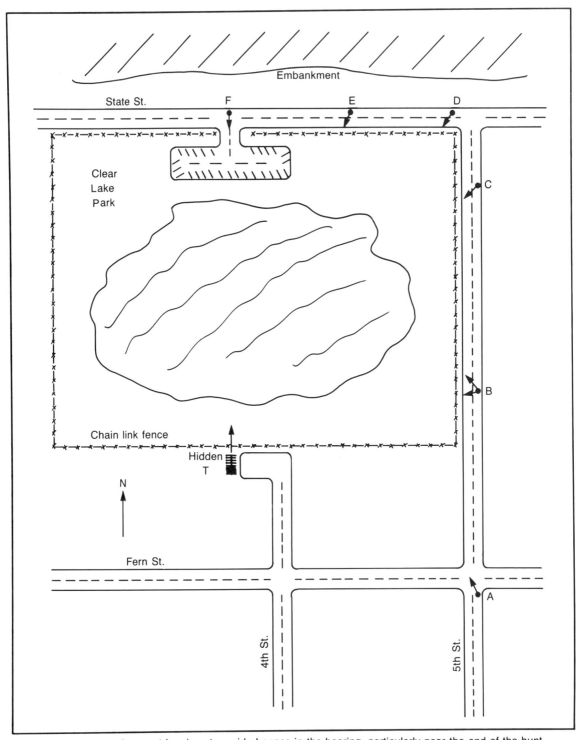

Fig. 3-7. Careful attention must be given to rapid changes in the bearing, particularly near the end of the hunt.

and in between, you would have concluded correctly that the bunny was south of the park, and you could have saved at least twenty minutes. So don't let heavy traffic and "No Stopping" signs keep you from doing a methodical job of triangulation. Pull into a parking lot and stop if necessary to get and triangulate those final bearings.

HUNTING LIKE A PRO

The most common mistake that new hunters make with a manually rotated quad or beam is to not swing it far enough to center the hidden T signal in the main lobe. They tend to swing it until they hear the signal, then stop immediately. You should swing across the signal, watching the meter go up and then down again. Then go back and forth a couple of times to center the signal carefully before reading the direction indicator.

Peaking the antenna can be a tedious task when the signal is fluctuating up and down due to vehicle movement or tricks of the fox. After quickly swinging the antenna all the way around to determine the approximate direction of the strongest signal, swing it slowly (about 10 rpm) across the peak and back again, using eyeball averaging to smooth out any flutter.

The same care is needed with a null reading antenna system. Find the deepest null by slowly and carefully sweeping across each dip in signal strength. Roll the car 20 feet or so further, retake the bearing again, and compare.

The second most common mistake of new DFers is not taking bearings continuously, and not checking the full 360 degree range. Their tendency is to stop swinging the mast for blocks at a time, so long as the signal is hearable. Figure 3-8 illustrates how this can be disastrous. The readings at the start (point A) and along the way at each of the mile road intersections (B, C, D, E) show only a weak reflected signal straight ahead toward the big hill, because the direct signal is obscured by the little hill.

If Mr. New Hunter isn't on the ball and continues to look for signal only from the north at points G and beyond, who knows how much farther out of the way he'll direct the driver to go! When he finally thinks to check other directions besides generally northerly, he's in for a big surprise. The signal from the southwest is far stronger. So remember to check consistently in all directions from start to finish.

HUNTING AS A TEAM

While it's true that some of the best hunters prefer to work alone, it's hard to top the performance of a well organized team, particularly on very long hunts. By properly dividing the tasks, you can shave off the minutes and miles from your score. Though usually not as efficient as a team of three, two person teams are most common. Map plotting is done by the antenna turner, and the driver assists in navigating.

For hunting with manually rotated antennas, a three person team is optimum. Here are the duties of each:

☐ The DFer is responsible for knowing the bearing to the transmitter at every moment during the hunt. He gets the blisters and sore wrist from swinging the quad or beam around constantly, making sure that the starting bearing wasn't just a reflection, that the hider didn't put it close in, and so on. The DFer makes receiver adjustments as required, and works with all the add-on meters and gimmicks. He operates the attenuator, too.

A good DFer continuously calls out readings relative to the vehicle, such as, "It's dead ahead . . . keep going," or, "It's 60 degrees left now . . . and swinging fast." The DFer must also be alert to conditions which could cause false bearings, such as electrical noise or a nearby embankment. He may also help identify roads for the navigator.

☐ The navigator should be seated where his large map and nighttime reading lights will not be a hindrance or hazard to the driver. His jobs are to know the location and heading of the vehicle at all times, to put vectors on the maps corresponding to bearings from the DFer, and to plot the minimum time and/or mileage route to the bunny in accordance with the scoring rules of the hunt.

The navigator is the one who gives instructions to the driver, such as, "Turn left at Maple Street or we'll be caught in an industrial complex," and, "Stop at Elm Street in the right lane to see if the bearing says 90 degrees there." The navigator constantly checks the map for one-way streets, dead ends, and private roads. He is always planning ahead for the driver.

☐ The driver concerns himself only with carrying out the instructions of the navigator, while watching traffic and moving ahead as rapidly as conditions allow on time factor hunts. Where possible he should call out important cross streets as they are approached, as well as important terrain features: "Overpass coming up, shall we stop on top for a bearing?"

The DFer position may be eliminated when an automatically rotated antenna system is used. The navigator

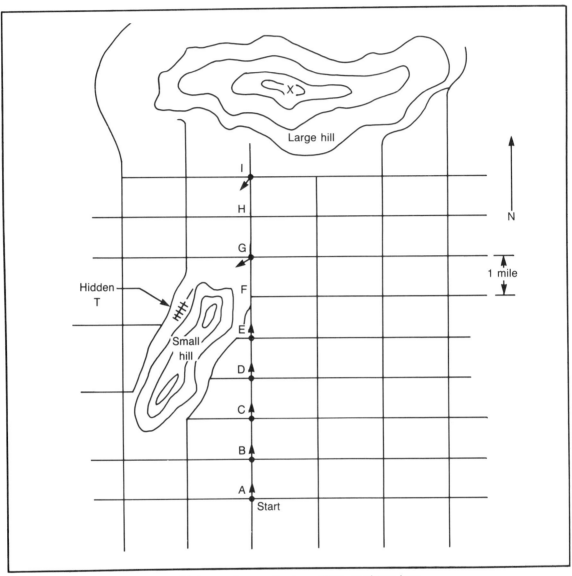

Fig. 3-8. Keep swinging the antenna! It is easy to pass the transmitter, as shown here.

watches the DF unit for trends and calls out the information to the driver. When time is not a factor, the driver can do it all, stopping frequently for critical bearings and for plotting.

COORDINATING A COOPERATIVE HUNT

Most of the advice in this book so far has been aimed at making you or your small team successful competitors. In contrast, when hunting a jammer, a keyed up radio,

or a station in distress, organization and cooperation between teams speed things along. You'll find that when time is short, a smooth-running group effort is the fastest way to find a signal source. If rules permit it, vehicle teams can help each other by triangulating their bearings, exchanging road information, and telling each other where the transmitter isn't.

All communications should be on a frequency other than that used by the target signal. This is particularly important in a jammer hunt for security reasons, but on any hunt the signals of reporting stations will interfere with DFing if they're both on the same frequency. We recommend that any communications among fixed stations which must be secure should not be done on any ham band. Use the telephone.

When cooperating mobiles are close to each other, they can use low power handhelds. But beware! Jammers have actually taped the conversations of hunters on their thought-to-be-secure frequencies and played them back through the jamming transmitters.

The more reporting stations, the better. It is important that they be widely dispersed, not all in the same area. A concentrated group of DF sites will probably all provide about the same bearings—not very useful for triangulation. If there's doubt about indicator accuracy, stations can give bearings relative to the repeater or another known fixed reference.

Stations without means for getting bearings should be encouraged to report strength readings. It's very important to know where the target signal is very strong, or not heard at all. Reports can be in the form of S-meter readings or amount of FM quieting. For full scale signals, an attenuator can give further information on just how strong the signal is. Since each station is different in some way, there is no true standardization of strength indications, but the information can still be useful. In the case of continuous carrier on a repeater input, knowing which stations can override the signal is also useful in determining the distance of the transmitter from the machine.

Some metropolitan repeaters have multiple receiver sites scattered through their coverage area with a sophisticated voting selection system. This can be very helpful for getting the general whereabouts of the source. Just check all the receivers for strongest signal and start there.

DFers must learn to give quick, precise indications of strength and azimuth to speed things up. They should also concisely pass along any other observations. Is there mobile flutter or any other distinguishing characteristic? The target's operation may change with time, so the times and dates of each observation should be recorded and considered.

One group that has worked hard to develop a cooperative DFing capability is the North Shore Repeater Association (NSRA) of Salem, Massachusetts. Users of the K1UGM repeater get training in how to report observations. Even non-licensed scanner listeners can telephone in reports.

A monthly DF drill is held on the repeater output frequency. After the hidden T transmits, all stations are invited to check in and report. The data is correlated, the bearings are plotted, and the hunt coordinators determine the suspected location or locations. The hidden operator then divulges his true position, so each reporter can check his own indications.

Occasionally a full scale cooperative practice hunt is done by NSRA. The fox is in a town or suburb in the repeater coverage area. Base stations collect and correlate data, then direct mobiles to the general area, where they then compete to find him.

Everyone benefits from these cooperative hunts and drills. The users learn how to take DF information and report it. The correlators learn who has the most accurate DFing and reporting capability, and whose reports not to trust. The mobile hunters get assistance and encouragement. Best of all, the whole effort is a great deterrent to malicious QRM.

Chapter 4

VHF Hunting with Directional Antennas

Despite the many specialized DF units now available, many expert vhf hunters still stay with their directional antennas and S-meters. What this system lacks in convenience is made up for by simplicity and the ability to hunt very weak signals with high gain antennas. In this chapter we'll cover a number of antenna schemes for vhf. We'll even show you how to get hunting today with a vhf loop. You'll see how easy and fun antenna experimentation is.

SIMPLE VHF ANTENNAS

Let's say that the local radio club is having its first 2 meter fun hunt on Saturday afternoon. It's Saturday morning now and you'd like to see if T-hunting is as fun as it sounds. Lots of work ahead? No. While you may not have the time or inclination to build a quad, attenuator, and mobile mount, you and a friend can still be part of the action, if you have a receiver with S-meter. It only takes a couple of hours to build and tune a loop antenna and become a transmitter hunter.

TWO METER LOOPS

This beginner's antenna was designed by Dick Reimer, W6ET, and passed along by John Gallegos,

W6EQ, who has used it to successfully find both hidden T's and a jammer. All you need to build it are a broomstick, a 10 picofarad trimmer capacitor, a small piece of perf board, six feet or so of coax with a connector for your rig, an alligator clip, and about three feet of solid AWG 12 copper house wire. A piston trimmer works best for the tuning capacitor, but an air or ceramic type can be used.

Mount the capacitor to the perf board. Form the solid wire into a loop and connect the ends to the capacitor terminals, as shown in Fig. 4-1. (Don't hook up the vertical sense antenna yet.) Secure the loop to the perf board for mechanical rigidity. Mount the loop to the broomstick mast. Five-minute epoxy glue is good for securing these parts. Connect the coax shield to the loop at the point exactly opposite the capacitor. Connect the center conductor to the gamma match as shown. It's done!

This antenna is about 0.3 wavelength around. It is much larger in terms of wavelength than the hf loops discussed elsewhere in this book, yet it acts like a small loop in its directivity. Figure 4-2 gives the pattern. Its sharp resonance gives it amazing sensitivity for its size. It has produced higher S-meter readings, when held above W6ET's car roof, than did a 5/8-wavelength magnetic mount whip on the rear deck.

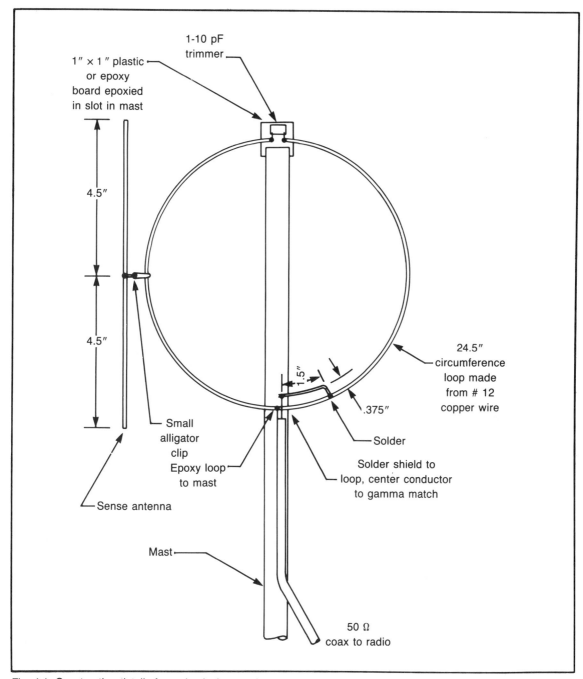

Fig. 4-1. Construction details for a simple 2 meter loop.

An electrostatic shield is not necessary or desirable on this particular loop. The higher you can get it above the car, the better it will work. Nearby objects won't have as bad an effect on pattern as they will with a loop on hf, because the objects are further away in terms of wavelengths. Still, it performs best when out in the clear.

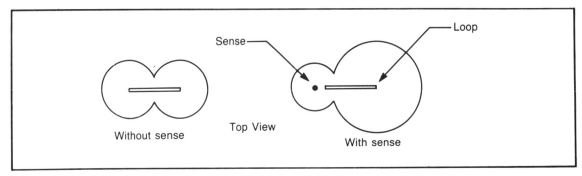

Fig. 4-2. Pattern of the simple 2 meter loop, both with and without the sense antenna, seen from above.

Use the local repeater as a signal source to tune the loop. Hold it end-on to the direction of the repeater antenna (one maximum lobe) and tune the capacitor with an insulated tuning tool for maximum signal. Tune slowly and remove your hand after each adjustment, as coupling to your body affects the adjustment. You may be able to improve performance by experimenting with the gamma match dimensions. Do not transmit into the loop. (Disconnecting the microphone prevents accidental transmissions.)

Though many experienced hunters prefer to work alone, you'll want to team up for this first hunt. The passenger holds the loop upright out the window and spins it slowly. For weak signals, hunt the peaks, which are in the plane of the loop (off the ends).

With stronger signals, hunt the nulls. They should be sharper and at exact right angles to the peak. They may be disguised or in the wrong direction due to interaction with the car and other antennas, however. Stay in the clear, away from fences and power lines when taking bearings, as they can upset nulls, too. Get some experience taking readings on the local repeater from several locations before setting out for the hunt.

One of the biggest problems with this simple loop is its bi-directionality. Is the signal in front or in back? One sneaky way for a beginner to tell at the starting point is to look at the direction all the quads and beams are pointing. Then drive a course that always keeps the bunny ahead of you, never letting him get more than 90 degrees left or right.

In the early days of hunting most hunters used these bi-directional loops. When they had a choice, they started the hunt from the edge of the boundary area instead of the center, so that the 180 degree ambiguity would not be a problem. Of course they were often a very long way from the fox because of starting at the boundary.

Fortunately there's a better way. The loop can be

made somewhat unidirectional. Clip the nine inch sense wire to the side of the loop as shown in Fig. 4-1. Now rotate the loop and notice that the signal peak on one side is much higher on the meter than on the one 180 degrees around. When oriented for the higher peak, the vertical rod will be on the side away from the hidden T.

Adding the vertical rod detunes the loop, decreasing its sensitivity by several dB. This is no problem unless the signal is very weak. The vertical sense antenna may not give proper unidirectional response if there is a high horizontal polarization component to the incoming signal.

Although this loop method will get you started in hunting (and maybe even win a hunt or two), it is admittedly crude. You'll soon be eager to improve your system. But don't throw your loop away then—keep it in the trunk. You may be glad you have it when your fancy wire quad gets mangled by a low hanging branch in the middle of a hunt.

OTHER INSTANT HUNTING IDEAS

Is the mini-loop the simplest directional antenna? Nope. Chuck Tavaris, N4FQ, reports that some hams in the Roanoke, Virginia, area have tried "The Garbage Can Emergency Antenna." They stick their magnetic mount whips in the middle of the inside of a galvanized iron garbage can lid. This contraption becomes a hand held DF antenna, held by the lid's handle like a knight of old would hold his shield.

Of course the polarization is wrong, the antenna pattern is wrong, and it can only be used outside of the car. But some say it works, sort of. It probably does best at hunting the null behind it. Keep it in mind as a last resort.

The idea of using a solid reflector to obtain directivity does have its merits, however. A proper metal shield behind a handheld radio can make a simple DF system. Try making a curved reflector out of corrugated cardboard as shown in Fig. 4-3. The radio, with rubber helix antenna, sits on a shelf to keep it centered.

Cover the inside of the shield with aluminum foil to make it reflect rf. Experiment with the shape for best directivity. To get fancy, mount a handle on the back of the shield. You should get a fairly good null off the back of this set-up. Very close-in hunting poses some problems, since there is no provision for rf attenuation.

ALUMINUM YAGIS

Commercial vhf beams, available at ham radio stores and mail order outlets, are a good way to get started hunting on 144, 220, and 420 MHz. You can transmit through them without fear. They are rugged and easy to put together. Larger yagis have a sharper pattern and more gain, but are heavier and have overhang problems when mounted on a vehicle.

A good choice for 2 meters is a four element yagi on a four foot boom, manufactured by Cushcraft (model A147-4), KLM (model 2M-4X), or others. The Cushcraft version uses a sliding gamma type match, while the KLM antenna comes with a 4:1 balun. Both match 50 ohm coax lines.

These antennas have the mounting clamps located behind the reflector for mounting on the side of a tower. Such an end mount should not be used on a car. Relocate the mounting point to the center of gravity, which can be found by balancing the antenna on your finger. For vertical polarization, use a wooden or other non-metallic mast, to avoid interaction. This mounting relocation is easiest to do when the center of gravity is between elements, not at an element—one reason why a four element 2 meter beam was recommended instead of one

Fig. 4-3. Sally Lucas, KA6SYT, demonstrates a reflector for use with a 2 meter handheld.

Fig. 4-4. WB6JPI hunts with a long KLM beam on a short car.

with three elements.

Loop the coax back behind the reflector to help preserve the pattern. Tune the matching circuit, if used, for best transmit SWR on the hunt frequency using your SWR bridge or wattmeter. Run a pattern check with a known signal, such as a repeater, to make sure that the signal peak is "on axis" along the boom.

Three or Four Elements on 2 Meters?

A three element wide spaced yagi on 2 meters has a typical forward gain of about 7.5 dB, and a four element yagi gives about 8.5 dB. Perhaps your thought is that one dB more gain isn't worth the added weight and wind load, particularly if the signal is never weak. Well, here are some additional factors to consider:

☐ The center of gravity of a beam with an odd number of elements is at or near an element, complicating the mounting on a mast. On a beam with an even number of elements, the center of gravity will be between elements.

☐ The higher the gain, the narrower the beamwidth, making it easier to take bearings in the presence of reflections and to separate the reflections.

The definition of beamwidth, and how to measure it, will be covered later. For now, it's important to note that the four element antenna, beam or quad, is noticeably narrower. A beam with more elements, if you can handle it mechanically, is even better in some cases. Figure 4-4 shows how Bob Thornburg, WB6JPI, gets ready for the All Day Hunt. He had better keep the signal ahead or behind him at all times, or he may get cited for illegal overhang (more on that later, too).

Choosing the Polarization

Using crossed yagis allows both horizontal and vertical polarization signals to be hunted without getting out of the car to switch the antenna mechanically. There are two ways to do this. The first is to put separate horizontal and vertical elements on the boom, with separate feedlines going to a coaxial switch selector. Check frequently throughout the hunt to see which switch setting gives strongest signals with fewest multipath lobes.

The other way is to use a circularly polarized (CP) antenna, which combines the signals from the horizontal and vertical driven elements. The driven elements are separated by a quarter wavelength, either physically along the boom or electrically with an extra electrical quarter wavelength of coax between one element and the combiner. Some hunters like CP yagis because horizontally and vertically polarized signals are each received with equal ease, with no switching, though at a 3 dB gain disadvantage.

For detailed information on the construction and properties of CP antennas, look in publications for amateur satellite enthusiasts. OSCAR operators use them to overcome signal fading caused by satellite tumbling. They often employ a switching scheme to select the sense of a CP antenna, which can be right-hand (RH) or left-hand (LH).

Reception of ordinary linearly polarized horizontal or vertical signals isn't affected by whether the receiving CP antenna is LH or RH. But two CP users in QSO must use the same sense, unless they are deliberately reflecting their signals off a terrain feature. Each bounce inverts the sense of the CP signal.

You probably won't need LH/RH sense switching for T-hunting. However, if the hidden transmitter operator happens to be using a CP antenna himself, and it's the opposite sense from yours, his direct signal will be rejected by your antenna, and reflections will be enhanced. (Hiders take note! This is how to foul up a CP hunter. The non-CP hunters won't have this trouble.)

QUAD ANTENNAS

A properly built and tuned quad is truly a high performance hunt antenna. It may be the only vhf hunt antenna you'll ever need. With a suitable mount it can hunt horizontally or vertically polarized signals. It excels for weak signal work, and does fine with an attenuator for strong signals. It's easy and inexpensive to build, and can cover more than one band. And of course you can transmit through it.

Hunters find that the quad has several advantages over the yagi at 2 meters:

□ For the same boom length, the quad has about 2 dB more gain than the yagi.

□ The quad seems to be less affected by proximity to the vehicle. It can therefore be mounted closer to the car roof, making it likely to hit trees.

□ For vertical polarization, it is only half as tall as a yagi (see Fig. 4-5). Again, this helps to keep it away from antenna-eating trees.

□ It can often be mounted from a right side window without exceeding the legal overhang requirements with either horizontal or vertical polarization.

Quad Building Details

You'll find that constructing quads is easy and fun. They aren't tricky, so build carefully and you can be assured of having good results. There are plenty of good designs around, and all will work. We think the following designs are better than most, because:

□ Wider spacing. Many designs call for only eight inches or so between elements on 2 meters. Though short spacing gives less weight and shorter turning radius than the following designs, it also gives significantly poorer gain and beamwidth characteristics. A short spaced four

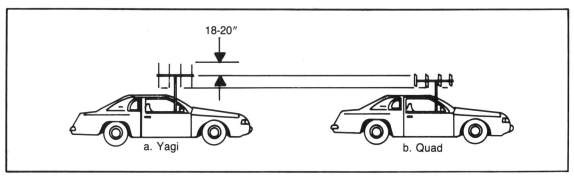

Fig. 4-5. A full sized 2 meter quad is much shorter than a similar Yagi.

element quad may be little better than a wide spaced three element quad of the same length. There is also less element interaction with wider spacing, which simplifies tuneup.

☐ Gamma match. The quad works with the coax directly connected to the driven element loop as is done in the shrunken sniffer quad in a later chapter. But the match is poor, giving more chance for feedline pickup, and transmitting SWR is less favorable. Time spent installing and tuning a gamma match is well spent.

Dimensions given here are for the upper end of the 2 meter ham band. Scale both the element sizes and spacings as required for other vhf/uhf frequencies, consulting one of the many ham antenna texts if necessary.

Before diving into the construction details, some words of caution for builders who aren't familiar with plastic pipe.

If you choose to glue any antenna elements together for a permanent setup, be careful with the glue. It sets up very rapidly once the pieces are pushed together. Coat both pieces with the glue and slide them together with a twisting motion. Be absolutely sure that the elements are aligned and level. The glue will set up in ten or fifteen seconds.

PVC pipe is easy to find at local building supply and hardware stores. Like lumber, the indicated sizes are far from exact. Tubing marked as 1/2 inch actually has an outside diameter of more than 13/16 inch, and so-called 3/4 inch tubing is closer to 1-1/16 inches on the outside. This is true for both Class 125 pipe which has thin walls and for the thick wall Schedule 40 pipe. Fittings are also marked 1/2 inch or 3/4 inch and can be used with either wall thickness. Read the markings carefully on both the pipe and the fittings when shopping. Do not buy threaded pipe or fittings.

The Strung Wire Quad

This 2 meter strung wire quad is based on a design by the late Clarence R. Mackay, K6OPS, who began experimenting with vhf quads in 1956. His work, along with the kitmaking and instructional efforts of WA6VQM, WA6TEY, KF6GQ and many others have made strung wire quads the most common 2 meter T-hunting antennas in southern California.

The mechanical configuration shown in Fig. 4-6 uses 1/4 inch fiberglass rod as the spreaders. Use 3/4 inch schedule 40 PVC pipe for the boom and mast. You will need a 3/4 inch TEE fitting and a can of PVC glue, along with the mast and boom pipe. Rod for the spreaders is available at plastic supply houses. Another source is masts from safety flags for bicycles, available in toy stores and bicycle shops. This material is great for small quad spreaders because it's very stiff and strong.

Cut each pair of spreaders about a half inch longer than the dimension S (see Table 4-1), and drill holes in each end for the loop wires. These holes should be spaced S inches apart. Set these pairs aside for the moment. The total spacing between the elements is 40.1 inches, so the boom length should be a little over 41 inches total. Trim the two boom pieces so that the boom is 41 inches or longer with these pieces installed in the mast TEE fitting. Mark the boom at each element position and remove the two boom pieces.

The actual hole positions on the boom must be offset each direction by .125 inch (assuming quarter-inch rods) so they clear each other (see detail in Fig. 4-6). Drilling of the boom is much easier if done with a drill press, but good results can be achieved with a hand drill if care is used. Drill one hole and push one of the spreaders through it. Clamp the boom section so that this spreader is parallel to the bench or drill press table. The second hole can then be drilled exactly perpendicular to the first.

After drilling all the holes, fit each pair of spreaders in the boom, making sure that the right pair is installed in the right holes and that they are centered in the boom. The loop lengths shown in the figure are exact for the spacing of the holes in the spreaders so when cutting the

Table 4-1. Element Lengths and Spacing for Strung Quad.

	LOOP LENGTH L	$\frac{L}{4}$	SPREADER LENGTH S	ELEMENT SPACING
REFLECTOR	84.1"	21.0"	29.7"	
				16.1"
DRIVEN ELEMENT	82.0"	20.5"	29.0"	
				12.0"
DIRECTOR 1	79.6"	19.9"	28.1"	
				12.0"
DIRECTOR 2	78.8"	19.7"	27.9"	

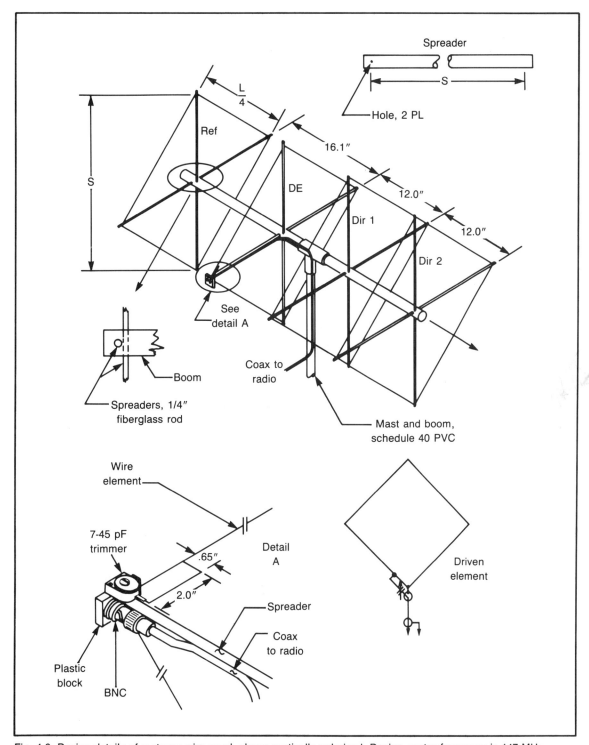

Fig. 4-6. Design details of a strung wire quad, shown vertically polarized. Design center frequency is 147 MHz.

wire allow about two inches extra to allow for twisting the ends together before soldering.

If you can find it, 22 AWG Copperweld™ wire is best for this antenna. Copperweld has a steel core with a very heavy layer of copper on the outside. It may stretch a little when you hit a tree but it's less likely to break than all-copper wire. Some hardware stores carry ordinary copper covered steel wire which may work fine at first, but the copper layer is thinner and it could eventually rust. If you use it, keep an eye on it for rust as it ages. Common 18 AWG enameled wire can be used if you can't find anything better and are willing to risk damage.

The gamma match, shown in the detail, is built on a small block of plastic. Drill a hole the size of the spreader on one side of the block and mount a BNC connector on the other. A small trimmer is mounted on the edge of the block. Glue the block on the end of the spreader and connect the center of the loop to the ground side of the connector and the center conductor of the BNC to the

trimmer. The other side of the trimmer goes to the gamma match stub. With the match on a side corner as shown, the antenna has vertical polarization.

Put the quad on its mast and get it up in the clear. Nearby objects can cause mistuning. Some hams point the antenna at the sky for SWR measurements to prevent any possible reflection effects. Connect the feed line to a SWR meter and transmitter and adjust the trimmer for lowest SWR. The shorting stub distance may need to be varied from the specified two inches to get the SWR to be 1:1. Move the stub and readjust the trimmer until you get a 1:1 SWR.

Figure 4-7 shows how to build a collapsible version of this quad. It comes apart readily for transport to and from the hunt. Each pair of spreaders is mounted on a PVC pipe coupler. Drill each coupler for spreaders as you would the regular pipe boom, and install the wire elements as before.

Cut the boom pipe into four pieces instead of two, and glue the two short pieces into the TEE connected

Fig. 4-7. Spreader detail for a collapsible version of the strung wire quad. Each pair of fiberglass spreaders is mounted in a PVC pipe coupler.

to the mast. The two other pieces go between the reflector and the driven element and between the two directors.

The notch keeps the elements lined up. Trim the boom pieces until the notches line up the elements and the element spacing is correct. File an extra set of notches 90 degrees away on the boom section connected to the mast TEE (the one that connects to the driven element). Then by pulling out the driven element, turning it 90 degrees, and pushing it back in you can easily change the antenna polarization from vertical to horizontal and back again.

If you don't trust friction to hold the quad together, use an elastic cord to hold the quad together. Sometimes called bungie cords, they have hooks on the end and are sold in hardware, bike, and automotive stores.

The Stiff-Wire Quad

Hunters who use strung thin-wire quads soon learn the most hated five words in T-hunting: "Look out for those (SCRAPE, TWANG, SNAP) trees!" The low-hanging willow is the quad's most feared enemy, and inspired the effort to find a tree-proof design.

This next model comes fairly close. While it may get bent or even mashed by a low obstruction, the damage is not likely to be permanent. After thirty seconds of straightening, you're back in the hunt again. The elements, made of AWG 10 bare solid wire, are simply bent back into shape after each mishap. Figure 4-8 shows a quad that's a veteran of years of hunts and still performs like new, even if it doesn't look new. It's just as easy to build as a strung wire quad, and has another advantage:

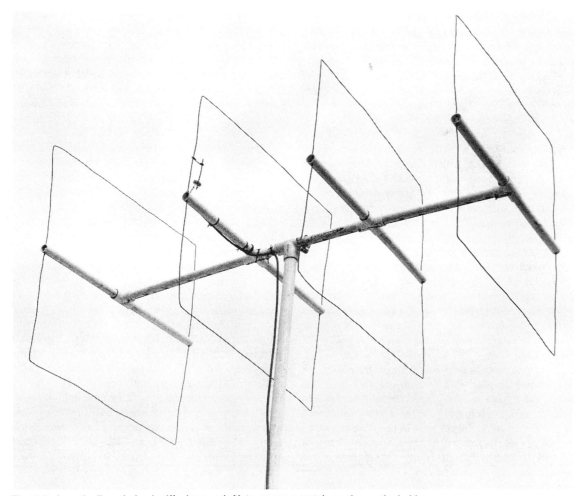

Fig. 4-8. A vertically polarized stiff wire quad. Note gamma match on the vertical side.

The wire diameter is a larger fraction of a wavelength, giving a somewhat more broadband antenna.

The stiff wire quad uses 1/2 inch Schedule 40 PVC pipe for the boom. Some weight is saved by making the spreaders from 1/2 inch Class 125 pipe. Two TEEs and two CROSSes for 1/2 inch PVC pipe are also needed, along with the special mast TEE to be described shortly. Though friction usually holds it all together, 1-1/4 inch 4-40 screws are placed through matched-drilled holes in the plastic TEE and CROSS flanges and held with nuts. This holds the sections together at any speed. The screws can be removed and the elements pulled apart for storage and transit if desired.

Construction is quite straightforward and shown in Fig. 4-9. Only the T-joint for connection to the mast is unusual. A 3/4 inch plastic TEE is split and reamed out to accept the half inch boom down the through arm (Fig. 4-10). It should be balanced for the exact center of gravity. One or two hose clamps are then tightened down enough to prevent sliding, while allowing rotation to select horizontal or vertical polarization. The drawing shows vertical polarization, but the entire boom can be rotated 90 degrees in the tee for horizontal polarization. Use care in routing the feed line to allow the rotation.

Make the spreader pieces for each element first. Cut and try them for size, allowing a quarter to a half inch beyond the wire at each end. (Alternately, you could make all of the spreaders 22″ long overall after cementing, and trim them after assembly of the boom.) After gluing the spreader pieces to the four pipe fittings, carefully measure the spacing needed for the wire loops. This is the distance L/4 shown in the figure. It is different for each element so be sure you have the right spreader/fitting assembly before the pipe is marked.

Though AWG 10 solid copper wire is called for, AWG 12 solid house wire can also be used. Take off any insulation, which would add weight and increase wind loading if left on. Drill holes through the spreader pipes that just clear the wire used for the quad elements. Make sure that the holes are drilled squarely so that the elements fit properly.

Cut each element loop an inch longer than called for in the figure (for a half inch overlap) and mark the points where the corners will be. Thread the wire through the holes in the spreaders and bend the wire at the marked spots. Before soldering the joint in the element (which should be inside a spreader), wrap small diameter bus wire around the joint to make it stronger. Wrap loops of bus wire around the element on each side of the spreaders and solder to hold the elements in place.

Cut the three boom pieces and try them for size. Keep trimming them until the element-to-element spacing is correct. Chances are the boom pieces will fit tightly into the cross fittings and friction will hold the elements together. One way to be sure and allow disassembly is to put screws through the cross fitting and mast at the six junctions, as mentioned earlier.

Make the gamma match as shown in the detail using another piece of #12 wire. The trimmer is supported by the stiff wire of the gamma match and is soldered to the center conductor of the coax. The shield of the coax is soldered to the center of the loop. Adjustment of the match is done with your SWR meter and transmitter, just as with the strung wire quad.

Helpful Hints

You will be spending a lot of time with your new quad, and you'll want trustworthy readings, so take the time to tune it up right and learn about its pattern. Practice with some known close-in targets to see what the normal beamwidth is, and where the minor lobes are. Mark the elements and radiation direction on the PVC with an indelible marker to ensure that it will always be assembled and put on the mast correctly in the rush before the hunt.

Boresight the antenna to verify that the pattern maximum is right along the boom axis. This can be done when receiving a known signal in the clear, or while transmitting, using a field strength meter 100 feet or so away, read with binoculars. In either case, swing the quad to peak the signal, then sight along the boom to check for off axis response.

If the response of the strung wire quad isn't on axis, try redistributing the wire in the driven element to correct it. For example, if the major lobe is biased to the right, slide about an inch of wire from the left side of the vertical spreader to the right side, and check it again. With a properly tuned gamma match, feed line pickup should not be a problem. If you suspect feed line pickup because of shallow nulls on the sides, try a simple sleeve balun and see if it helps. Such a balun, sometimes called a bazooka, is an electrical quarter wavelength long and is shown in the Shrunken Quad diagram in Chapter 12.

A mast of 3/4 inch Schedule 40 PVC pipe will take a lot of abuse, but it can flop around a lot. A more sturdy mast can be made from 3/4 inch Class 125 PVC pipe with a broom handle slid down inside. It's a tight fit but it works well, making a nice stiff non-metallic mast, with no chance of splinters.

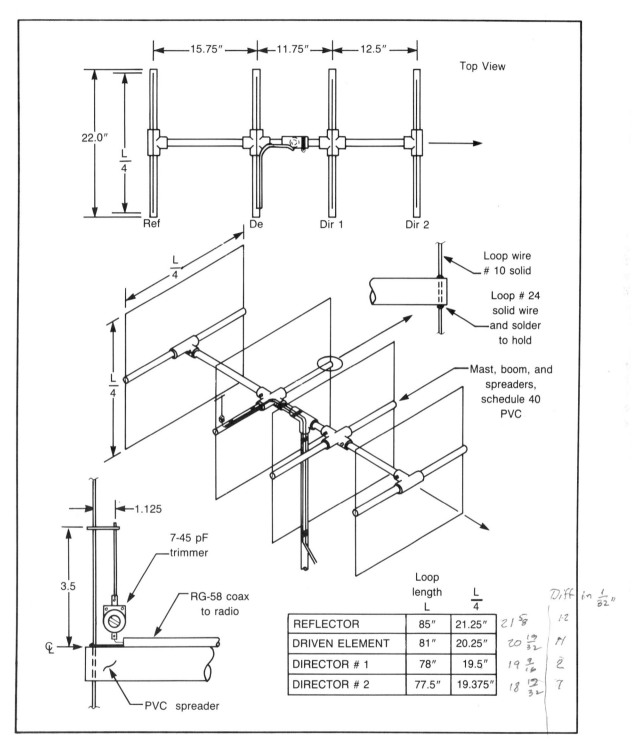

Fig. 4-9. Construction details of the stiff wire quad, shown vertically polarized.

The table within the figure:

	Loop length L	L/4		Diff in 1/32"
REFLECTOR	85"	21.25"	21 ⅝	12
DRIVEN ELEMENT	81"	20.25"	20 19/32	11
DIRECTOR # 1	78"	19.5"	19 7/16	2
DIRECTOR # 2	77.5"	19.375"	18 12/32	7

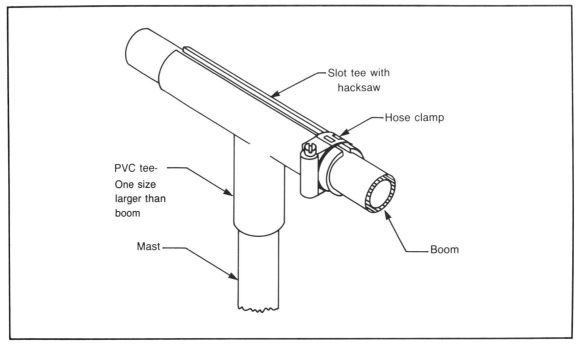

Fig. 4-10. Detail of the center mounting TEE on the stiff wire quad.

At highway speeds, a beam or quad can be hard to control, particularly if the mount has been designed to be easy to turn. Most of the time the signal will be straight ahead, so the antenna should be made to "home" in that direction. To do that, put a vertical fin of plexiglas sheet on the boom at the reflector end. The antenna will then "weathervane" to the forward-facing direction. The fin should be no bigger than necessary. Try 4″ × 4″ for starters.

Creative Configurations

Several avid hunters have built dual-band strung quads, with both 144 MHz and 220 megahertz band elements on the same spreaders. The result is a short spaced 2 meter antenna with near optimum spacing on the 1-1/4 meter portion. Separate feed lines, with sleeve baluns, are used for each driven element. The feed line for the driven element not in use should be terminated with a 50-ohm resistive load. (This presumes you only hunt on one band at a time!) There is still interaction of course, but it is tolerable.

Performance won't be affected if the elements are in a square, as in the Fig. 4-11A, or a circle, as in Fig. 4-11B, so long as the feed point is in the right place for correct polarization. Furthermore, there is nothing sacred about the square element shapes used on the quads in this chapter. A few hams have had success with circular elements. A pentagon or hexagon format is just as feasible technically, but offers no particular advantages. Go ahead and experiment.

Remember that the circumference of each element determines its resonant frequency, not the diameter. You will notice marked feed point impedance changes if you "squish down" the loops into a rectangular shape. The extreme case of this is a folded dipole element (Fig. 4-11D), which has an impedance of about 300 ohms, and is used as the driven element of some yagis.

One really eye-catching configuration is the triangular shape, properly called the delta loop or delta quad. Antenna experimenter Walt Brackmann, WA6SJA, really likes this type. He mounted the elements on the side of a bar and fed it for vertical polarization. Since it looks like something you'd cook Thanksgiving dinner in, he is known all over southern California for his "turkey rack."

MEASURING BEAM AND QUAD PERFORMANCE

Forward gain is the specification that is always compared when hams judge various gain antennas against each other. Gain has a lot of importance for DXers, but it's only of secondary importance in judging a T-hunt antenna. For hunting, other electrical specs that count heavily are:

☐ Beamwidth
☐ Sidelobe level
☐ Front-to-back ratio

For complete information about the meaning of these specs, and how to optimize them in a particular design, consult a good book on amateur radio antennas. For this discussion, just remember that the beamwidth is the angle in azimuth between the 3 dB points of the pattern (see Fig. 4-12). Signal voltage is 70.7 percent of maximum at the 3 dB points. The term half-beamwidth refers to the angle to the 3 dB point with respect to the direction of maximum radiation. The sidelobe and front-to-back ratios are the levels in dB with respect to the main lobe to which signals from the sides and rear respectively are suppressed.

It turns out that beamwidth is directly related to gain. That's easy to understand when one realizes that an antenna doesn't amplify rf signals to give gain. What it really does is concentrate the rf energy in a specific direction or directions. Given an equal amount of energy, the more concentrated the signal becomes, and hence the narrower the beamwidth, then the higher the gain.

Hams find it tedious to measure gain directly, because it requires special reference antennas and some sort of pattern range. Antenna pioneer John Kraus, W8JK, has derived a math formula that directly relates gain to beamwidth. Measuring beamwidth is easier, but requires an accurate indicator of the 3 dB points on the main lobe. The calibration of S-meters and attenuators most hams use isn't good enough to measure 3 dB points with a great deal of accuracy.

Some clever hams (Gunter Hoch, DL6WU, and Joe Reisert, W1JR) have made things easy by determining that a good estimate of beamwidth can be gotten by measuring the nulls at the edges of the forward lobe, instead of the 3 dB points. The nulls are easy to find and this method requires no special calibration. It turns out that the 3 dB point in each direction is at about half (47.5 percent, to be exact) the angle to the null.

All we have to do is find the azimuth swing between

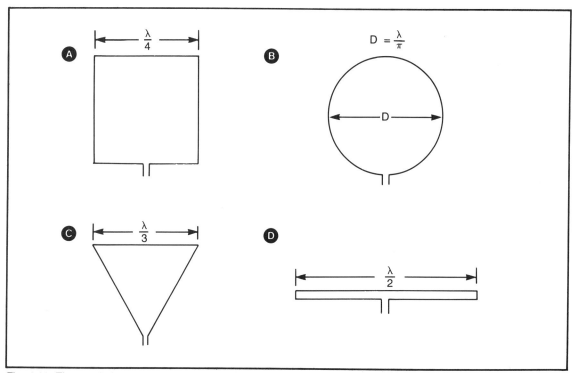

Fig. 4-11. The quad elements can be configured in many ways. These driven elements are all configured and fed for horizontal polarization. Each is a full wave around. Note that the feed impedance changes with the different shapes. No matching networks are shown.

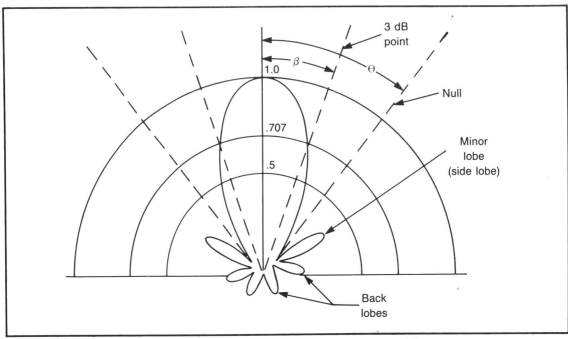

Fig. 4-12. Typical antenna pattern showing 3 dB points, nulls, sidelobes, and back lobes.

the nulls, and simple calculations give both the beamwidth and the gain. With that in mind, here's the procedure for evaluating your vhf T-hunt antenna.

1. Find a clear area, with no large nearby buildings or other reflections. A local repeater or a friend's station can be used as a signal source as long as signal is good and the path between source and receiver is unobstructed. A circularly polarized test signal is best, but if that's unavailable, reconfigure antennas as required so that the source and receive antennas have the same polarization.

2. Aim the antenna directly at the source. Note your azimuth indicator reading. This is the "boresight" direction. Use an attenuator before the receiver to reduce the signal level if the receiver S-meter is pinned.

3. With both antennas set for vertical polarization, rotate the antenna first to the left until the signal is in a null. Note the number of degrees moved in azimuth. Now find the null on the right side, and note the number of degrees.

4. The left null and the right null should be equidistant from the boresight direction. If not, the antenna is asymmetrical. A few degrees of asymmetry is not worth worrying about, but if the asymmetry is large, look into it.

5. Take the average of the left and right swings to null. This will be a_V in the formula.

6. Reconfigure both the test and source antenna for horizontal polarization. Repeat steps 3 and 5. Average the swings to find a_H.

7. Sidelobes and front-to-back ratio can be measured using your attenuator to estimate sidelobes and rear pickup level. Swing the antenna to find the worst sidelobe and set the attenuator for half scale S-meter reading on it. Now swing the antenna to boresight and put in additional attenuation until the S-meter reads the same as it did on the sidelobe. The added attenuation is the same as the antenna's sidelobe suppression. Rear lobes can be evaluated in the same way.

8. Assuming that the highest sidelobes are 20 dB down, the following formula can be used to estimate the antenna gain:

$$G = 10 \log \frac{38781}{a_H \times a_V}$$

where G = gain in dB
a_H = average angle of null each side, horizontal polarization.
a_V = average angle of null each side, vertical polarization

As an example, assume a hypothetical four element

40

$$G = 10 \log \frac{38781}{80 \times 75} = 10 \log 6.4635$$
$$G = 10 \times .8105 =$$
$$G = 8.1$$

antenna. The data on the angle of the nulls is as follows:

POLARIZATION	DIRECTION	ANGLE TO NULL
Vertical	Easterly	79
Vertical	Westerly	81
Horizontal	Easterly	77
Horizontal	Westerly	73

So a_V is 80 degrees and a_H is 75 degrees. Putting these values into the formula gives a gain of 8.1 dB.

For worst sidelobe levels other than 20 dB, add a correction factor to the computed gain figure as follows:

SIDELOBE LEVEL	CORRECTION
− 30 dB	+ 0.71 dB
− 25 dB	+ 0.40 dB
− 15 dB	− 0.31 dB
− 10 dB	− 0.67 dB

If our sample antenna had − 25 dB sidelobes, its gain would be 8.5 dB.

Be warned that the results of sidelobe and back lobe measurements are highly dependent on the accuracy of your attenuator, the impedance match to the line at both the antenna and the receiver, and any reflecting objects in the near and far field of the antenna. The formula also does not account for any inefficiency in the feed system. Take all these measurements with more than a grain of salt.

It is possible that your antenna may have such bad asymmetry that the peak response is not in the boresight direction. Here are some suggestions to help solve this problem:

☐ Change the matching system. Some gamma matches and other unbalanced feed systems can skew the pattern.

☐ Add a balun to decouple the feedline from the antenna.

☐ Re-route the coax path from the antenna. It should not follow close to an element, but instead depart from the driven element at a right angle.

PHASED ARRAYS

All of the gain antennas so far have used only one driven element. Directivity has been achieved by parasitic elements—elements not connected to the feed line.

Another useful way to achieve gain (and nulls) in a multi-element antenna is to control the phase of the radiated wave in the far field from two or more elements. This is done by adjusting both the phase of the rf signal to each element and the spacing between elements.

The prime example of a phased array is the Adcock. Two dipoles fed opposite polarity (180 degrees) will have a sharp null looking between them, and some gain off the ends. The spacing is not critical for null depth. At one half wavelength spacing, the antenna has the classic figure 8 pattern.

As spacing is increased, the pattern is distorted and more lobes appear. At one wavelength spacing, there is a second set of nulls appearing in the line joining the two elements. At two wavelengths spacing there are null about every 45 degrees around, and lobes between. At three wavelengths there are 12 lobes and 12 nulls, at four wavelengths there are 16, and so on . . .

An antenna with a single lobe and a single null is usually of most interest to DFers. A unidirectional phased system is shown in Fig. 4-13 for 2 meters. Each half of the dipole is 19-1/2 inches, the spacing is 20 inches, and the coax line length is 40 inches. Dimensions are somewhat more critical for a deep null. The antenna could be built with ground planes instead of dipoles.

This unidirectional antenna will not present an ex-

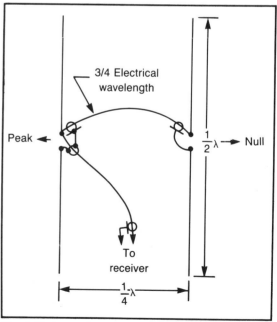

Fig. 4-13. A two element phased array. This antenna has a single null with a broad peak 180 degrees away.

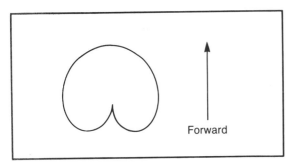

Fig. 4-14. Typical pattern of the ZL Special.

act match to 50 ohm coax due to the paralleling of elements. It will work for receiving, but isn't intended for transmitting. Some more advanced uses of phased antennas, both receiving and transmitting, are covered elsewhere in this book.

THE ZL SPECIAL

For those who want a 2 meter directional gain antenna that is smaller and easier to turn than a four element quad, this two element phased array may be of interest. It's called a ZL Special, and has been popular with many hams as a simple DX band beam.

Besides its smaller size, the ZL Special has the advantage of excellent front-to-back ratio (about 20 dB). The null in back is quite sharp, and is the best way to get a highly accurate bearing with it. The forward lobe is very broad, as seen in Fig. 4-14, but it can be used for hunting when the signal is too weak to hunt with the null.

Users report an apparent sharpening of the forward pattern when receiving weak signals. The word "apparent" should be stressed here, because the pattern doesn't actually change with signal strength. It appears to sharpen up on weak bunnies due to threshold effects in the FM receiver.

The classic ZL Special design is two folded dipoles spaced 0.1 wavelength apart and driven 135 degrees out of phase, as shown in Fig. 4-15. Perhaps the easiest way to make a mobile ZL Special was the method used by Jane Rice, AD6Z. TV type 300-ohm twin lead was used for both the folded dipoles and the phasing line. The twin lead was mounted to a wooden frame for light weight.

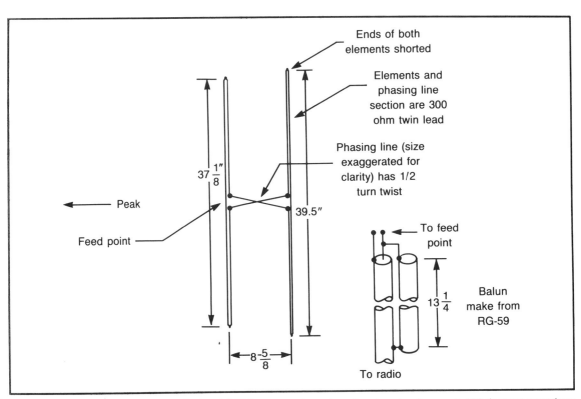

Fig. 4-15. Electrical information for a ZL Special. This antenna has a broad peak and a sharp null, 180 degrees away from each other.

If you build a ZL Special or other phased array system using this method, here are some suggestions:

□ If at all possible, get good quality twin lead with AWG 20 or 22 conductors. Most hardware store twin lead uses 24 gauge wire, which has more loss. You needn't use the foam type, however.

□ A balun of some sort is vital for good null characteristics. A simple sleeve balun like the one in the diagram is suitable. Route the coax carefully at right angles to the driven element.

□ Put a dual direction pointer on the mast which readily indicates to the antenna turner which side of the pointer corresponds to the peak and which side to the null.

□ Despite its low gain, an attenuator will still be needed for close-in hunting, as strong signals will fill in the null if there's no attenuation.

Null-hunting antennas such as these are more prone to proximity effects than ordinary gain antennas. The effect of the car body, the coax feed, and other nearby metal is to partially fill in and skew the nulls. AD6Z and Don Barrett, KA6DJK, have found that tilting the antenna about 30 degrees from vertical appears to minimize the proximity effects, and is a good compromise for hunting both horizontally and vertically polarized signals. Turn-ing radius is increased about 50 percent, but is still much less than that of a four element quad. The tilt angle may also have an effect on the feed point impedance.

Don't assume that just because the SWR is fairly constant over the band, the directional properties are too. Check to be sure. Before hunting a fox with your phased array system, hunt some known signal sources. Be sure that the null is exactly in the expected direction, and that the peak is exactly opposite, at your frequencies of interest.

The vhf ZL Special antenna may be built for bands other than 2 meters, but some experimentation is necessary to get the feed point impedance to be optimum. At 1-1/4 meters, it can be small enough to be an outstanding sniffer antenna for foot work, and even at 2 meters is comparable in size to switched antenna units for foot hunting.

This ZL Special and other phased antennas are easy to build, and their light weight and small turning radius when mounted vertically or diagonally make them easy to install on a passenger car. Of course the small size does not come free. The price is reduced gain—only 3 dB or so forward gain compared to about 8.5 for a good four element beam or quad—and reduced sharpness of the peak. These are not the antennas to use if you know the signal is going to be very weak.

Chapter 5

All About S-Meters

Good signal strength indications are vital for consistently successful T-hunting. It's true that you may be able to find a fox out in the open with just a directional indicator and no S-meter. But you'll find it much sooner if you have strength information and use it properly. A vhf hidden T concealed in an area of multiple nearby reflections may be impossible to find with azimuth indicators alone.

A good S-meter serves two purposes:

☐ With a rotating directional antenna, the S-meter indicates the pattern peaks and nulls, and thus the direction to the signal.

☐ It tells, in a relative way, that you're getting close to the hidden T. Carefully used, it may even predict the distance.

On FM, it's very difficult to judge the strength of a signal by ear once it's strong enough to quiet the receiver. You need a meter or other indicator. This chapter will cover the use of S-meters and the addition of metering circuits to rigs that are meterless.

A good S-meter is:

☐ Sensitive. Don't expect meter indications on very weak signals, but a 90% quieting signal on FM should move the meter off zero. Hf S-meters should be sensi-

tive enough to see atmospheric noise.

☐ Linear. For good hunting, the peak to null ratio of the antenna pattern should represent a sizeable meter deflection. Easy pinning is not likely to be a problem, as it can be eliminated with an attenuator. Some AM and SSB rig meters have a logarithmic scale which expands the dynamic range as much as 80 dB. That's fine for a base station, but hard to hunt with in a car.

☐ Consistent. The meter reading should not be drastically affected by car battery voltage variations and temperature changes. Extra voltage regulation may be desirable.

☐ Properly damped. An overdamped meter circuit responds too slowly to allow rapidly swinging the antenna. At the other extreme, a meter which flickers and bounces wildly may also give poor results. A meter which sticks mechanically is worthless and should be replaced.

☐ Easy to see. The meter should have a large scale, be well-lit, and be mounted so that it is easily seen by driver and antenna-turner alike. (If the rig is under the dash, consider a remotely located meter, described later.)

VHF-FM S-METER CIRCUITS

Figure 5-1 shows a typical S-meter circuit found in

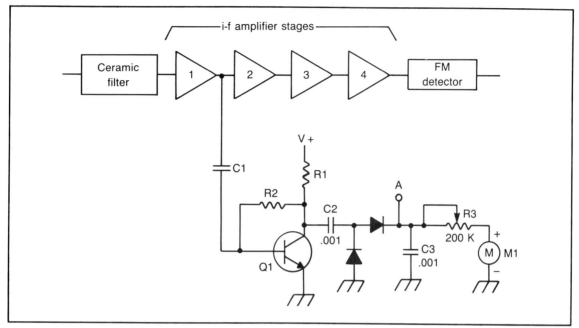

Fig. 5-1. Schematic of the classic VHF-FM receiver S-meter circuit showing tap points.

many solid state vhf-FM rigs. The i-f signal is sampled after the first or second stage, at a transistor collector or drain. Q1 provides some amplification, but is also important because it keeps the diodes from loading the i-f stage. R2 provides bias to set the collector voltage of Q1 at about half the supply voltage, with about 1 mA of collector current. Typical values are 5.6 k for R1 and 200 k for R2.

Such a circuit is simple to add to a meterless rig. The meter should be a sensitive one—get a 50 microampere movement if you can. Use germanium diodes such as 1N60, and a silicon rf transistor such as 2N3563, 2N2222A, or 2N3904. R2 may have to be changed to get half the supply voltage at the collector.

Why is the takeoff point after the first i-f stage? Of course it's important for it to be after the crystal filter to avoid reading other strong signals in the band, but why not put it at the i-f output for greater sensitivity? Try it and see. Chances are you'll find that the gain is so high that receiver front-end noise will cause significant upscale reading on the meter. Disconnect the receiver antenna, put a 50-ohm resistor across the receiver input, and move the tap point back a stage at a time until the noise no longer moves the needle. It will probably be at the point shown.

Potentiometer R3 is the full-scale adjustment. The FM i-f stages successively saturate on strong signals (this

is called limiting), and the saturation point of the stage ahead of the tap point determines the maximum S-meter reading, which we want to be full scale without pinning.

Connect a high impedance dc voltmeter to point A. Put an on-channel signal into the receiver, from an adjustable signal generator if possible. Increase the level until the voltage reading stops increasing. Use caution, because further increases in signal level may make the voltage decrease somewhat. Read the voltage at maximum. The voltage should be approximately equal to the dc collector voltage of Q1.

The target value of R3 is given by this voltage divided by the full scale current of the meter movement. For 6 volts and a 50 microampere movement, R3 is 120 k ohms. Use a 200 k potentiometer and adjust for full scale on the S-meter.

Purists will note that this biasing arrangement of Q1 may not be optimum for constant gain over a wide temperature and voltage range. Chances are the variations in the gain of the receiver rf and i-f stages are even greater, and mask any variation in Q1. If not, try putting in a small emitter resistor and using fixed bias per standard practice for the design of common emitter amplifiers.

EXTERNAL S-METERS

When the rig is under the dash, it's hard for the driver and all passengers to keep an eye on it. It's easy to add an additional meter with a large scale up on the dash in plain view. Check the local surplus shops and flea markets for a suitable meter, the larger the better.

The best movements are 100 microamperes or less for full scale deflection. The added meter should have sensitivity equal to or greater than the internal meter. A back-lighted meter would be truly first class, but any dash-mounted meter can be illuminated with a light mounted on the meter box, or even by the car's map light or dome light.

Some receivers have a "limiter test point" or "i-f test point" used for alignment. This may be a good place to directly connect a sensitive meter. Check for good sensitivity and range at that point before making a permanent installation.

If the receiver already has an S-meter, the quick-and-dirty method of connecting an external meter is to parallel the outside one (M2) and its dropping resistor (R4) across the internal meter and resistor, as shown in Fig. 5-2. Bring the lead out of the rig via the accessory jack, if supplied. Use shielded cable to help keep strong rf signals from sneaking in this way. The rf choke and capacitor help keep rf out, too.

The use of separate full-scale setting potentiometers allows using non-identical meters inside and outside with the same deflections. Note that the external meter will not properly display the transmitter's metering information, but that's not important during hunting anyway.

There may be some loading of the internal meter circuit by the external meter. With the strong signal source set for maximum reading as before, R3 and R4 should each be readjusted for full scale on the respective meters.

If you plan to disconnect the external meter when not hunting, it's probably better not to change the R3 setting, as the internal meter would then read too high when unloaded between hunts. Alternately, you can use a miniature phone jack for the external meter input, wiring it so that when the external meter is plugged in, the internal meter is disconnected.

THE AMPLIFIED EXTERNAL METER

Perhaps you found the perfect meter, but it has a one milliampere movement. How can it be used? With a little additional work, you can whip up a custom meter amplifier circuit that can drive it, and has several other features:

☐ It doesn't load the rig's circuits. You can use it and the meter or bar graph display in the rig simultaneously.
☐ It can display both transmit and receive readings.
☐ Damping can be controlled.
☐ It can drive an audible indicator, as described later.

The heart of this deluxe metering system is a CMOS operational amplifier (U1) used as a dc amplifier. The CA81E is shown, but most other field effect transistor (FET) input types will work. Try the TL081 (RS 276-1716). The circuit of Fig. 5-3 shows components selected for a real find — a surplus 200 microampere long-scale meter. This movement rotates 250 degrees instead of the usual 90 degrees, resulting in a scale almost three

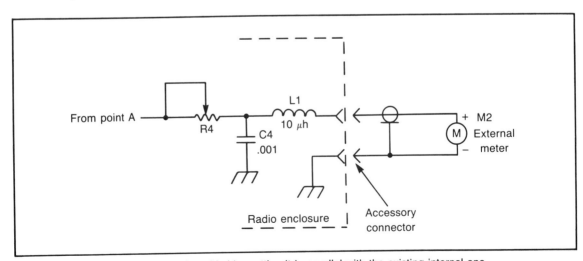

Fig. 5-2. An additional S-meter can be added by putting it in parallel with the existing internal one.

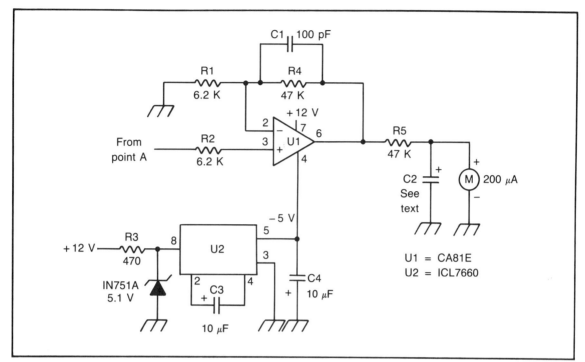

Fig. 5-3. An op-amp can be used as a meter driver.

times the normal length for a 3-1/4 inch meter. There's a photo of this meter box in Chapter 11.

This movement had such high inductance and back EMF that it could not be connected directly across the internal meter without the action being so sluggish that it took several seconds to rise from zero to full scale. That's unacceptable for rapidly swinging the quad!

With the amplifier, the large series resistor (R5) allows very fast response. We prefer it this way, to be able to spin the quad around. You may find your meter overshoots or flickers too fast and that some added damping is desirable. If so, there are two tricks to try. You can add capacitance across the meter. Do not add it at the amplifier output. Put it at the meter at C2, as shown in the schematic. A large value may be required. The other way is to add a resistor in parallel with the meter movement (in place of C2) and decrease the value of R5. With a little trial and error, you can have just the amount of damping you like.

Some customizing is desirable as you build a meter amplifier for your particular receiver. R1 and R4 set the gain and are selected according to standard operational amplifier design procedure such that the voltage at U1 pin 6 is about +10 volts when maximum signal (representing full scale) is present at the circuit input. R1 and R2

are made equal to minimize the effect of the IC input offset. Maximum value of R5 is a function of the meter full scale current (I_m).

$$R4 = \frac{10}{I_m}$$

For example, if M1 is a 1 milliampere movement, R5 becomes 10 kilohms maximum. R5 can be made variable for an adjustable full scale setting to match the internal meter. The Intersil ICL7660 (RS 276-2335) used at U2 is a remarkable little 8-pin IC. Give it +5 Vdc, and out comes −5 Vdc, at up to 10 mA. Just be sure not to exceed +6 volts at the input when connected as shown. It's used for a number of projects in this book. A 1N4733 (RS 276-565) can be substituted for the 1N751A 5 volt zener diode.

LINEARITY EVALUATION

If you have access to good rf test equipment, make a calibration curve of your receiver/S-meter calibration. Set the receiver at mid-band, and the generator to that frequency. Then increase the level and note the rf input level for each scale mark on the meter face.

47

S-meter performance of four typical 2 meter FM receivers and one AM/SSB/FM receiver has been graphed in Fig. 3-4. Measurements were made with a Hewlett Packard model 608D signal generator. The meter scale markings are 0 through 10 in all cases except the IC-25A, which has a seven segment LED display. Low-scale sensitivity is about the same for each of the receivers. Any of the FM-only receivers works well for quad hunting, but the TR-33C has the most linear response (and the tiniest meter).

The VHF-1 S-meter circuit we tested saturated before full scale was reached. Its scale was very compressed above "7." Note that this can happen on other receivers, and also may not be representative of all VHF-1's. There are three ways to deal with this problem, any one of which may suit your needs:

☐ Use caution to keep the meter below "7" at all times when hunting, using the attenuator.

☐ Readjust the S-METER ADJUST pot (R125 in the VHF-1) in the receiver for full scale reading at 5 microvolts. (It will pin above this.)

☐ Add an external meter, adjusted for greater sensitivity.

Due to misalignment at the factory, improper servicing, or tweaking by a previous owner, your receiver S-meter adjustment may also be incorrectly set. Take the next opportunity to check it on a good calibrated source.

LED METERS

The sluggishness and bounce problems of mechanical meter movements can be eliminated with an all-electronic indicator. "Tuning eye" tubes were once commonplace on ham gear. Made in both circular and bar configurations, they were inexpensive, instantaneously responsive, and easy to see in the dark. With the advent of solid state low voltage circuitry, they have almost disappeared.

Today's most common electronic indicator is the light emitting diode (LED) bar graph display. It has become competitive in cost to mechanical meters. It is easy to see in the dark, though it may be washed out in direct sunlight. With care in selection of circuit time constants, it has a very fast response time.

As signal strength increases, more LEDs in the display light in succession from left to right. The circuitry required to accomplish this successive illumination is rather complex. In effect, each LED must have its own threshold circuit. Fortunately, such a complex scheme is easy to implement in an integrated circuit. Many imported rigs use the TA7612 IC, which drives seven LEDs at less cost than a mechanical meter.

Most hams find that their LED meters are usable for T-hunting. The biggest problem is lack of resolution. It may be difficult to determine the exact signal peak or null direction with only seven LEDs, the typical number provided.

Adding an LED Meter

Perhaps you'd like to try a bar graph meter on your receiver. It can easily be read at night out of the corner of your eye. It responds instantaneously as you swing the beam. It can replace the present meter, or be added on externally while leaving the existing meter in operation.

The LM3914N IC, made by National Semiconductor, is perfect for this application. This 18-pin DIP packaged part has all of the comparators and drive circuitry for a 10-LED display. No current limit resistors for the LEDs are needed, as the outputs are current limited. LED brightness is easily controlled.

The LM3914N (RS 276-1707) and a 10-LED bar graph display (RS 276-081) should be readily available and inexpensive. You can also make your own display from individual LEDs. You might try using more than one color to make reading the display easier, as is done on some commercial rigs.

The high input impedance of the LM3914 allows it to be used in parallel with the present meter in your rig. If the rig has no meter at all now, just add a meter rectifier circuit to the i-f stages as in Fig. 5-1, ahead of the LM3914.

If 10-LED resolution isn't enough to satisfy you, more LM3914's can be chained together. A 100-LED display is easy! Display linearity is no problem. With the internal comparators and precision reference, the IC is far more linear than the signal you'll drive into it.

Figure 5-4 shows how the bar graph LED display was installed in place of the panel meter in a Drake UV-3 FM rig. The existing meter was sticking and thus useless for hunting. The bar graph indicator supplements the external long-scale meter described earlier.

Figure 5-5 is the schematic of the bar graph add-on, which can be built on a small piece of perf board to fit inside the radio. The value of R3 determines the current in each LED, which is 21.5 mA for the value given. This much current is necessary for the chosen 10-LED display to be visible in daylight through the smoked plastic bezel. A home brew display of super-bright LEDs would reduce this current requirement.

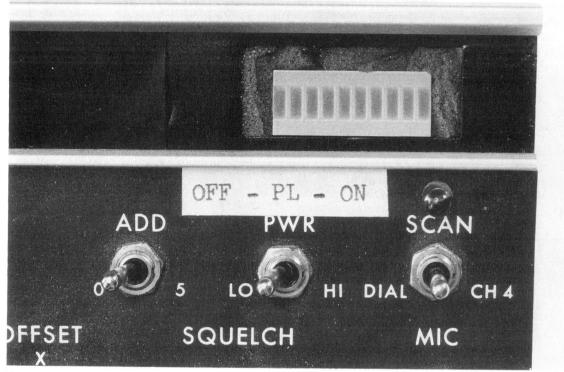

Fig. 5-4. An LED bar graph S-meter installed in a Drake UV-3 control head. The new S-meter was installed in place of the old mechanical meter.

To prevent oscillations, keep all leads as short as possible. Don't omit supply bypass capacitors C1 and C2, and utilize a single-point grounding technique. By connecting all circuit return wires to a single point near pin 2 of the IC, high ground currents don't develop signal voltages to couple back into the input and cause feedback. Current limiting for the LEDs is provided within the LM3914N.

The regulator (U1) and voltage dropping diodes (D1 and D2) keep the U2 power dissipation from becoming excessive. Reducing the LED voltage source reduces voltage drop across each LED driver. R1 reduces the dissipation in U1 and allows the TO-5 package version (LM309H) to be used, provided a clip-on heat sink is attached. Both R1 and U1 get quite warm in a vehicle, where the supply voltage can approach 15 volts.

The voltage range required at the U2 input for zero to ten LEDs lit is 0 to 1.25 volts. Potentiometer R2 is used to adjust the IC input to this level. The input of the board is connected to the meter rectifier output (Point A in Fig. 5-1). There should be no effect on the operation of the original meter by tapping there. If the original meter is removed, an equivalent resistance should be substituted for it.

Setting R1 is simple. Find a signal that's full scale on the internal meter. Set R1 to the extreme counterclockwise point (no LEDs lit), then bring it up slowly until the tenth LED just lights.

Using an LED Meter

Even if you already have a large mechanical meter, the bar graph makes a worthwhile addition. It is instantly responsive without overshoot and can grab signal levels on very short transmissions. It also tells a lot about signal quality. Fast flutter, for example, shows up in flickering of the bar.

The LED indicator really shines (oops, pardon the pun) in its ability to identify high noise levels. In hunting a weak signal, a mechanical meter may bounce upscale when passing through an area of high pulsed electrical noise, even though the hidden signal may not be strong enough to move the meter by itself. The pulsed noise could be from a power line or a neon sign, for example.

Swinging the antenna toward the noise will increase the reading, giving a false bearing. An experienced hunter might be suspicious, because the hidden signal will not quiet when the antenna points to the noise.

The bar graph display diagnoses this condition immediately. The hidden T's carrier will make the LEDs light in succession as the signal gets stronger. Only the right-most light in the bar may be dimmer than the ones on the left. Pulsed noise, on the other hand, gives a display with varying intensity across the bar, as Fig. 5-6 illustrates. This happens because the noise pulses are above the IC comparator thresholds for varying lengths of time. The lower the threshold, the longer the time. By checking the bar graph display regularly, you'll keep from being snookered into hunting QRN instead of the bunny.

The LED display does not show noise in this way if the input is damped or rolled off. Some commercial ham rigs have large electrolytic capacitors at the input of the bar graph IC. Experiment with reducing the value of this capacitor to speed up response.

By proper connections to pin 9 of the LM3914, the display becomes a moving dot (one LED at a time, with soft transitions) instead of a growing and shrinking bar. The National Semiconductor application note should be consulted before trying this, particularly if you plan to chain the ICs. The dot display will have lower power dissipation, but it's hard to read a dot display at night in a dark vehicle. You'll be constantly trying to figure out, "Where's the dot, to the left or right?"

AN AUDIBLE SIGNAL STRENGTH INDICATOR

For safer and more efficient hunting, put an audible signal indicator after the meter amplifier just described.

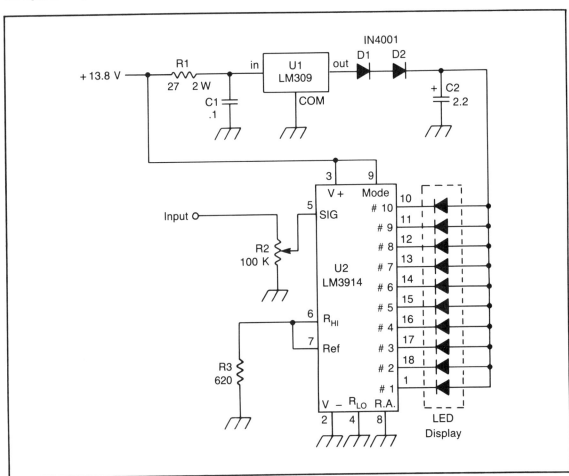

Fig. 5-5. The LED bar graph S-meter. Bar graph and driver are built on a small piece of perf board.

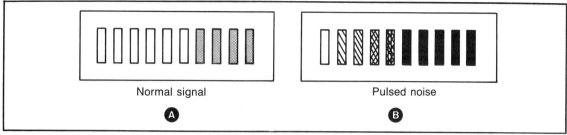

Fig. 5-6. The LED S-meter can distinguish between normal signal (A) and high ambient noise (B).

The driver can keep eyes on the road and the navigator can keep eyes on the mast pointer and the maps. Figure 5-7 is the schematic of this simple add-on. The drive voltage comes from pin 6 of the operational amplifier in Fig. 5-3.

With no signal and the meter at zero, no sound is heard. As the meter moves up a tone begins, rising from a low buzz to a high whine at full scale. The 555 is a universal timer IC, widely available (RS 276-1723).

S-METERS FOR HF AM/SSB RECEIVERS

Unlike FM receivers, AM and sideband sets cannot use limiting techniques. The signal level to the detector stage is optimized by an automatic gain control (AGC) circuit. While the presence of AGC makes the i-f S-meter

tap-off impractical, the AGC voltage itself makes a convenient strength indicator.

AGC techniques and metering vary widely from receiver to receiver, making an individual design almost mandatory for added or external S-meters on hf receivers. Use a variable level signal generator to drive the receiver input while measuring the voltage on the AGC line. Note the polarity, the voltage with no signal input, and the voltage with a very strong signal. You may then be able to come up with a simple circuit that works over the range in your receiver. AGC conditions in receivers can be described as follows:

A. Negative AGC, voltage becomes more negative with increasing signal.

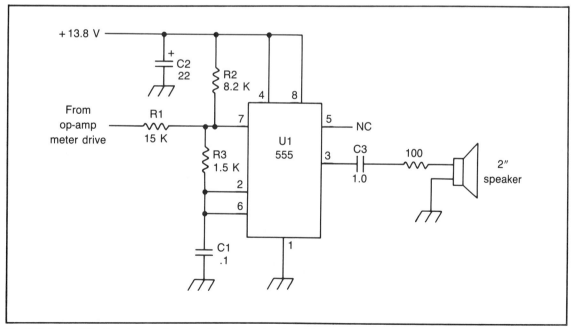

Fig. 5-7. An audible S-meter can be added in place of or in addition to the existing external S-meter.

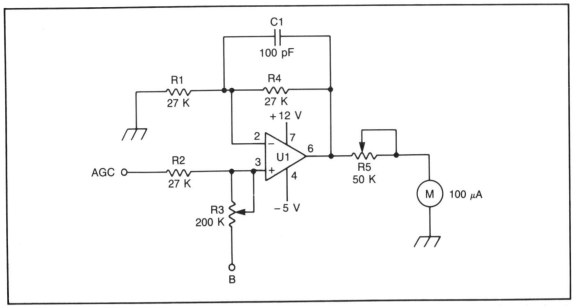

Fig. 5-8. A universal S-meter circuit for hf receivers. This circuit accommodates either a positive or negative going input. See text for proper connection of point B.

B. Negative AGC, voltage becomes more positive with increasing signal.

C. Positive AGC, voltage becomes more negative with increasing signal.

D. Positive AGC, voltage becomes more positive with increasing signal.

Figure 5-8 gives a universal op-amp circuit that should be adaptable to most receivers. The circuit is quite similar to Fig. 5-3, and uses the same source of negative voltage. Component values must be modified for the particular voltage range encountered in the receiver. To accommodate the four cases, wire the circuit as follows:

A. Point B to +5 Vdc; Positive meter terminal to ground

B. Point B to +5 Vdc; Negative meter terminal to ground

C. Point B to −5 Vdc; Positive meter terminal to ground

D. Point B to −5 Vdc; Negative meter terminal to ground

In each case, R4 and R5 control sensitivity, and R3 controls threshold or zeroing. The +/− volts to point B must be regulated. Use the ICL7660 circuit in Fig. 5-3. Get a +5 V from the 1N751A cathode (U2-8) and −5 V from the voltage converter output (U2-3).

To set up the circuit, first set R5 to maximum resistance. With no signal input, set R3 for zero on the meter. Now input an on-frequency signal to the receiver of about 50 microvolts (S9). If the meter is pinned at full-scale, decrease the value of R4 until it comes off the pin. When the meter is off the pin, set R5 for exactly full scale on the meter. With this setting, any signal above S9 will pin the meter. This lowered dynamic range is good for mobile hunting, but if you want more range, just use a stronger signal (say 5000 microvolts for 40 dB over S9) for full scale adjustments.

These are merely suggested starting point circuits, and may require a lot of optimization for your particular receiver. Meter response may be very non-linear, and you may wish to try non-linear feedback circuits to compensate using logarithmic amplifier techniques.

THE BRIDGE CIRCUIT

The classic bridge circuit (Fig. 5-9) may be useful for adding S-meters to some receivers. It is capable of very high sensitivity. The example shows a junction field effect transistor (JFET) as the control element, but other devices can be used with appropriate polarity, component value, and biasing changes. With no signal, R5 is set such that the current through path ABD equals the current through path ACD. The bridge is balanced, voltages at B and C are equal, and the meter reads zero.

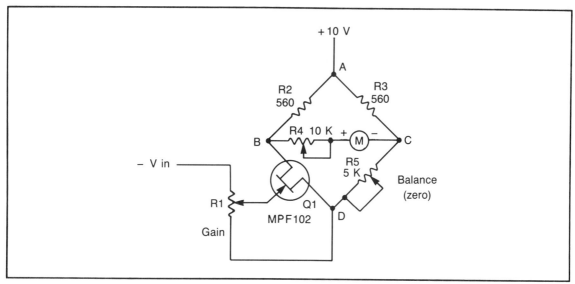

Fig. 5-9. The bridge circuit offers a possible high sensitivity S-meter driver.

As the input voltage rises (becomes more negative, in this case), Q1 current decreases, the voltage at point B increases, and the meter reading increases. The meter deflection follows the input until the active device saturates or cuts off. In the case of the MPF102 JFET, pinch-off occurs at about − 8 volts on the gate. The value of R4 is chosen such that pinch-off results in a reading at or slightly above full scale on the meter used. R1 sets the sensitivity, and may need to be a front panel control in your installation. The MPF102 is readily available (RS 276-2062) but other small signal N-channel types can be substituted.

The bridge circuit will respond to changes in supply voltage (V +). It should therefore be operated from a regulated source instead of directly from a vehicle supply. The device's equivalent resistance may be thermally sensitive, resulting in the need to re-zero the meter if a large change in temperature takes place. If you are meticulous (and very patient), you can add temperature compensating components, such as thermistors, in the R2 leg of the bridge to eliminate thermal effects.

FM GAIN BLOCK ICs

A signal meter circuit with a choice of greater dynamic range may be desirable in some cases. Such a meter can be added to an FM receiver by tapping the i-f circuits ahead of the limiting stages. It's good for many AM and SSB receivers, too. The RCA CA3089E integrated circuit has an internal meter driver function that

is ideal for this application. The LM3089 is an identical part made by National Semiconductor.

The 3089 contains a three stage i-f amplifier with peak detectors on each stage, which are summed by the meter driver. The output at pin 13 goes from zero with i-f input below 5 microvolts to 5 volts when input is above 50 millivolts. The IC also contains a quadrature FM detector, a squelch circuit, audio output and AFC output. For S-meter use only we can ignore these other sections and leave out the external components associated with them. This leaves the very simple circuit of Fig. 5-10.

The 3089 has very high gain. Precautions must be taken to prevent oscillation. Build the circuit on a board with a ground plane. Keep leads, particularly the ground and bypass capacitor leads, very short. Attach the capacitors right to the IC pins. Lowering the value of R1 may stop oscillations. Reduce it to as low as 50 ohms as required.

Connect the 3089 input to the output of the first i-f stage as shown in Fig. 5-11. The IC will work with any first i-f frequency from 2 to 20 MHz. It is optimized for the common FM i-f of 10.7 MHz. It will not work properly with 455 kHz i-fs. If the circuit loads down the receiver i-f too much, reduce the value of C1.

This circuit, with up to 80 dB dynamic range, is ideal as shown for telemetry from a remote site, where attenuation cannot be readily switched into the antenna lead. For mobile hunts, this range should be reduced and the sensitivity at the low end should be increased. Use a 50 microampere full scale meter at M1 to reduce dynamic

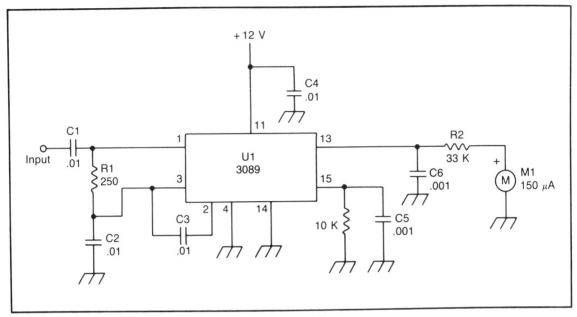

Fig. 5-10. The 3089 IC provides a simple, wide-range S-meter driver.

range to about 43 dB. Full scale on the meter will be about 700 microvolts into the 3089.

The 3089 can also be used to add a wide range S-meter to an AM or SSB set that does not have one. The AGC may have to be disabled for wide range response. If you have studied Chapter 9's coverage of Doppler DFs, you may be wondering if the 3089 makes a good add-on FM detector to drive a Doppler DF unit from an AM-only receiver such as those used on the aircraft band. That idea works, but special care must be taken with the external detector components. High Q, low drift parts are needed to get sufficient audio from the low deviation Doppler signal at high i-f frequencies. Consult the manufacturer's application notes for more information.

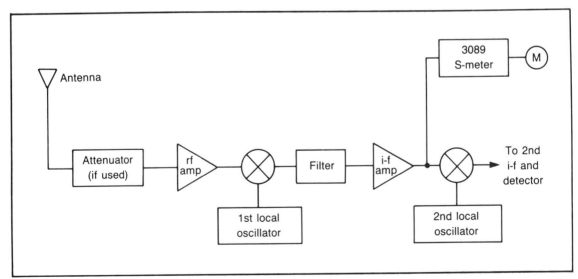

Fig. 5-11. Tapoff point for the single IC S-meter. Note that this IC works from around 2 to 20 MHz and is optimized for 10.7 MHz i-fs.

Chapter 6

Knocking Down the Signal

Manufacturers take great care to give modern receivers good sensitivity. Signals on the order of a few tenths of a microvolt will open the squelch on most FM sets on the market today. When hunting weak signals this level of performance is great. But once the signal gets strong, new problems start.

Some of the characteristics of a good S-meter also tend to make it worthless on strong signals. As shown earlier, a meter suitable for T-hunting should have good sensitivity and a relatively small dynamic range. So as soon as the signal gets 20 dB or so above the noise, a good meter is near or above full scale. Without something to reduce the signal level, it's not possible to move closer without bearings becoming difficult or impossible to read accurately.

The problem is generally worst with FM receivers. They are designed to have the successive stages saturate as the signal gets stronger. This is fine for eliminating noise on the FM signal but makes it impossible for the T-hunter to use the S-meter on signals above the i-f saturation level. There are two solutions to the problem: decrease the signal strength at the receiver input or decrease the sensitivity of the radio.

EXTERNAL ATTENUATORS

Most new hunters start out with external attenuators. They can be used with any receiver and their calibration gives a good indication of the proximity of the bunny. A well-calibrated attenuator has many other uses around the shack; for example, measuring the gain of an amplifier or receiving antenna.

Switched Resistive Attenuators

It is doubtful that you can justify to yourself spending $100 to $500 for a new commercial switched attenuator. But if there are electronics surplus outlets or industrial salvage sales in the area, give them a good look. You might find a real bargain. There are several possibilities for attenuators, beginning with individual cylindrical units that have fixed values and must be chained together as you close in. These pads, as they are called, are made by dozens of companies, including Weinschel, Hewlett Packard, and Narda. They have been made with every conceivable type of coax connector.

It's appropriate to digress at this point and put in a word about the choice of connectors for your setup. The good old uhf connector (PL-259, SO-239) is not great at either uhf or vhf. Type N fittings are best from the standpoint of being waterproof and rf tight, but are inconvenient. The BNC type is best for hunting. Interconnecting cables with molded BNC connectors are available ready-made from suppliers such as Pomona Electronics, so you don't have to assemble your own. The bayonet mating feature makes possible rapid assembly and reconfiguration of the system as you substitute antennas or get out of the car to go portable.

A commercial step attenuator with one or more rotary switches is wonderful for hunting if you can find it surplus. It should be adjustable in small steps, but have total attenuation of 100 dB or more. The Hewlett Packard HP-355D will go to 120 dB in 10 dB steps with only about 0.25 dB loss at 150 megahertz when set at zero. Special switches are used in these units to achieve isolation between the steps. Home constructors will find this isolation can't be achieved with ordinary rotary switches. That's why there isn't a rotary switch attenuator project in this book.

Step attenuators using toggle, slide, or rocker switches are more common in the surplus market than the rotary switch variety (Fig. 6-1). Look for one with enough steps to total 100 dB or more. The ultimate attenuation figure, which is the loss with all sections switched in, equals the sum of the steps in a top quality unit. In others, the ultimate attenuation may be less than the total due to leakage.

A Slide Switch Attenuator

Cellular switch type attenuators are the easiest for experimenters to build successfully. Refer to Fig. 6-2 for construction details of a simple unit with slide switches. The enclosure and partitions are made of double sided unetched copper clad circuit board. Shields of the same material, with a small hole for an insulated wire, are placed between stages for isolation. All seams are soldered together and the outside of the single-hole BNC connector (RS 278-105) is soldered to the box before parts are installed. More sections can be included if desired.

Slide switches are shown in this version. They should be standard size switches of high quality. The switch body *Radio Shack*

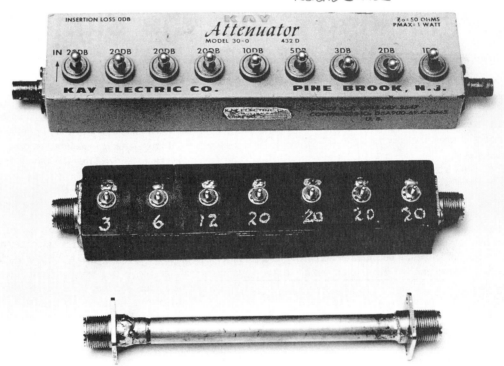

Fig. 6-1. Typical attenuators used by amateurs for T-Hunting: a commercial switched unit, a home brew switched unit, and a home brew waveguide-below-cutoff unit. Due to the poor vhf/uhf performance of the uhf connectors, BNC connectors are suggested.

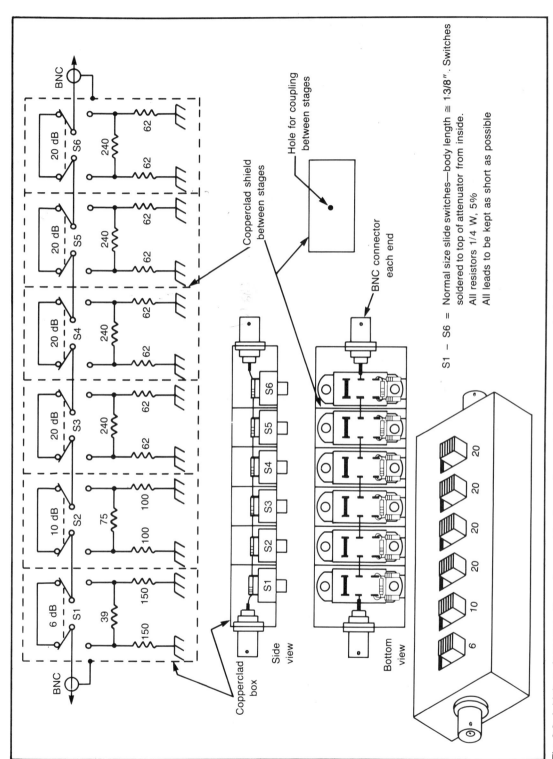

Fig. 6-2. A high performance attenuator capable of operation up into the uhf region can be built from commonly available components. This unit uses copper-clad PC board material and slide switches.

57

is about 7/8 inch long. Don't use miniature slide switches. For best uhf performance, thoroughly degrease the switches, preferably using a non-flammable solvent such as trichoroethylene or freon. Solvents such as acetone or lacquer thinner can also be used, but make sure that the solvent won't dissolve switch parts. Use good ventilation with any solvent and be especially careful with flammable solvents. Acetone and lacquer thinner are highly flammable. Use tongs to hold the parts, and don't come in contact with or breathe the fumes of any degreaser you use.

Cut the board sections carefully so that everything fits together with square edges. It's easiest to solder up the inside seams before installing the switches, but make constant fit checks of the switches while installing the shields, to prevent an unpleasant surprise. Tack solder the shields, fit check, then run a solder bead along the entire shield length. Then the flanges of the slide switches can be soldered to the back of the front panel. File away excess solder fillets if necessary so the switches sit flat. When the unit is completed, carefully clean out excess solder flux. Coating or painting the outside isn't necessary for performance, but can be done for appearance.

Cut the switch lugs down to half size before wiring in the resistors and bus wires. Use carbon composition or film resistors for good vhf performance. Keep the resistor leads very short and direct, but be careful not to overheat the resistors when soldering. Half watt resistors can be used instead of quarter watt ones if the box is made slightly larger to accommodate them. Using 5 percent resistors will give better accuracy than ones with 10 percent or greater tolerance. If step accuracy is important,

measure the resistors with an accurate ohmmeter before installing, particularly if you're using surplus resistors. Resistance value changes with time aren't unusual with composition resistors.

The smallest step value in this design is 6 dB. Smaller steps aren't useful in hunting. Other step values can be used to suit your particular needs. Refer to Table 6-1 for resistances to give other values of attenuation per section. Stage losses of greater than 20 dB aren't practical, due to leakage.

The rear panel should have a tight fit, with good metal connection all around to prevent strong signal leakage into the unit. If the rear panel is made slightly oversize, it can be soldered in place all around. Copper foil tape (if you can find it) is another way to get a good rf seal. Rear panel grounding straps, as described later, are recommended.

The ARRL Attenuator

Circuit Board Specialists of Denver, Colorado, offers a complete kit of parts for an eight-step attenuator, using slide switches, at an attractive price. It was developed in cooperation with the American Radio Relay League (ARRL). Be sure to ask for assembly instructions (a copy of the original *QST* article) when ordering. It was designed more for measurement purposes than for DFing, with 1, 2, 3, 5, 10, and three 20 dB steps. The circuit board material is pre-cut and fits together readily with only some slight touch up filing necessary. All parts are supplied.

We measured the performance of a carefully constructed ARRL attenuator on a Hewlett Packard model

Table 6-1. Table of PI Attenuator Resistor Values.

ATTENUATION (dB)	R1	R2	
1	886.939	5.88457	
2	444.936	11.8472	
3	298.25	17.9671	
4	225.391	24.3254	
5	182.058	31.0064	
6	153.486	38.0989	
7	133.343	45.697	
8	118.465	53.9014	
9	107.094	62.821	
10	98.1725	72.5743	
15	73.0616	138.862	
20	62.3333	252.45	
25	57.0777	452.027	
30	54.3309	805.574	

8505A network analyzer. This instrument provides a calibrated CRT plot of attenuation versus frequency. Figure 6-3 shows the characteristics of the first four cells. The horizontal scale goes from 2 to 500 megahertz at about 50 megahertz per division. The diamond shaped markers are in the center of the vhf/uhf ham bands: 6, 2, 1-1/4, and 3/4 meters. The vertical scale is expanded to 1 dB per division.

The straight line near the top is the 0 dB reference. The top curved trace is the attenuation of the unit with all steps switched out. This is the measured insertion loss, which is less than 1 dB up to 330 MHz. Traces below the insertion loss trace are 1, 2, 3, and 5 dB, respectively. Note how well they track the insertion loss trace.

For performance of the higher steps, look at Fig. 6-4. The vertical scale has been changed to 5 dB per division. The top trace is the insertion loss again, and below it are 10, 20, and 40 dB (two 20 dB sections) traces. These measurements show good vhf performance and good step accuracy. SWR presented to the transmitter with no attenuation switched in is about 1.6:1, which will keep almost any transmitter happy.

The biggest problem in using the ARRL attenuator for transmitter hunting is its 81 dB maximum setting. A good solution would be to change the resistors in the first steps to give higher attenuation values, following the suggestions given earlier in this chapter. The rear panel is held in place with screws in the four corners, which is not optimum for highest ultimate attenuation. Ground-

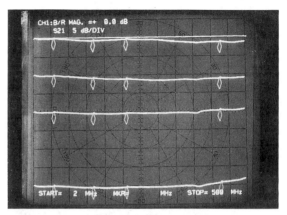

Fig. 6-4. Network analyzer response of the Circuit Board Specialists attenuator. Horizontal scale is from 2 to 500 MHz. Vertical scales are 5 dB per division. The straight line on the top is the 0 dB reference point. The other lines are the 0, 10, 20, and 40 dB attenuator settings. Small diamonds mark the 50, 144, 220, and 450 MHz amateur bands.

ing straps should be added in each cell as detailed in the next section, especially if the maximum attenuation is increased. Without the straps, the stock ARRL attenuator had ultimate attenuation of only 66 dB.

Using Toggle Switches

It's tricky to make the rectangular holes for slide switches with hand tools. Slide switches may also not take lots of abuse. They won't stand up to coffee or cola spills very well, for example. You might be thinking that toggle switches would be easier to mount, more durable, and make for a more rf-tight home-brew attenuator. Well, go ahead. Many other hunters are successfully using toggle switch attenuators. Electrical performance at vhf/uhf isn't as good, and the result isn't a piece of laboratory test equipment. But it works fine for hunting foxes. Figure 6-5 shows the inside of a unit made with miniature toggle switches by Clarke Harris, WB6ADC. (Don't use the full sized ones—they have too much inductance.)

Figure 6-6 shows the performance of a toggle switch attenuator as measured on the network analyzer. The horizontal scale is the same as before, and the vertical scale is 5 dB per division. The top trace is insertion loss and the traces below are 3, 6, 10, and 20 dB individual steps. Insertion loss at 2 and 1-1/4 meters is about 2 dB, rising to 5 dB at 3/4 meters. The unit gives good attenuation accuracy up to about 300 MHz, but above that the 10 and 20 dB sections begin to both approach about 15 dB attenuation. SWR looking into the unit isn't nearly as good as the ARRL attenuator, but it's OK for most

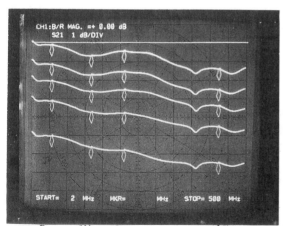

Fig. 6-3. Network analyzer response of the Circuit Board Specialists attenuator. Horizontal scale is from 2 to 500 MHz. Vertical scales are 1 dB per division. The straight line on the top is the 0 dB reference point. The other lines are the 0, 1, 2, 3, and 5 dB attenuator settings. Small diamonds mark the 50, 144, 220, and 450 MHz amateur bands.

use braid
make shorter
in each section

Fig. 6-5. Inside of toggle switch attenuator.

applications. It is about 4:1 with all sections switched out, dropping to a much better match when the section nearest the transmitter is switched in.

To improve the SWR at minimum attenuation, try replacing the insulated wires between sections with straight pieces of 0.141 inch semi-rigid coax (RG-402). The outer shield of the coax should be soldered to the enlarged hole in the shield on both sides. No other shield connections are necessary.

Notice the wires connecting the rear panel to the inside ground of most attenuator cells in Fig. 6-5. We highly recommend adding these, but they should be shorter, made of braid, and put into each section. The straps will markedly increase the ultimate attenuation and reduce the strong signal pickup. Without them, the ultimate attenuation may be 20 to 30 dB less than the sum of all sections, which can spell disaster in hunting a high power transmitter. If you use the straps, you'll find that carefully soldering or bolting the back panel on isn't as critical for strong signal performance.

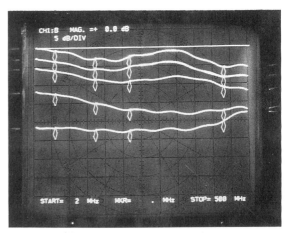

Fig. 6-6. Network analyzer response of a home brew toggle switch attenuator. Horizontal scale is from 2 to 500 MHz. Vertical scales are 5 dB per division. The straight line on top is the 0 dB reference point. The other lines are the 0, 3, 6, 10, and 20 dB attenuator settings. Small diamonds mark the 50, 144, 220, and 450 MHz amateur bands.

The Waveguide Attenuator

As if all that weren't enough, there's one more external attenuator type to consider. The following design gives a unit with smoothly adjustable loss from 8 dB to almost 100 dB at two meters. Designer Russ Andrews, K6BMG, calls it an indestructible attenuator because it can't be damaged by transmitting through it. We call it a waveguide attenuator because it works like a piece of circular waveguide operated below its characteristic cutoff frequency. Figure 6-1 lets you see how small it is.

Before telling you how to build it, we should give you a warning. Electrically, the unit is nearly indestructible. Your transmitter final stage may not be. When set for minimum attenuation (center conductor making contact), it presents a nice 1.35:1 SWR to a two meter transmitter, assuming a perfectly matched load following. But when the sections are pulled apart enough to break the center conductor contact, the SWR suddenly becomes extremely high. Transmitters without protective shutdown circuitry may be damaged quickly by this poor match.

This type of attenuator is called reflective, because the power not passed to the antenna is reflected back to the transmitter. In contrast, the switch type units described previously are called absorptive, because the power not passed to the antenna is absorbed in the resistors of the various stages. In order for a reflective attenuator to present a low SWR to the transmitter, another device, such as an isolator or circulator, must be placed between them. Unfortunately, a good Phelps Dodge isolator for the purpose will cost over 250 dollars—not too practical just for T-hunting. So before building this attenuator with the idea of tranmsmitting through it, make sure your transmitter isn't going to be damaged by transmitting into a bad match.

Figure 6-7 gives construction details for the slide attenuator. Brass tubing can be purchased in one foot lengths at some hardware stores and hobby shops. This tubing comes in diameters that increase by .03125 inch (1/32 inch) each size. Since the tubing wall thickness is 1/64 inch, each size tubing will telescope with the next smaller and larger size.

First cut the tubing sections to the proper lengths. If a tubing cutter is used, you may have problems getting the tubing to slide together unless you file off the cut end until it is square. If you use a saw, use the finest blade you can find, that is, the most teeth per inch. A jeweler's saw is great if you can find one. Remember that this type of tubing has very thin walls and will bend if clamped too tight.

Clean up the cut ends with a fine file and verify that the telescoping sections fit together. Polish the sections with fine steel wool so that they will solder easily. Use a 75 to 100 watt iron with a 3/16 to 1/4 inch tip to solder the larger pieces.

Starting with the left half, slide section B into section A until the ends are flush. Heat the outside near the end until solder will melt on the tubing. Slowly feed solder to the joint between the two pieces, letting it flow in before feeding more to the joint. Use care to keep the ends of the two sections flush. Feed solder all around the joint. While the tubing is hot, tin the inside and outside of the tubing about a sixteenth of an inch back from the end.

Using a fine file, bevel the outside of section D at point "Y" so that this section will easily mate with the center section of the other half of the attenuator. Slide sections C, D, E, and F together. Sections C, D, and E should have the ends flush while F (the center section) should stick out about an eighth of an inch. Solder these sections together as in the previous paragraph.

After soldering, carefully tap the assembled sections at point "X" with a center punch, distorting the metal so that when fully mated, there is a snug fit with the center section of the other half of the attenuator. Solder this assembly to the center contact pin of one of the BNC connectors, making sure that it is straight. Glue two of the fiberglass washers onto the assembled center section. When the glue is dry, carefully file the outside edge of the washers, a little at a time, until the large outer tube (section "A") will slide over the inner coax line section. Set this assembly aside for the time being.

Assemble the right half by sliding section H over section G as the figure illustrates. Solder these two sections together as was done with the other half of the attenuator. Tin the end of the tubing. Assemble sections I, J, and K and adjust the length of the assembly so that the end of section I is flush with the end of section G when everything is soldered. Solder the center sections together and solder this assembly to the center pin of the BNC fitting. Glue the other two fiber washers on the center assembly and file the outer diameter until they just fit the inside of section G, the outside tube.

Remove the outside sections from both halves of the attenuator and slide the two inside sections together fully. These two sections should just fit inside the two outer sections when they are fully mated. If not, adjust the lengths as necessary. If the lengths are correct, then solder the outside tubing sections to the rear of the BNC connectors.

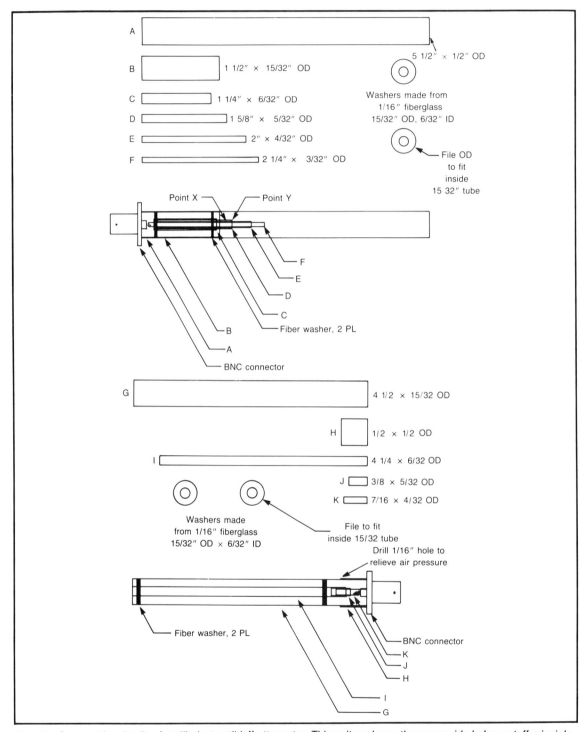

A   5 1/2″ × 1/2″ OD

B   1 1/2″ × 15/32″ OD

C   1 1/4″ × 6/32″ OD

D   1 5/8″ × 5/32″ OD

E   2″ × 4/32″ OD

F   2 1/4″ × 3/32″ OD

Washers made from
1/16″ fiberglass
15/32″ OD, 6/32″ ID

File OD
to fit
inside
15 32″ tube

Point X   Point Y

F
E
D
C
Fiber washer, 2 PL
B
A
BNC connector

G   4 1/2 × 15/32 OD

H   1/2 × 1/2 OD

I   4 1/4 × 6/32 OD

J   3/8 × 5/32 OD

K   7/16 × 4/32 OD

Washers made
from 1/16″ fiberglass
15/32″ OD × 6/32″ ID

File to fit
inside 15/32 tube

Drill 1/16″ hole to
relieve air pressure

Fiber washer, 2 PL

BNC connector
K
J
H
I
G

Fig. 6-7. Construction details of an "indestructible" attenuator. This unit works on the waveguide-below-cutoff principle.

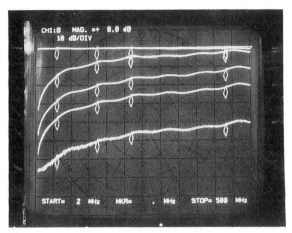

Fig. 6-8. Network analyzer response of a home brew waveguide-below-cutoff attenuator. Horizontal scale is from 2 to 500 MHz. Vertical scales are 10 dB per division. The straight line on top is the 0 dB reference point. The other lines show the increase in attenuation when the unit is pulled apart about 1 cm for each step. Small diamonds mark the 50, 144, 220, and 450 MHz amateur bands.

Measured performance of the waveguide attenuator is illustrated by Fig. 6-8. The top trace is made with the tubes pushed fully together, showing that insertion loss is very small, even at 500 MHz. It might have been even less if it had been built with N or BNC fittings instead of uhf types. Each successive lower trace represents about one centimeter of extension of the tubes. The first step is very sharp, and occurs when the center conductor connection is broken. You can see that it is not possible to set this unit for attenuation values between 0 and 15 dB on six meters or 0 and 8 dB on two meters.

Note that the first loss step becomes very steep below 25 MHz. You should never consider using this design on a 75 meter hunt! After the first step, the increase in attenuation is smooth out to the maximum of 100 dB or so. Attenuation is not linear with extension, so the traces don't have the same vertical separation.

Pot Attenuators

Perhaps you have wondered if ordinary potentiometers can be used for rf attenuation. They can, making a quick-and-dirty way to get signal reduction. Figure 6-9 shows an example. More than one pot can be cascaded for high attenuation if shields are put between each section. Each section can achieve about 30 dB maximum.

Build the unit in a sealed, shielded box as was done with the switched attenuator earlier. Rotary carbon com-

position pots must be used, and the lugs should be cut as short as practical. Keep all leads very short. Ground leads should be soldered directly to the case.

Inductance L can be added to cancel some of the stray capacitance in the pot. The value must be found by empirical methods (fancy words for "trial and error"). A different value of L will be required for each frequency band to form a parallel resonant circuit with the stray capacitance. You'll probably be disappointed in the performance of this attenuator at vhf and uhf, but it may be quite satisfactory for hunts on 75 or 160 meters.

THE MAGIC ANTENNA SWITCHER

Here's a sure-fire way to be able to use any kind of external attenuator or antenna safely with any transceiver. You'll never have to worry about accidentally damaging the transmitter or antenna. This antenna selection box can be built up with connectors to fit all your radios so that any can be set up on short notice for DFing.

A toggle switch (S1) selects either NORMAL operation or DF operation. In the NORMAL mode, both transmitting and receiving takes place with the regular vertical vehicle antenna. In the DF mode, receiving is done through the DF antenna and attenuator, but when the mike push-to-talk (PTT) button is pressed, the set is switched over automatically to the normal vehicle antenna for the duration of the transmission.

The circuit (Fig. 6-10) is an adaptation of a switching method used by some marine vhf radio/DF combination sets. It is not limited to quad, yagi, or loop hunting—an add-on dual antenna or Doppler DF could also be switched in and out with it. The control box can be built separate from the relay and attached to the radio.

The key component is a vhf coaxial relay (K1) with a 12 volt coil and SPDT auxiliary contacts. These relays are fully shielded and most feature rf fittings (BNC, N, or uhf) on them. They are often found in good condition on the surplus market, or can be salvaged from junked commercial vhf/uhf two-way mobile radios. Dow Key (a division of Kilovac Corporation) is one manufacturer.

The auxiliary contacts are needed to insure that the relay has closed before the transmitter is keyed. This way, rf won't be transmitted into the DF antenna system, even for an instant. Wiring is shown for the standard grounding PTT scheme used on most transceivers.

INTERNAL ATTENUATORS

While most hams use external attenuators quite successfully, they are not trouble-free. Resistive units can

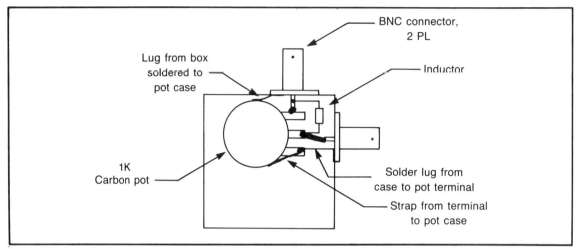

Fig. 6-9. For hf and non critical vhf needs, a pot attenuator may work well. Use a carbon composition pot as wire wound types have too much inductance.

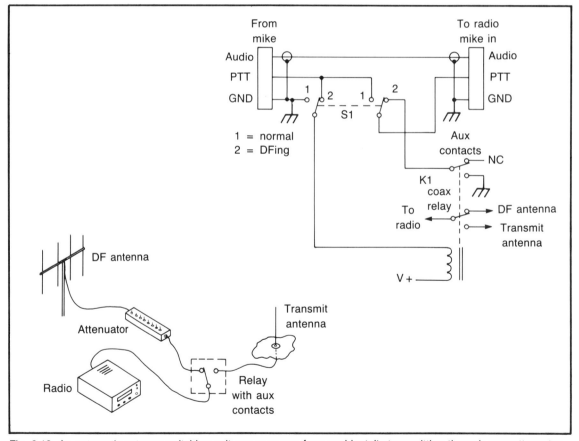

Fig. 6-10. An automatic antenna switching unit can save you from accidentally transmitting through your attenuator.

be fried when transmitted through. In most cases 100 dB or so of attenuation is the most that is practical. Above that, signal leakage around the attenuator and through the case still results in pinning of the S-meter. The shielded enclosure and double-shielded coax described in the chapter on sniffing help. So does using double-shielded or semi-rigid coax between the attenuator and the receiver. A simpler solution, though, is to decrease the rf gain of the receiver.

Internal gain reduction is an attractive alternative to an add-on attenuator box. The attenuation system is always ready for hunting. It eliminates the problem of rf leakage around an external box and into the receiver case. The attenuator dial can be calibrated in dB by using a calibrated vhf signal generator. You need never fear transmitting through your attenuator box, burning it out or damaging the transmitter.

A disadvantage of an internal attenuation system is that it must be specifically designed and installed in each receiver, while an external attenuator box can be used right away on any handy rig. Adding an internal attenuation scheme is not a good project for the novice at construction and repair, or for those squeamish about modifying their commercial gear.

There are wide variations in the circuits and devices used in receivers. The rf gain controls that we will describe here are given as examples. They work for specific radios and may not work exactly as shown for others. We don't recommend that you simply build these by rote. Carefully examine your own radio and see if any of the following circuits can be used as is or must be modified to your specific needs.

Supply Voltage Control of Gain

If you use an external FET preamp ahead of your receiver, try varying the supply voltage to it and noting the effect on gain. (Hint: If you don't have a calibrated vhf signal generator, you can use a steady on-the-air signal and your external type attenuator to make this gain measurement, provided the steps are reasonably accurate.) You'll find voltage control is far from linear, but it's effective. With the preamp described in the weak signal chapter later in this book, varying the supply voltage changes its gain from +13 dB to −8 dB, effectively turning it into a 21 dB attenuator.

Just 21 dB is not sufficient attenuation for most purposes. More attenuation can be obtained by varying the supply voltage to internal stages of your T-hunt receiver. The circuit shown in Fig. 6-11 is used by Ken Diekman,

WA6JQN, with his Icom IC-22U. He found a jumper wire on the PC board that connects the Vcc voltage to the receiver preamp, first mixer, and first i-f amp stages.

By cutting this jumper and connecting it to the wiper of the pot, the sensitivity of the radio can be varied over a wide range. An ordinary 5 kohm pot works fine because the stages draw little current. Ken says that with this modification he can get meter readings on almost any signal that is likely to be presented, even a 70 watt signal at five feet. Ten watts at ten feet is easy to hunt. His tests show 3000 microvolts into the antenna input of the radio can be brought down to mid scale on the meter.

Reduction of voltage to rf and i-f stages will worsen the cross modulation performance of a receiver. This usually doesn't matter because the signal being DFed will be strong enough to override the cross-mod products when the attenuation is cranked in. The voltage reduction may also reduce dynamic range of the S-meter by causing limiting in early stages below the limiting level of the last i-f stage. One suggestion for overcoming this problem when it occurs comes from John Moore, NJ7E, who has had success with reducing the level of just the first local oscillator. This results in direct reduction of signal level at the i-f in well designed receivers. Try it!

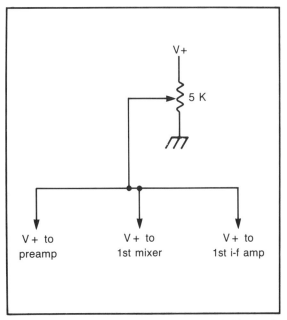

Fig. 6-11. Ken Diekman, WA6JQN, developed this circuit for the ICOM IC-22U. Using this circuit, the radio can give an on-scale S-meter reading with almost any imaginable signal.

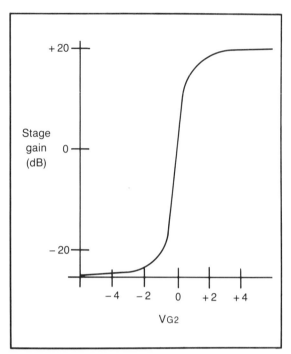

Fig. 6-12. The gain of a MOSFET stage varies as a function of the gate 2 voltage.

Altering the Bias

The gain of some devices is very sensitive to biasing. Stages with field effect transistors are easily controlled with gate voltage changes. Dual gate Metal Oxide Semiconductor (MOSFET) devices have a second gate which makes an excellent gain control point. The effect of varying the second (non-rf) gate voltage on stage gain is illustrated by Fig. 6-12. This performance is typical of the common dual gate MOSFETs such as the 3N140 and 40673.

WA6JQN developed a simple gain reduction circuit using this principle for the Heathkit HW-2036 two meter FM transceiver (Fig. 6-13). S1, B1, and R1 are added to the first rf stage. Closing S1 changes the dc voltage at gate 2 of Q201 from +2.5 to −2.0 volts, cutting it off and giving about 40dB of attenuation. Current drain from the battery is very low, so type N cells (about half the size of AA), or even smaller ones, work fine.

Some MOSFETs may require a higher negative voltage to cut them off, in which case more cells can be stacked. If you never want to worry about replacing the battery, use an ICL-7660 converter IC, described in the last chapter, to generate the negative voltage. Attenuation can be adjusted by varying R1, but the very sharp cutoff characteristic of gate 2 will usually not allow a

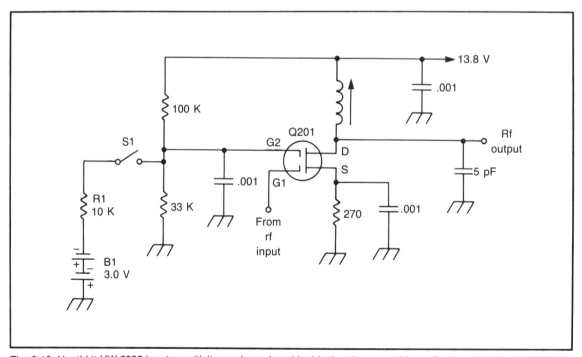

Fig. 6-13. Heathkit HW-2036 input sensitivity can be reduced by biasing the second (non-rf) gate of the input MOSFET.

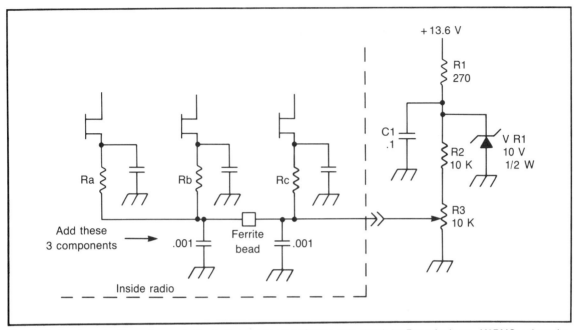

Fig. 6-14. A gain reduction scheme similar to the one for the IC-22U, this circuit by Russ Andrews, K6BMG, raises the FET sources above ground.

smooth, linear attenuation control. The exact cutoff voltage is also sensitive to temperature variations.

Another bias changing scheme, shown in Fig. 6-14, was developed by Russ Andrews. He raises the circuit return connection of each of several stages up above ground. This changes the operating point in a similar way—the stage current is reduced. The stages in this case use MOS devices, but bipolar transistors can frequently be controlled this way, too. Ra, Rb, and Rc are the source resistors presently used in these stages. The ground end of each resistor is disconnected from ground, then tied together and to the bias source. Note the importance of Zener regulation of the bias voltage supply.

THE AUTOMATIC ATTENUATION CONTROL

In all of these circuits the rf gain/loss settings must be changed manually as the signal gets stronger or weaker. This may not be a big problem in open, flat terrain. However, on winding roads in the hills the signal can fluctuate wildly, leaving you to madly working the attenuator trying to keep the meter on scale. Why not control the attenuator setting automatically?

In addition to the rf/i-f stage gain reduction circuit already described, K6BMG has also developed a driver circuit to do just that. The circuit, shown in Fig. 6-15,

uses only two ICs—one 555 timer to generate the negative supply voltage, and one LM324N quad operational amp. The result is automatic gain control (AGC) of the receiver, but since there's no additional gain involved, we'll call it an Automatic Attenuation Control (AAC) instead.

This system was originally designed for the Yaesu FT-227R two meter mobile transceiver. The internal meter drive circuit is tapped and fed out of the radio to a voltage follower/buffer, U1A. The output of the buffer is used to drive an external S-meter through R1 and R2. This meter is mounted in a box in easy view on the dash of your T-hunt vehicle.

U1B amplifies the meter voltage by the ratio of R10 to R5. The voltage divider chain composed of R6, R7, and R8 determines the level at which the output of U1B is offset above ground. Adjusting R7 sets the level at which the AAC action starts. This is an inverting amplifier, so as this level is set further above ground, the voltage at the output of U1A (the signal strength) must be higher to forward bias diode D1.

With no signal, the output of U1B is offset positive and D1 is reverse biased. The external AAC meter reads zero volts. Also, zero volts applied to the input of the inverting log amp (U1D) gives zero volts out, effectively grounding the sources of the three FETs in the modified

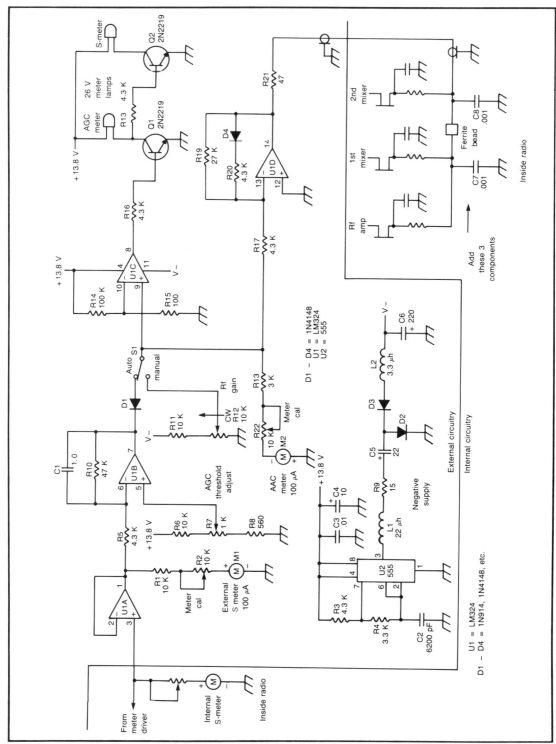

Fig. 6-15. Combined with the circuit of Fig. 6-14, this driver circuit automatically reduces the signal as you close in on the transmitter.

radio. This gives the stages maximum gain for greatest receiver sensitivity.

As the signal strength increases, the voltage at the output of U1A also increases. This voltage is amplified by the inverting amplifier U1B. With increasing signal strength the output of U1B decreases to the point where D1 is forward biased. This is set to take place when the internal S-meter is around 85% of full scale. At this point two things happen. The external AAC meter starts to indicate and the output of U1D starts to go positive.

Amplifier U1D is called a log amp because the presence of diode D4 in its feedback path makes the amplitude response of the amplifier become an approximate function of the logarithm of the voltage input. A log amp is used here to complement the nearly exponential response of the internal attenuation scheme shown.

The output of this log amp is connected to the source leads of the FETs in the modified radio. When the sources are raised above ground the sensitivity of the radio is decreased. Further increases of signal strength will further decrease the sensitivity of the radio. When this is happening the S-meter only moves a slight amount above the 85% level while the AAC meter indicates the increasing signal strength.

U1C is used as a voltage comparator. With low signal strength the output of the comparator is low, turning off transistor Q1. Transistor Q2 is turned on with bias through the "AAC" lamp and R13. When Q2 is on the "S-meter" lamp is on. When D1 starts to forward bias the output of U1C switches, turning on Q1 and turning off Q2.

With Q1 saturated, the "AAC" lamp is on and the "S-meter" lamp is off. These lamps use 26 V bulbs mounted inside the respective meters. When either meter is active the proper lamp is on, illuminating that meter. The higher voltage bulbs have very long life, but plenty of illumination for night hunting. If your meters do not provide for illumination, use LED indicators instead, placed next to the appropriate meter.

A 555 timer IC (U2) is used in the K6BMG design to invert the +12 volt source and provide −7 V to −11 V to the operational amps. Exact negative voltage is not critical. The values of R3, R4, and C2 give an oscillator frequency of around 20 kilohertz at pin 3. Inductors L1 and L2 are included in the circuit to keep glitches generated by the 555 out of the rest of the system. For the same reason keep the ground connections to the inverter circuit short and direct, using heavy or multiple leads. Braid from RG-174 coax makes a good ground lead. C3 and C4 should be as close to the 555 as practical. Alternately,

the ICL7660 voltage converter IC negative supply described in Chapter 5 could be used in place of the 555 circuit.

Build the unit in a box that fits on your dash, using meters that are easy to read. Wide scale edgewise meters are excellent for M1 and M2. Two edgewise meters can be stacked. Quad operational amplifier U1 is readily available (RS 276-1711). Transistors Q1 and Q2 can be any silicon NPN device that handles the lamp current. Diodes are general purpose silicon types such as 1N4148.

When modifying the radio keep leads as short as possible and away from all other circuitry. Feed the signal from the AAC circuit to the radio with RG-174, using connectors suitable for your set. When everything is hooked up, set the AUTO/MANUAL switch to the MANUAL position and turn the rf GAIN (R12) control fully counterclockwise. The wiper voltage will be zero. The radio should work normally. If not, check the outputs at U1-1, U1-7, and U1-14. With no signal into the receiver, the voltages at U1-1 and U1-14 should both be zero, and the output at U1-7 should be equal to the voltage at U1-5. If these voltages appear to be correct, check your modification of the radio. If one of these voltages is incorrect, start at the first stage (U1A) and troubleshoot until you find the problem.

A well-shielded vhf signal generator with adjustable output is recommended to set up the automatic attenuator. Connect it to the receiver input, set it to the receive frequency and increase the output until the receiver's internal meter reaches full scale. Adjust R2 to set the external S-meter to full scale. The values of R1 and/or R2 may need to be changed to suit your specific radio or meter. With the meter set, turn the rf GAIN control (R12) slowly clockwise and note that the meter reading decreases. This means that the rf gain of your radio is being reduced. Now adjust R12 back to full counterclockwise.

Adjust the signal strength into the radio so that the S-meter reads about 85% of full scale. Set the AUTO/MANUAL switch to the AUTO position and connect an oscilloscope or meter to U1-8. Adjust R7 until the output of U1-8 just goes to logic LOW, and then slowly turn the pot the other way until the level just goes HIGH. This sets the point at which the AAC circuit starts taking effect. If this adjustment can't be made, make sure that U1C is wired properly and is a good part, then try changing the values of R6, R7, or R8.

At this point U1-8 should switch when the signal level exceeds 85% of full scale on the S-meter. Below this point the AAC meter should read zero. As the signal level is

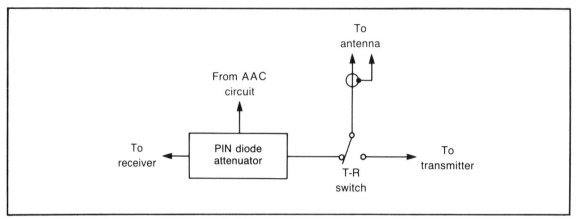

Fig. 6-16. A PIN diode attenuator can be installed between the antenna relay and the first rf stage. The driver circuit in Fig. 6-15 can be modified to control the attenuator.

varied around this point, the meter illumination lamps should switch back and forth. When the signal level reaches 85% the S-meter should stop moving and the AAC meter should start increasing.

Set AAC meter calibration control R22 so that the M2 reads full scale without pinning on the very strongest signals. Russ says the AAC meter tops out about ten feet away from a 10 watt hidden T. If unexplained fluctuations or oscillations occur on strong signals, try increasing the value of C1.

Advanced constructors may want to try building this driver to control a PIN diode type rf attenuator instead of using the internal stage gain reduction method. The PIN attenuator could be in the radio's antenna lead, making the system portable and adjustable to any FM radio. Or, to prevent possible burnout of the PIN diodes, it could be mounted inside the radio between the antenna relay and the first rf stage (Fig. 6-16).

Chapter 7

Equipping Your Vehicle

No matter what kind of gain antenna system you decide on, you'll eventually want some way to mount it on your vehicle. Perhaps for your first hunt you just stopped regularly and pulled the quad or beam out of the trunk to get bearings. Or maybe you hunted like J. Scott Bovitz, N6MI, (Fig. 7-1). Having to stop for each bearing is time consuming, and not knowing what the signal is doing (direction-wise) between bearings can put you past the hidden T before you realize what's happening. It's a great way to lose hunts, not to mention getting soaked if it rains.

Antenna mounts can be anything from a piece of PVC pipe stuck out of the window and tied with rope to the Rube Goldberg contraption shown in Fig. 7-2. Serious and devil-may-care hunters will simply bore a hole in the middle of the roof and turn the antenna mast directly. Our wives claim to be not that serious yet, at least not with *their* cars.

The type of mount you choose depends on the physical layout of your vehicle, what kind of DF equipment you plan to hunt with, and how elaborate a mount you want to build. The mount for hunting once or twice a year on a summer afternoon may be quite different from one used two or three times a month year around in all kinds of weather. Maybe you'll be happy with John Shockley's (WA6UVS) scheme (Fig. 7-3). Or maybe you'll want to

be totally enclosed and weatherproof. In this chapter you'll get lots of ideas of how to set up your vehicle for quick and convenient hunting.

Every vehicle is different. The details of mounting an antenna are quite diverse on different makes and models of mid-size passenger cars. Trucks, VW bugs, vans, and convertibles all present their own special quirks. The following pages provide a variety of ways that antennas can be mounted, but don't be afraid to experiment. Dig around in your garage and in hardware and automotive stores, and try out your own ideas. It's the best way to find the perfect way to mount your specific antenna system. Part of the satisfaction of T-hunting is setting up your vehicle to suit yourself and your needs with hardware that you have found and made.

THROUGH-THE-WINDOW MOUNTS

Probably the simplest mast holder is shown in Fig. 7-4. The bottom bearing is a PVC pipe end cap one or two sizes larger than your mast. In this installation, it is wedged into the armrest for hunting and is removed at other times. The broom handle mast rests inside and turns freely. It is shown raised up out of the cap in the photo for clarity.

Fig. 7-1. No antenna mount yet, but lots of enthusiasm. J. Scott Bovitz, N6MI, stands in the bed of his pickup to take bearings.

Instead of the cap, some hunters mount a piece of tubing, a piece of PVC pipe, or a fruit juice can to the inside door panel. When the screws are removed, the holes are hardly visible, particularly if the door panel is covered with fabric. If you want to remove it between hunts, but don't want to remove the door panel to get to the nuts inside, use expansion anchors such as Molly bolts. Your local hardware store stocks these and many other interesting styles of fasteners. Tell them what you are trying to do and see what they can provide.

The mast can be secured at the top of the window frame with thin nylon cord, particularly if there is a wing window post to hold it laterally. There's a good chance of slippage this way, though. A better solution is shown in Fig. 7-5. Drill a hole from inside the window channel up through the door frame. Use a nylon cable tie to secure the mast to the door frame. Leave just enough slack

to allow free turning, but not enough to let it flop around. It probably would be a good idea to seal the top hole with silicone caulking material, but the hole is shielded by the frame of the car. We haven't had any leakage problems as is.

For a bare bones mount this is hard to beat, but a few refinements can be made. A ball bearing or large marble dropped into the bottom support will make the mast turn more smoothly. Also a dry lube, either a spray or graphite powder, will help.

For better holding and turning at freeway speeds, add a support bracket to the car frame. The bracket is bent from .062 inch sheet steel and is mounted with self-tapping screws just behind the mast. Place the bracket so that the mast just clears it on opening or closing the door. Use dry lube here, too. The nylon cable tie and bracket make the antenna easier to turn when driving, since the antenna is kept from blowing around.

Some car doors don't have a solid frame around the windows. The bottom mount can be the same as above, but the top of the antenna mast must be mounted to the frame of the car, secured to the rain gutter. Drilling a small hole in the rain gutter and using a cable tie to secure the mast is the easiest method. A better way would be the bracket in Fig. 7-6. Cut the bracket from .062 sheet metal and mount it on the rain gutter with 4-40 or 6-32 screws.

The main problem with frameless window mounts is that to open the door, the antenna mast must be raised up out of the bottom mount, or the gutter mount must be detached. The antenna-turner can't exit the car at the ending point until someone helps him by removing the mast. If he forgets, the mount could be damaged.

MIRROR MOUNTS

For those who own a vehicle that has oversize side view mirrors, the bracket for the mirror can be used as an antenna mount. One way is to mount a piece of pipe or tubing vertically on the mirror bracket. Use a piece that your antenna mast will smoothly slip inside. PVC plastic pipe is a good material since it is lightweight and easy to work. Glue an end cap on the pipe and mount it on the mirror either with brackets or, for a temporary mount, with duct tape.

Another method is illustrated in Fig. 7-7. Two small brackets are cut from aluminum or steel and mounted at the top and bottom of the mirror assembly. The bottom bracket has a small dowel or rod screwed to it that sticks up into the antenna mast. The screw shown in the middle of the mast is used to indicate the direction that the

Fig. 7-2. Rube Goldberg looking, perhaps, but this roof mount puts the antenna solidly in the optimum spot, right in the center of the roof. The antenna can be turned by either the driver or the navigator, with protractors mounted on both sides of the car. The quad boom rotates, allowing the polarization to be changed with only a pull of a cord.

Fig. 7-3. John Schockley, WA6UVS, seats the navigator and beam turner in the rear of the pickup. The radio is mounted on the table, with attenuator and mast in easy reach. Note the headset to communicate with the driver.

antenna is pointed. This pointer is removed to install or remove the mast.

The nice feature of the mirror mount is that if your T-hunt partner can't make it and you decide to hunt alone, the mount can easily be switched from passenger's side to driver's side. This assumes you have mirrors on both sides. The bad part is that you may have to reach well out of the window to turn the antenna in a pouring rain.

DOOR MOUNTS

Figure 7-8 shows a mount that can be used on almost any vehicle except doorless jeeps. The frame is made from approximately 6.5 feet of 1 × 3 pine, cut into four pieces, three 24 inches, and one 6 inches long. Trim the two di-

agonal pieces as shown in the illustration and glue the pieces together using small nails to hold everything tight.

Bend the two window brackets from sheet steel or aluminum. The size of these brackets will depend on the thickness of your particular door. With the mount installed on the door, these brackets should just hook down inside the window well on the door. To keep the mount solidly on the door, and to prevent damage to your vehicle's paint, install three suction cups on the rear of the frame. Suction cups can usually be found at better hardware and auto parts stores.

To hold the bottom of the antenna mast, use a piece of PVC pipe large enough for the mast to drop inside. Use heavy-walled pipe, Schedule 40 or thicker. Referring

Fig. 7-4. The bottom of the mast can be secured with a PVC pipe cap pushed into the armrest of the car door.

Fig. 7-5. A simple cable tie is used to secure the mast to the door frame.

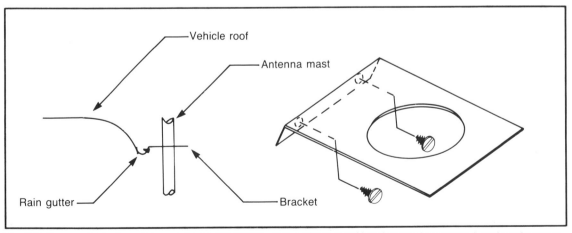

Fig. 7-6. If your vehicle doesn't have a frame around the door, a sheet metal bracket can be mounted on the rain gutter.

Fig. 7-7. Sheet metal brackets can be mounted on truck mirror mounts to hold the antenna mast. The bottom bracket has a protruding pin that fits up inside the mast. No modifications to the mirror are required.

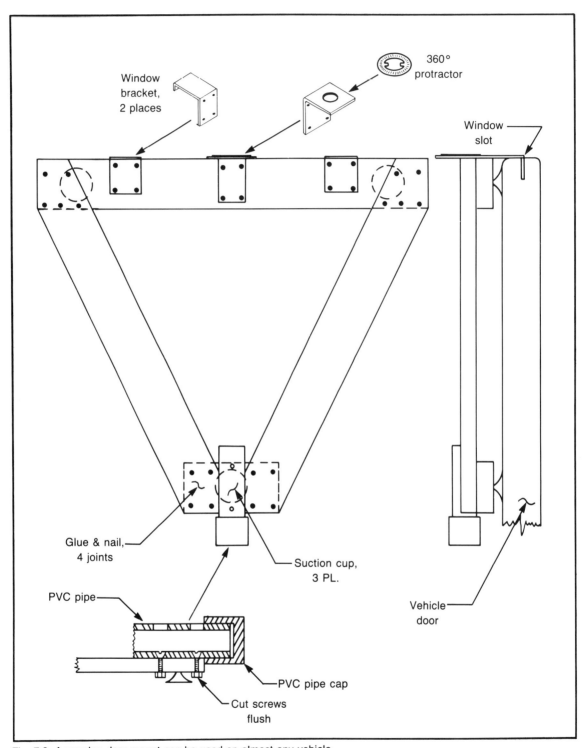

Window
bracket,
2 places

360°
protractor

Window
slot

Glue & nail,
4 joints

Suction cup,
3 PL.

PVC pipe

Vehicle
door

PVC pipe cap

Cut screws
flush

Fig. 7-8. A wooden door mount can be used on almost any vehicle.

to the drawing detail, drill two holes through both sides of the pipe. These holes should be large enough to clear the threaded portion of the flathead mounting screws (#8 or #10 should be OK).

Next, drill out the holes, on one side only, to clear the heads of the mounting screws. Use the same drill to bevel the inside of the smaller holes so that the flathead screws will be almost flush with the inside of the pipe. Glue a pipe cap on the pipe and mount it on the wood framework. Make sure that the screws do not stick out beyond the suction cups or you'll have scratched paint.

Bend the top mast bracket from sheet metal. By varying the horizontal position of the hole in the bracket, the mounted mast can be made to be vertical, even if your vehicle door isn't. If the outside of your vehicle door isn't vertical (most aren't), you may need to put a shim between the bottom pipe and the wood frame near the top mounting screw, or the antenna mast may bind. Using a larger diameter pipe helps eliminate the problem, but the antenna wiggles around more.

The actual dimensions of the door mount are not given since every door is different. You may find a better way to mount the mast or the assembly on your door. Dig around in your own junk box and see what you can find. Don't be afraid to experiment and improvise.

SUN ROOFS AND CONVERTIBLES

A slick way to keep the wind out of the car, but still be able to turn the antenna from the inside, is to run the mast down through the sun roof. If your vehicle has a sun roof that is opened and closed with a crank, one mounting method is shown in Fig. 7-9.

Cut a piece of quarter-inch plywood about two inches wider than the side to side opening of the sun roof. The length of the plywood will depend on where you want the antenna mast to come through the sun roof. Decide how far back from the front of the sun roof opening you want the mast and add another four to six inches back from there. About one inch in from the front and back of the plywood sheet attach a piece of 2×2. These pieces should be short enough so that they will fit inside the sun roof opening.

Carefully set the mount over the sun roof with the front 2×2 touching the front lip of the opening. Slowly crank the sun roof closed until it is just touching the rear 2×2. Now, closely examine all around the mount to see all the places where wood is touching painted surfaces. To keep from scratching the paint, glue either old carpet or foam rubber to the wood. An easy way to do this is to use press-on foam rubber weather stripping.

One way to secure the bottom of the mast is also shown in Fig. 7-9. Use a pipe cap larger than the mast and mount it on a block of wood. Attach the wood block to the floor of the vehicle, either by screws or using Velcro™ strips. This material is available in yardage stores and comes in matching strips that stick to each other. Glue one side of the bottom of the wood block and sew the other to the floorboard carpet. The nice thing about mounting the block with Velcro is that it can be mounted without tools and removed with only a pull.

Another method for securing the mast bottom is to modify an automotive beverage holder. Any type of holder that mounts on the hump or between the seats will work as long as it doesn't slip around. One kind commonly found has lead shot bags that drape down the side of the hump on the floor.

The lack of a solid upper frame limits the type of mount that can be used on a convertible. One solution is to use a door mount as previously described. Another way is to mount a bracket for the mast just above the rear view mirror. Make the bracket from sheet metal and mount it to the window frame just above or around the rear view mirror. The bottom of the mast can be held in the same manner as with the sun roof.

THE HOLE IN THE ROOF

Having a through-the-roof mast is about the nicest way to hunt, especially in bad weather. The mast can be turned by either the driver or navigator with equal ease, and direction indicators are easily mounted. Perhaps you've thought about boring a hole in the car's roof. You've stopped worrying about the resale value, but:

- ☐ "I don't have a machine shop handy to make a bushing."
- ☐ "I don't want to stuff a rag in the hole."
- ☐ "I don't have a cork that big."

Excuses, excuses! Relax. Making a bushing that seals out the weather between hunts is simple and cheap. A mount of this kind can be made totally waterproof very easily.

Three-quarter inch PVC pipe is a common mast material, so we'll describe a bushing and weather seal for this pipe. It is made of commonly available PVC pipe fittings. Total cost is about three dollars. The only tools needed for the bushing are a saw and assorted files. For other mast sizes, take a piece down to your local hardware store and try parts for proper fit.

Figure 7-10 shows the details of the bushing assem-

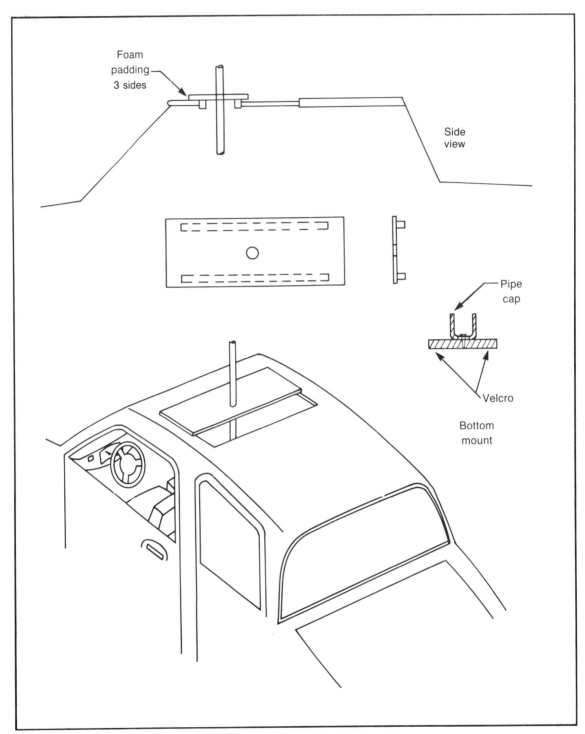

Fig. 7-9. If your vehicle has a sunroof, a through-the-roof mount is easy to build and very nice to hunt with. The antenna mast can easily be turned by both the driver and navigator.

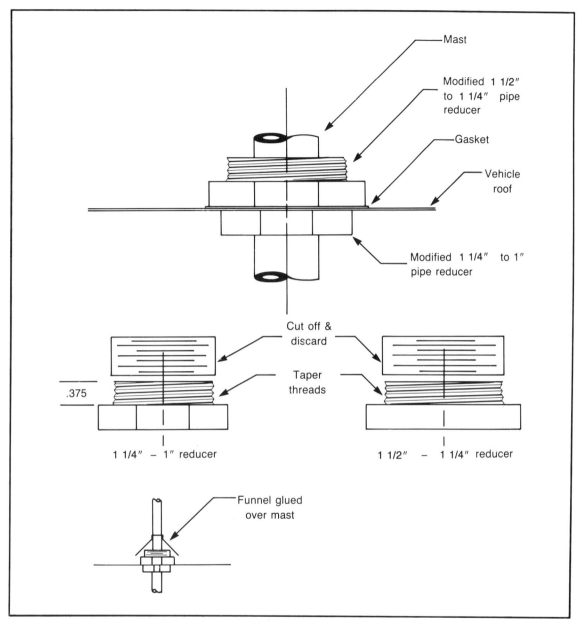

Fig. 7-10. For only a few dollars and a few hours work, a very neat and serviceable roof bushing can be made. When the bushing is not being used, the hole can be sealed by screwing on a pipe cap.

bly. The two bushing halves are modified PVC pipe reducing adapters. The smaller one is 1″ to 1-1/4″, and the larger is 1-1/4 to 1-1/2″. This is the size marked on the adapters; measured diameters will be different. Be sure to buy the threaded variety. Cut off part of the threads as shown, leaving 3/8 inch.

The adapters have tapered pipe threads, meaning that the further they are screwed together, the tighter they get. Since the section cut off is the smaller diameter portion, the remainder will fit tighter than desired. Carefully file a taper on both sets of threads, cleaning up the grooves with a small triangular file or pocket knife.

81

The fit of the two reducing adapters should be fairly tight, but if a lot of wrench torque is needed for the final mounting on the roof, they may crack. Keep trying the fit until they will screw together as far as needed. After cutting a hole in the roof, mount the bushing using a gasket made from gasket material or a piece of inner tube. Before screwing the pieces together, coat both sides of the gasket with gasket compound, such as Permatex™, or silicone bathtub caulk.

Used as is, the bushing gives about one tenth inch of clearance with the fittings and pipe we had. For a tighter fit, get an extra pipe cap and bore a hole that will just match the mast diameter. If your mast is slightly larger than the inside of the bushing, the unused threads may be easily filed out to the proper diameter.

Even with a tight fitting mast, some rain water will dribble through the hole. Rather than tying a rag around the mast or building a complex waterproof bearing, try the simple solution shown in the detail. Cut off the spout of a plastic funnel and widen the opening until you can just force your mast through it. With the mast mounted on the vehicle in the hunt position, slide the funnel down so that it is just over, but not touching, the bushing assembly. Glue it in place onto the mast, taking care that it stays centered. A plastic spray can top can also be used, bored out and glued on.

After the hunt is over, just unscrew the bored out pipe cap, if used, pull out the mast, and screw on an unmodified pipe cap. Your roof is now weatherproof, and unless someone looks very closely, it looks like an antenna base minus the whip.

WINDOW BRACKETS

To many hunters, the thought of having a hole in the car roof conjures up visions of the salesman saying, "Well, your car is in fine shape, but because of this small hole, we'll cut $1000 off the trade-in allowance." Even though this isn't a realistic fear, most folks don't like to cut a 1-1/4 inch hole in the family vehicle. Hams with vinyl-covered car roofs are particularly reluctant. The good news is that you can make a mount that attaches to the window frame without any clamps, screws or holes.

The mount, shown in Fig. 7-11, is made from three pieces of quarter inch plywood and two pieces of 2 × 4 or 2 × 6. Cut the plywood pieces about three quarters of an inch shorter than the height of the window opening. Glue and nail the pieces together as shown with the center piece sticking out about a half inch.

Be sure to mount the antenna mounting brackets on the outside plywood before gluing the plywood pieces to-

gether. If not, the mounting for the bottom bracket may be in the slot for the window when it is rolled up into the slot. Taper the center piece so that it will slide up inside the window well on your door. Just roll down your window, stick the mount into the slot in the top of the window frame, and roll the window up into the slot in the bottom of the mount.

Except for trucks and vans, almost all vehicles have windows that aren't exactly vertical. The mounting brackets must be offset to hold the mast vertical. The brackets can be made from metal, but a quicker method is to use blocks of wood mounted on the wood frame. After you have mounted the blocks, attach the whole antenna mount on the window and determine at what angle the holes for the mast need to be drilled.

If you want a deluxe easy-to-spin mount (your navigator will thank you), put bearings at the top and bottom as shown in Fig. 7-12. Automotive parts stores have bearings. Find a store that specializes in bearings for an even better selection.

WINDOW COVERINGS

Without our wives as navigators, turning the quad mounted in the right side window, we simply wouldn't be competitive hunters. (At least that's what we always tell them!). However, in the winter months our concentration was often interrupted by their complaints about blue and stiff fingers, wet clothes, windblown hair, and general discomfort. Obviously not in the true T-hunt competitive spirit, right?

Our compassion (and their threats of nonparticipation) moved us to develop some ways (short of a hole in the roof) to keep the car reasonably sealed up while allowing the mast to be easily turned. One way to keep the weather outside is to build a window box. Shown in Fig. 7-13, the box is made from sheet Plexiglas.

Look for plastic suppliers in the local phone book. Cadillac Plastics is one nationwide plastic supplier. That company has a booklet on gluing and cutting Plexiglas.

Select a piece of plastic sheet that is slightly larger than the window opening and is slightly thicker than the window. Make a template from cardboard the size and shape of your window. Cut it so that the template sticks up inside the window channel on the top and front, is flush on the rear, and clears the bottom by one inch or so. Use this template to cut the sheet of Plexiglas.

Clean up the cut edges with a file and try it for size. You may need to bevel the plastic where it fits inside the window well. With the plastic installed in the window opening it should look something like the window except

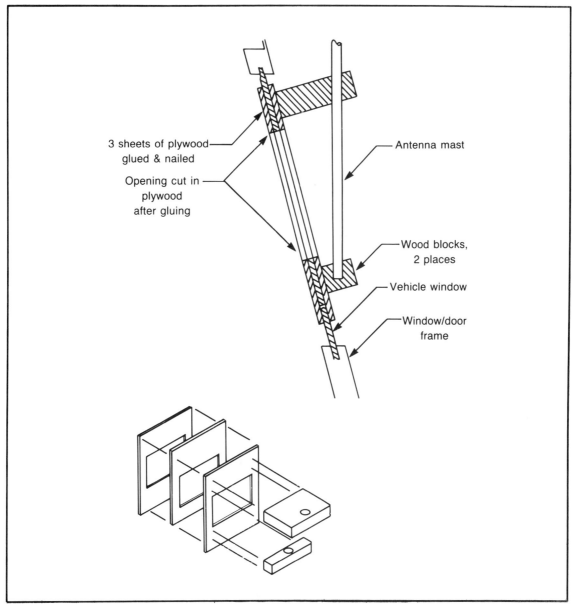

3 sheets of plywood
glued & nailed

Opening cut in
plywood
after gluing

Antenna mast

Wood blocks,
2 places

Vehicle window

Window/door
frame

Fig. 7-11. This simple window mount can be thrown together in a few hours. Remember to install the mast mounting blocks before assembling the plywood pieces.

for the one-inch gap in the bottom.

Cut two strips of plastic the length of the window opening and about 1.5 inches in width. These pieces are glued on each side of the bottom of the plastic window with each piece sticking down below the edge about a half an inch. A solvent type of glue is very easy to use with Plexiglas, and a two ounce bottle will last through a number of projects of this scope.

Lightly clamp the assembly and, using a medicine dropper, apply the solvent along the joint of the overlapping pieces. You can watch the solvent flow between the two pieces by capillary action. Let the joint dry for a few hours (preferably overnight) before removing the clamps.

After the glue is thoroughly dry, try the panel for fit.

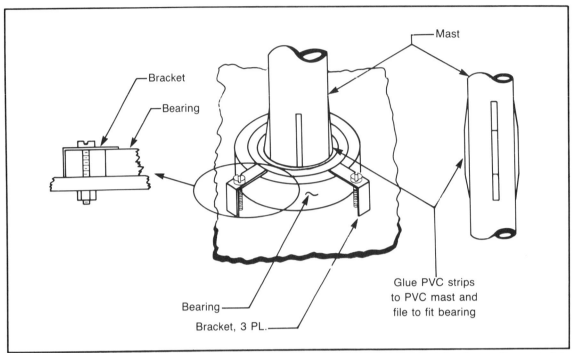

Fig. 7-12. Installing bearings at top and bottom will greatly ease turning the antenna, especially when on the move. Automotive parts suppliers and bearing specialty houses have a wide selection of appropriate parts. Mounting of the bearing can be varied from the figure to suit your own setup and available hardware.

After sticking it in the slot at the top and front of the window frame, the window should roll up into the slot formed by the pieces glued on the bottom. You may need to file the edges of this slot so that the window will slide up smoothly inside it.

With the sheet attached to the window, determine how big the cutout needs to be for your antenna setup. Because the window opening is usually somewhere around shoulder height, your hand is up fairly high when turning the antenna. The opening should be as low as practical. Determine the dimensions of the cutout and lay them out on the plastic sheet.

A rectangular piece can be cut out to form a hole, but the sharp corners of the hole form high stress points. Though there will probably be no problem with this, a better way is to drill large (about a half inch diameter) holes at each corner before cutting the piece out. The hole will look the same, except that the corners will have a quarter inch radius instead of a sharp corner. This reduces the chances of cracking the main plastic sheet.

Determine how deep the box portion needs to be to hold the mast and provide clearance for your knuckles when turning the antenna. The pieces must be cut at an angle so that the top and bottom pieces are parallel to the ground, to hold the mast vertical. You will end up with two trapezoid shaped pieces, a small rectangular piece for the bottom, a larger one for the top, and a rectangular piece about the size of the hole cut out of the main plastic sheet. The top and bottom sheets need to have the inside edges cut at an angle to match the angle of the window from vertical.

Smooth the cut edges with a file and tape the whole assembly together to check for proper fit. Keep filing until all of the joints fit smoothly. When everything fits, glue all the joints with the solvent. Leave the unit taped together and undisturbed overnight to thoroughly dry.

If the joints were fitted very carefully, the assembled box should be sufficiently strong. For even more strength, either drill and tap all the joints for assembly with screws before gluing, or bolt in small angle strips or brackets in the joints.

After the box is fully dry, try it for fit in the window. With the box installed on the window, sit in the car and figure out approximately where the mast should be. Bore a hole through the top for the mast, and make a bearing assembly to hold the bottom of the mast. As before, a

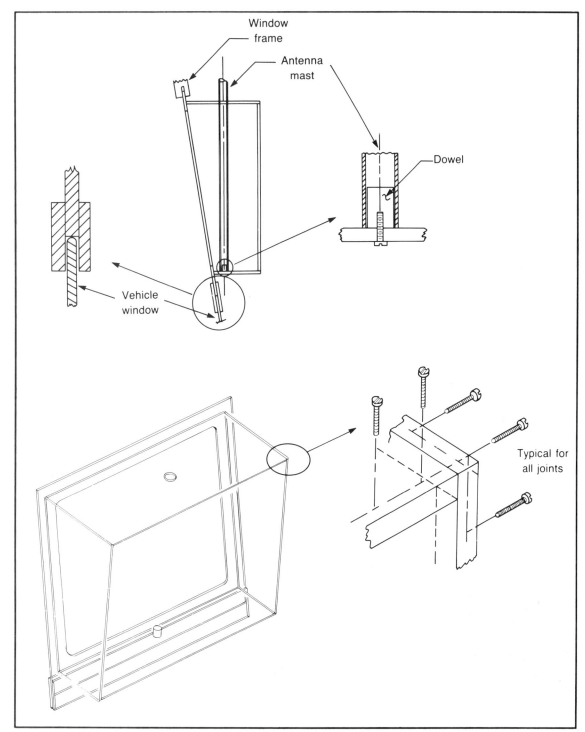

Fig. 7-13. A window box allows hunting in the worst weather, keeping the antenna turner warm and dry. Using Plexiglas for all the sides gives excellent visibility.

bearing assembly can be installed at both top and bottom.

Not only does a box mount of this type keep the weather outside, but the ledge inside provides space to mount the attenuator, direction indicator and other DF accessories. A similar window box, but made partly from plywood, is pictured in Fig. 7-14, complete with plumber's helper rain deflector. Note the area around the mast has a protractor direction indicator.

WINDOW INSERTS

Weather coverings for use with through-the-window mounts described earlier can be made very easily. Referring to the previous section on the window box, make a template and cut a piece of Plexiglas as before. After gluing on the bottom strips, mount the plastic sheet in the window. Carefully note where the mast would go

through the plastic. Cut or drill an undersize hole in the plastic and, using a large round file, bevel the edges until the mast slides through and match up with the top and bottom mounts.

When finished, the hole should look as if it has been drilled at a sharp angle and will be noticeably oval shaped. With some windows and mast installations it may be necessary to split the plastic into two pieces. Cut it down through the mast hole. Figure 7-15 shows the completed installation.

DIRECTION INDICATORS

No antenna mounting system is complete without an easy-to-read direction indicator. Of course you can look up at the antenna when hunting in a convertible or car with a sun roof, but it's dangerous for the driver, and is

Fig. 7-14. A rather unusual window box, but it works. Note the "plumbers' helper" rain deflector and the large protractor around the base of the mast.

Fig. 7-15. When using a mount such as those shown in Fig. 7-5 or 7-6, a Plexiglas insert is an easy way to enclose the window. In certain installations it may need to be split and installed in two sections. Except for the hole and the split, the window is completely sealed. An insert of this type can easily be built in an afternoon.

Fig. 7-16. A straight-forward handle and direction indicator combination. Note the top sheet metal bracket, held in place by closing the door on it. This is an outstanding example of a simple and cheap antenna mount that can be put together in a hour or two.

sure to produce a sore neck after a while. A good direction indicator minimizes the distraction to the driver's attention and provides readout accuracy to within a degree or so.

WB6UZZ admits that for his first few hunts the only method he used to tell bearing direction was to remember that the antenna feed line was taped to the rear (reflector end) of the mast. As he held the mast, he could feel the position of the coax. It wasn't highly accurate, but it worked.

Some hunters are happy with just a long screw through the antenna mast. Others take a cheap screwdriver, cut off most of the shank, and stick it through a hole drilled in the mast, pointing in the direction of the beam lobe. It forms both a pointer and a convenient handle for turning the antenna. The handle of Fig. 7-16 is more elegant. Figure 7-17 shows a more straightforward (dare we say crude?) means.

Accuracy By Degrees

For truly competitive starting bearings and for triangulation, readout to the nearest degree or so is a must. More accurate bearing readouts can be mechanical, electro-mechanical, or electronic. We'll describe several methods, going from austere to elegant. Keep in mind that none of them are far and away more accurate than the others, but the more complex designs add convenience. Best accuracy results from doing a lot of careful boresighting and eliminating all mechanical "slop" in the mounting.

If your installation permits, mount a 360 degree protractor at the base of the antenna mast as shown in Fig. 7-18. Attach a wire or plastic pointer on the mast and adjust it to read zero degrees when the antenna is pointed straight ahead. Look around in your local military surplus store. You may find all sorts of strange degree wheels and scales that can be adapted to antenna readouts.

Fig. 7-17. One way to get a firm grip on the mast to turn it. Just be sure you clamp the pliers in the same direction each time.

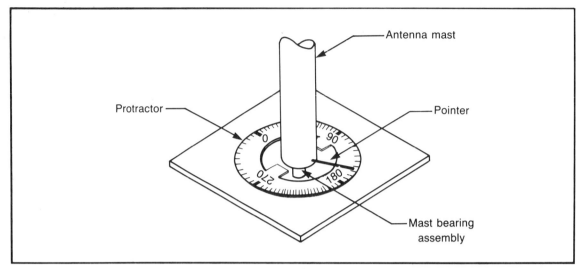

Fig. 7-18. For accurate bearings, some type of protractor readout is needed on the antenna mast. The easiest, if your installation will allow it, is to mount it directly under the mast.

One idea for a readout is shown in Fig. 7-19. The circular wheel is made from the cover for a 50-foot reel of Super-8 movie film, glued to the PVC pipe mast. The pointer is attached to the window sill with double-sticky foam. The long screw through the mast gives a general indication and allows eyes-off direction checks when in close.

Selsyns and Synchros

Selsyn transmitter-receiver pairs have been popular with hams as antenna readouts since they first became available on the surplus market. A selsyn (sometimes called a synchro transformer) is a specially wound AC motor. When two matching units are connected together and excited from an ac source, their positions track perfectly, even when stationary. Turn one 84 degrees, and the other follows by turning almost exactly 84 degrees. The error is less than one degree.

Acquiring a pair can be a challenge nowadays. WB6UZZ uses a transmitter-receiver pair from a Bendix aircraft radio compass (Fig. 7-20). You might try looking around your local airport. Some radio shops may have some junked equipment of this type lying around. Selsyns have also been used in ground, airborne, and shipboard radars. They can sometimes be found in surplus stores and ham flea markets. Before you buy, try to determine the design frequency and voltage. This is usually, but not always, marked on the side.

Hook up is usually straightforward. The terminals are typically marked S1, S2, S3, R1, and R2. Connect wires from transmitter S1 to receiver S1, transmitter S2 to receiver S2, and so forth. If they are unmarked, and position of the terminals isn't a clue, remove the rear cover. Look for color coding on the internal wires and connect wires between matching colors. If you aren't sure, don't be afraid to experiment. Swap the wires around until they work. A typical hookup is shown in Fig. 7-21. A method of mounting the selsyn to the mast is shown in Fig. 7-22.

One drawback of selsyns is that they require ac for excitation, usually 60 or 400 Hz. A dc/ac inverter is required for mobile operation. Ideally, the inverter should generate a sine wave. But a square wave is much easier to generate, and the selsyns don't seem to mind.

Two types of inverters are practical for driving selsyns, self excited and externally excited (clocked). A self excited inverter is shown schematically in Fig. 7-23. The circuit dates back to the 1960s and was popular when it appeared in the *ARRL Antenna Book* edition of that era. It's a good use for those old round-can germanium PNP power transistors in your junk box. Silicon power types work fine too. Depending on transformer and transistor characteristics, the values of R1, C1, and C2 may have to be changed to ensure steady oscillation on the correct frequency.

The clocked inverter circuit of Fig. 7-24 is much closer to the state of the art. Frequency is determined by a separate oscillator stage, and will not vary with load or transformer characteristics. The flip-flop stage, which

Fig. 7-19. For a window mount, the protractor can be mounted on the mast itself, with the pointer fixed.

Fig. 7-20. Using a selsyn pair is an excellent way to get a remote readout of the antenna direction. In WB6UZZ's installation, the selsyn is part of a radio compass (Automatic Direction Finder, ADF) panel indicator. The other two meters are a remote S-meter and a noise meter (see Chapters 5 and 11).

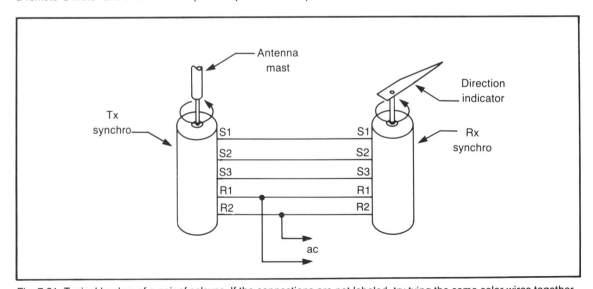

Fig. 7-21. Typical hookup of a pair of selsyns. If the connections are not labeled, try tying the same color wires together.

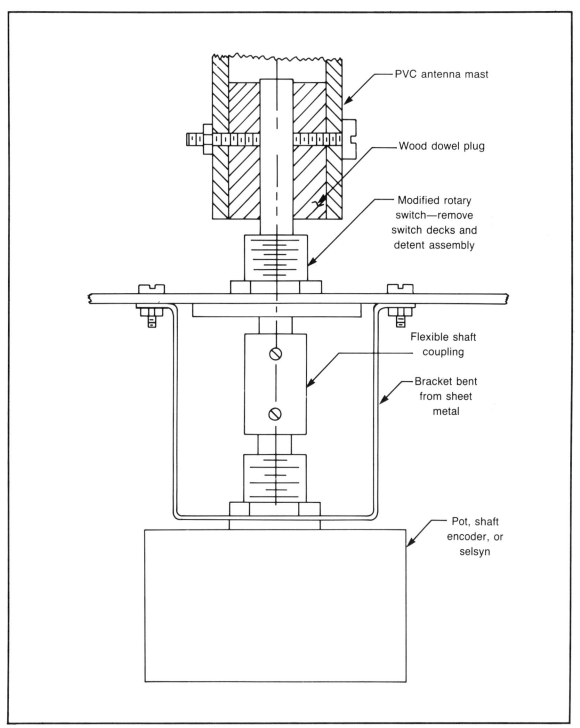

PVC antenna mast

Wood dowel plug

Modified rotary switch—remove switch decks and detent assembly

Flexible shaft coupling

Bracket bent from sheet metal

Pot, shaft encoder, or selsyn

Fig. 7-22. One method of mounting a selsyn (or pot or shaft encoder) to your antenna mast. The bushing itself is a modified rotary switch.

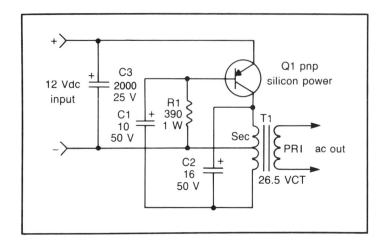

Fig. 7-23. A self-excited dc to ac inverter can be built from a transformer and a single transistor. Resistor and capacitor values may need to be changed to get the desired frequency. You may find that the output frequency changes with load.

divides the oscillator frequency by two, gives perfect square wave drive to the Vertical Metal Oxide Semiconductor (VMOS) chopper transistors.

A small power transformer with a 24 volt center tapped secondary connected backwards gives approximately 120 Vac output with 13.6 Vdc input to the inverter. The output voltage can be changed to suit your selsyn requirements by appropriate transformer selection. For 26 V selsyns, try a 110V/110VCT transformer.

The VMOS transistors must be handled carefully to prevent electrostatic discharge (ESD) damage. Use a grounded soldering iron and a grounded work bench. The devices may be packaged with the leads in conductive foam. Wrap a piece of fine wire around the leads of the

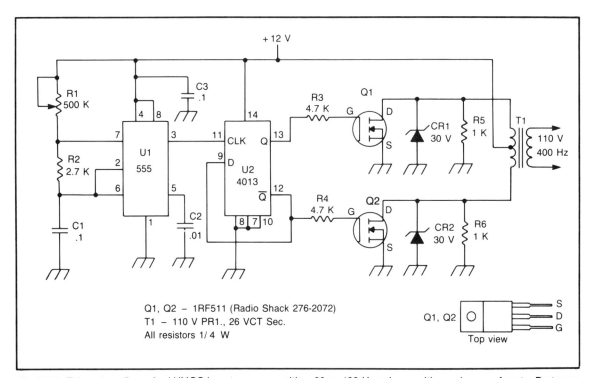

Fig. 7-24. This externally-excited VMOS inverter powers either 60 or 400 Hz selsyns with no change of parts. Parts are commonly available and construction is not critical.

transistor directly adjacent to the body before removing it from the conductive foam. Remove this shorting wire after all circuit wiring is completed.

Mount the transistors on a heat sink or chassis using insulating hardware. The sink should be about the size and surface area of the prototype in Fig. 7-25. Use thermal compound (heat sink grease) on both sides of the insulating washer.

The resistor and diode across Q1 and Q2 are soldered directly to the transistor leads after they are mounted on the heat sink. The ICs and associated components can be mounted on a small piece of perforated board. Mount the board and transformer as close to each other as practical. The wire used for Q1 and Q2 source and drain leads should be AWG 22 or larger to minimize voltage drop.

After completing the wiring and checking it over, temporarily disconnect the transformer center tap from the supply line and apply power to the remainder of the circuit. Set the oscillator frequency at U1-3 to twice the desired inverter output frequency with R1. Range of R1 should go from 60 to 400 Hz. The outputs of U2, on pins 1 and 2, are complementary square waves at the inverter frequency.

Now reconnect the center tap of the transformer. The waveforms on Q1/Q2 drains will be square waves with peak to peak voltage of twice the supply. Voltages are about 26 to 28 V peak to peak when used in a car.

The LED Ring

Remember the inexpensive bar-graph LED driver used earlier for a rapidly responding S-meter? It can also form the basis of a direction indicator that's easy to see in the dark, even out of the corner of the driver's eye. The indicator can have as few as ten or as many as a hundred LEDs, depending on your desire for accuracy and your patience in drilling mounting holes.

The light-emitting diodes are arranged in a circle and

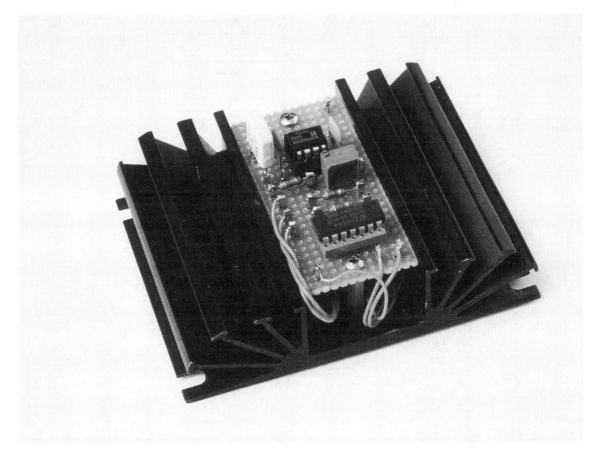

Fig. 7-25. The VMOS inverter as built on a small heat sink.

calibrated in degrees or direction. The circuit of Fig. 7-26 has 20 LEDs giving 18 degrees per step. By adding more drivers, more diodes can be driven in groups of ten. How about a ring of 36 for ten degrees per step? Just use four ICs, offset with resistors, and ignore the last four outputs.

The National Semiconductor Linear Data Book contains detailed information on daisy-chaining LM3914s. Other drivers work as well but the specifics of hooking them up differ. Consult the manufacturer's literature before substituting. Make sure the part is a linear driver—some are logarithmic units for audio applications and cannot be used here.

The most critical component of this indicator is the mechanical transducer, a continuously rotatable potentiometer. Such a pot doesn't have the normal stops, and the resistance element covers the full 360 degrees of shaft rotation instead of the usual 300 degrees or so. It must also have a linear taper. It is mounted under the mast in the same way as the selsyn shaft in Fig. 7-22.

Pots such as these are very expensive when purchased new. The best places to find one are surplus stores or salvage sales where electronic instrumentation equipment is featured. The exact value isn't really important as the circuit can be modified to accept many values. The 1 k to 5 kohm range is ideal.

If you have a choice of pots, select the one with the

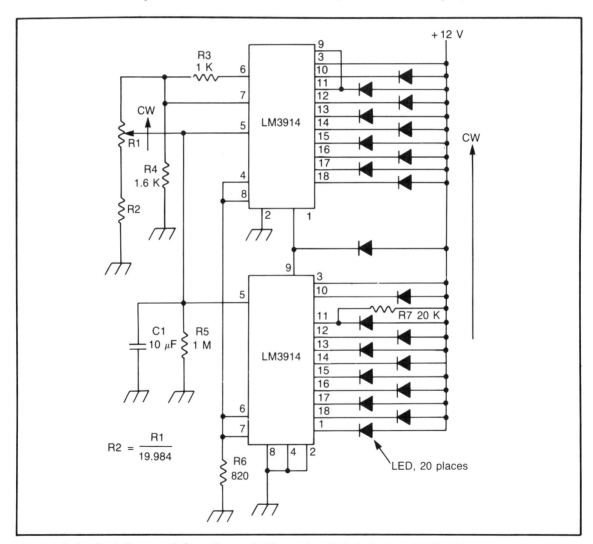

Fig. 7-26. A direction indicator made from a bar graph driver and pot. This design has 20 LEDs giving 18 degree resolution.

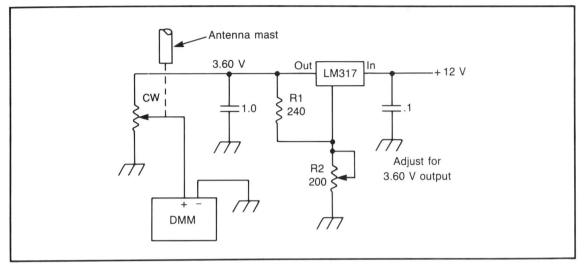

Fig. 7-27. Digital direction indicator using a digital multimeter (DMM).

smallest dead space. Dead space is the rotational area between maximum resistance and zero ohms when measuring from center tap to one end. Take along your VOM when you shop so you can check this. Instrumentation pots often have only an undecipherable part number instead of a marked resistance value, making a VOM check a necessity.

To ensure that one LED is on when the pot is at zero ohms, the voltage output from the pot must be offset by one step of the bar graph driver. This is accomplished by R2 between the bottom of the pot and ground. The value of R2 will vary depending on the value of the pot (R1) chosen. Assuming that a 20-LED display is used, the value of R2 is given by:

$$R2 = R1 / 19.984$$

For example, if a 1 kohm pot is used at R1, then R2 is 50 ohms. For more than 20 LEDs, consult the data book. If the calculated value falls between two standard resistance values, use the next higher value instead of the lower one.

Digital Direction Readouts

The continuous rotation pot can also be used to make a digital direction readout, as shown in Fig. 7-27. The heart of this indicator scheme is a digital voltmeter (DVM) module. Panel mount DVM modules can be purchased new in the $30 to $50 range and can be found surplus for less. Use of the DVM eliminates the need to home

brew the analog to digital conversion circuitry, latches, and digital readout. The regulator provides a precise 3.60 volt reference which of course corresponds to 360 degrees.

Another method is to use a digital multimeter (DMM). Radio Shack and other outlets sell DMMs for under forty dollars. When not being used for T-hunting, the DMM will find many uses in testing and troubleshooting other projects. As before, the value of the pot isn't critical so long as there is a minimum amount of dead space.

Here's a final idea to think about. Larry Starkweather, WD6EJN, mounted an ordinary TV rotor and mast solidly to his rear bumper, to turn an aluminum yagi. It turns a bit slowly, but the digital readout on the rotor control gives great accuracy, and he stays dry in the rain. The best way to do this is to replace the ac rotor motor with a dc unit and rewire the control box.

Keep Your Setup Legal

Before you complete your antenna and mounting plans, give some thought to the legal aspects of your installation. The three requirements pointed out here are based on the laws of California. You can be sure there are similar limitations in other states. They may be explained in a Driver's Handbook, used for preparation for the written driver's exam, or you may have to check your state's Vehicle Code at the local library. Some municipalities may have additional limitations.

☐ Overhang. In California, you must not carry any-

Fig. 7-28. Many states do not allow any overhang on the side of the vehicle. There is no overhang with this dual antenna DF.

thing in or on a passenger vehicle which protrudes beyond the line of the fenders on the left side, or more than six inches beyond the line of the fenders on the right side. This may be just the excuse you need to punch that hole in the car top, particularly if you hunt alone. A properly spaced 2 meter 4-element quad probably can't be mounted on the left side of your car and meet this requirement, although it may squeak by on the right side. The through-the-window mount on the thick door of an American car may allow the legal use of a dual antenna DF unit on the left side, as shown in Fig. 7-28. Measure before you build. There is no legal limit stated on front/back overhang,

which allows some cross-country hunters to use a very long vertical vhf yagi, keeping the boom secured along the length of the car or van while driving and stopping off the road to swing it for bearings.

☐ Obstruction. It is unlawful to add T-hunting equipment which obstructs the driver's view in front or to the sides. If you put in a window box to hold the rotating mast and keep out the cold, make sure you can still see out the window enough to be safe. Use lots of clear Plexiglas where possible.

☐ Headsets. Earphones are useful for hearing weak signals, but one ear must be kept uncovered while you're driving.

Patrol officers are very familiar with these rules, and while they may not stop you for an extra inch or two of overhang, they will not hesitate to pull you over if you appear to create a hazard. And they will be unimpressed with your plea to let you go quickly because you're on a timed hunt! Avoid trouble—make sure your installation conforms with your local laws.

Besides the legalities, be sure to think of the safety of your setup. Might you be liable if a personal injury occurs? Check your gear for the following:

☐ Is the front seat full of cables and wires which could tangle in the steering wheel?

☐ Could you or a passenger be injured by flying gear if you have to brake suddenly?

☐ Could your antenna or mount be a danger to passers-by? How about tip protection on those Doppler radials?

☐ Is your quad or beam top less than 13 feet, 8 inches above ground to prevent it striking an overpass?

Finally, remember to give some human engineering consideration to equipment placement. You may not be able to have a "heads up" display like a fighter pilot, but equipment should be placed on the dash in such a manner that you won't have to take your eyes off the road or look down at the seat for long periods.

Chapter 8

Homing DF Units

Despite its advantages for weak signals and simplicity, a gain antenna such as a beam or quad has a disadvantage—it is not a very sharp indicator of azimuth. When the source is far away across open country, a very sharp directional indication is a great help. Also, a quad or beam is not appropriate for airborne DFing.

It was with these needs in mind that experimenters developed DF units that take advantage of a receiver's capability to process the information from more than one antenna, when the antennas are switched back and forth at a rapid rate. These units are called homing DFs because they home in by giving left/right and front/back indications of which way to turn the antenna, vehicle, or aircraft in order to be aimed at the source.

The common features of the DF units described in this chapter include:

☐ Electronic switching in the antenna system at an audio rate.

☐ A single sharp indication of apparent signal direction, which may be rendered inaccurate by high multipath levels.

☐ Indication of which way to turn, left or right, to aim toward the signal source (except on the DDDF).

Some homing DFs use AM receivers and some use FM receivers, but none use both. Some are kits and some come complete and ready to hunt. In operation, all the switched antenna units described here except the DDDF are used in the same way. They have two vertical antennas, a left/right indicator, and a DF buzz in the receiver audio. In this chapter we'll look at some of them in detail and tell what to expect in performance and accuracy.

The left/right indication of a homing type DF makes it possible to find signals of very short duration (often called "kerchunks") because the indicator tells which way to turn the antenna. It only takes a few short transmissions to find crossover by moving the antenna according to the indication after each transmission. The first kerchunk left/right indication tells which 180 degree hemisphere it's in. Turn the antenna 90 degrees in that direction. The next burst tells which half of the hemisphere. Move the antenna 45 degrees in that direction, and so on . . .

The flip side of this is that these DFs are more affected by reflections, and they require time averaging in motion for best bearings. This means that it's important to minimize reflections when getting bearings on intermittent signals by getting as high and in the clear as possible.

It would seem that aircraft and homing type DFs were made for each other. They are easy to install, require only two vertical antennas, and give a direction-to-turn readout. Almost every popular airborne DF unit for vhf/uhf use is in this family. For the same reasons, they are well suited for use on boats, though most boat owners seem to prefer Doppler DFs.

Wouldn't it be nice if cars could be maneuvered as easily as planes or boats? Unfortunately the authorities take a dim view of 360 degree turns on streets and freeways, so fixed mounting of the two antennas on a car isn't good enough. Many years ago, one of the authors rode along with a hunter who was proudly trying out his new secret weapon, a Happy Flyers unit with the two antennas mounted solidly on a ski rack on the roof. It was a rainy night, and the signal was weak. Countless times the car was stopped and someone got out, pulled the quad out of the back of the wagon, and spun it to get a good bearing on the weak signal. We never did arrive at the T in the allotted time.

To take best advantage of the sharp crossover characteristic of a homing type DF in a car, a rotatable mount, such as those described in Chapter 7, is necessary. Some hunters report that the car body can affect the bearing when the antenna is rotated. Using full dipoles as far above the car as practical instead of ground planes avoids this effect. However, there is a method for using multiple fixed antennas that does work, and is described later.

When used handheld in close, a homing DF can be turned sideways and used to measure elevation of the signal. It can often tell, for example, that the transmitter is on the roof of a building, near the ground, or on a middle floor. Cross polarization adversely affects attempts at vertical angle measurements in some cases.

THE DOUBLE-DUCKY DIRECTION FINDER

Hams who like to home brew their gear may be interested in a simple handheld DF unit for use with FM receivers. Called the Double-Ducky Direction Finder (DDDF), this unit was designed by David Geiser, WA2ANU. Its main features are its small size and simplicity, and the ability to work over the entire band covered by the radio it's used with. In operation it is quite different from the other DFs in this chapter, but its switched antennas make it a homing unit.

Standard short helix (rubber duckie) antennas are mounted on an oval ground plane about 20 by 10 inches. The square wave used for switching comes from a 567 tone decoder IC. A switch position puts a fixed bias on the antenna diodes to allow transmitting at low power

through the DF without disconnecting it.

The only connection to the receiver is the antenna. In the DF mode, the DDDF superimposes an audio tone on the received signal audio. The tone is at a minimum, or null, when the antennas are equidistant from the transmitter. Thus at null, the bunny is along a line which is perpendicular to the line connecting the two antennas.

The biggest disadvantage of this unit is that it is bidirectional, with no provision to tell front from back. WA2ANU suggests experimenting with reflectors to make the pattern asymmetrical and resolve the ambiguity. However, an effective reflector would probably be so large as to take away much of the advantage of the unit's small size.

Another problem is that finding the exact null point, particularly in the presence of multipath, takes some skill and practice. As with the null-seeking antennas discussed in the antenna chapter, some locations provide no null at all. Audio frequency harmonics may be generated at the null, causing timbre changes in the DF tone that require some interpretation.

If the signal being hunted has heavy tone modulation, as often happens on fun hunts, it can be difficult to use a unit of this type. It is hard for an inexperienced user to hear and resolve nulls on the DF tone in the presence of the other tones on the audio. Take time for plenty of practice before using the unit in competition.

The construction plans for the DDDF have been reprinted in the *Handbook for Local Interference Committees*, available from the American Radio Relay League. A complete kit of parts, including enclosure, is available from Circuit Board Specialists of Pueblo, Colorado.

SWITCHED CARDIOID PATTERN HOMING UNITS

There are many varieties in the designs of switched pattern direction finders. We won't try to point them all out; instead, we'll discuss the general principles that make them work. Phased antennas that produce cardioid patterns at vhf were discussed in Chapter 4. The antenna set of Fig. 8-1 is just a two element phased array with switched selection of directivity. Each element is a half wavelength long. The electrical distance between the antennas (D1) is the same as the electrical length of the delay line L, because the velocity factor of the coax is accounted for. Feed lines can be any reasonable length (D2), but must be equal.

Spend a few minutes analyzing the geometry of the system (this is what your textbooks used to call an "exercise for the reader"), and you'll conclude that when the switch is in position #1, the antenna has a cardioid pat-

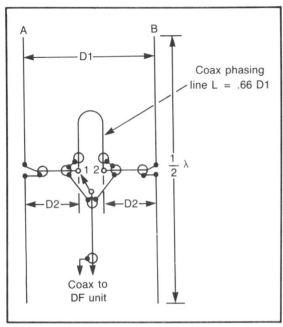

Fig. 8-1. Example of the basic switched pattern antenna.

tern with the null to the left, and for position #2, the null is toward the right. The resultant patterns are given in Fig. 8-2. Best cancellation occurs when D1 equals one quarter wavelength, but a perfect null is not necessary for the DF to function.

Notice what happens when receiving a signal from 45 degrees azimuth. Position #1 produces a much greater amplitude response than position #2. The opposite is true for the signal from 315 degrees. When the signal is from zero degrees, the patterns produce exactly the same output. All that needs to be done to make a DF using this principle is to use the position #1 signal amplitude information to move a meter to the right, and the position #2 amplitude to move the same meter left. At crossover, the equal amplitudes make the meter read zero.

Switching circuits with rf type diodes can do just this. Replacing the mechanical switch with the circuit of Fig. 8-3 and putting a matching switching circuit in the display makes the same receiver process position #1 and position #2 information alternately, at any practical rate. A rate in the audio frequency range is usually used.

DF experimenters have found that more sensitivity at crossover can be achieved by using the phase relation-

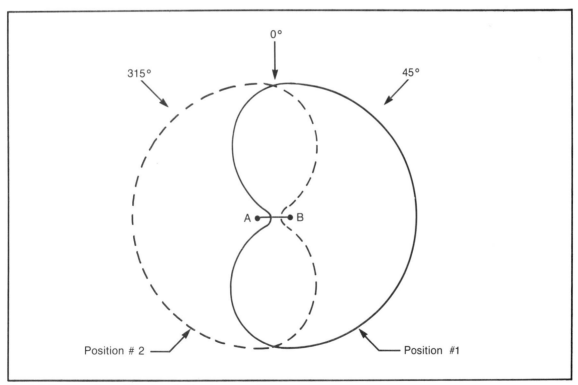

Fig. 8-2. Pattern of a switched cardioid antenna system.

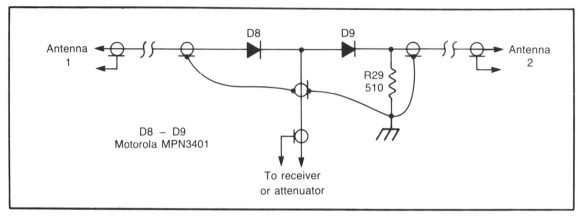

Fig. 8-3. Circuit of the antenna switching portion of the Happy Flyers DF.

ship of the signals from the patterns than obtained from the amplitudes. Some DF units are both phase and amplitude sensitive, and a strong signal will cause more needle deflection than a weak one. Others respond only to phase, and have meter swing corresponding only to the number of degrees away from crossover, regardless of amplitude.

THE LITTLE L-PER DIRECTION FINDER

First offered in the summer of 1975, the Little L-Per LH series by L-Tronics rapidly became the most widely used piece of commercial gear for amateur radio DFing, and has found wide acceptance in all forms of search and rescue work.

L-Tronics provides a complete integrated system, with a sensitive double-conversion superheterodyne receiver and folding dual dipole antenna system, which can be put to work by even an inexperienced DFer with only a few minutes of training. It weighs less than 2 1/2 pounds, including integral batteries, and folds up for easy storage and carrying.

The L-Per is a switched pattern DF system. In the DF mode the two antennas form an array with cardioid patterns alternately looking left and then right. The meter circuit compares the signal from the two patterns, which are being selected back and forth about 120 times a second, and determines the pattern of maximum signal.

The meter swings to the left to tell the user that the signal is to the left, and vice versa. The user need only remember to turn himself and the unit in the direction of the meter indication until the needle is at center scale. This crossover point indicates that the L-Per is pointing to the origin of the signal. Unlike the lobes of a quad, the L-Per's crossover point in the DF mode is quite sensitive, and only a few degrees of turning result in a wide swing of the meter.

With a reflectionless signal, there are two crossover

Table 8-1. Performance of the L-Per DF in the Presence of Controlled Reflections in NASAR Test.

REFLECTION ANGLE (θ) DEGREES	REFLECTION AMPLITUDE					
	− 20 dB		− 10 dB		− 6 dB	
	−	+	−	+	−	+
10	0	2	4	4	10	6
30	4	4	12	10	24	14
50	4	6	18	14	30	22
70	6	6	18	18	30	30
90	8	6	18	22	26	38

points 180 degrees apart. Does this result in ambiguity? Not if care is used. At the correct crossover, turning the unit left results in the needle moving to the right, as if to tell the user to turn back to the crossover. If the needle goes the other way, the L-Per is pointing away from the signal. It's easy to resolve a bearing to within two degrees of arc.

There is some loss of receiver sensitivity in the switched pattern DF mode. To aid in hearing weak signals, a receiver (REC) mode is provided, which stops the switching with the antennas configured to have maximum gain to the left of the unit. The meter becomes an ordinary S-meter in this mode.

When using the dual antennas in the REC mode, the response lobe is very broad, and highly accurate bearings cannot be taken. Similarly, the nulls do not occur at predictable points. For better bearings on very weak signal hunts, a beam or quad should be connected to the L-Per and used with the REC mode.

L-Tronics claims that the L-Per tracks AM, FM, pulsed signals, and random noise. DFing sensitivity is specified at 0.15 microvolt in the DF mode for vhf frequencies and 0.3 microvolt for uhf. Since an AM detector is employed, the receiver gain must be reduced for strong signals to prevent overloading. A sensitivity control and automatic gain control (AGC) circuit do this. The claimed 120 dB range allows DFing up to a foot away from a 10-watt rig. The stock i-f filter may be too broad for use on 2 meters in urban areas with 15 kHz channel spacing. L-Tronics offers an optional 6-pole filter if this is a problem.

Models and Frequency Ranges

The basic L-Per for ELT searches is the LH-10, which works on the vhf aircraft band only and comes with a crystal for 121.5 MHz. Other models have tuners (receiver rf stages) for the 2 meter ham band, the vhf marine band, the vhf commercial and business bands, the uhf aircraft band, and the 1-1/4 meter ham band.

A second tuner can be incorporated into the L-Per receiver for dual band coverage, at about a 20% increase in price. Popular combinations are vhf plus uhf aircraft bands, vhf aircraft plus marine, and vhf aircraft plus amateur 2 meters. The four crystal positions can be divided among the tuners as desired, but in no case is it possible to have more than four total channels.

This receiver frequency selection scheme is fine for search and rescue, but is the single biggest disadvantage of the L-Per for ham radio operators. Hams in populated areas will find the four channel limitation a big handicap.

There is no provision for frequency synthesis or use with an outboard receiver. If the operating frequencies cover a span of more than 500 kHz in one band, sensitivity on some will be reduced.

For portability, the L-Per is powered by two 9 volt transistor radio batteries in parallel, good for a maximum of 50 hours. A single battery lasts less than half as long as two in parallel. Both batteries must be changed at the same time and be of the same freshness. No provision is made for a switchable spare pair. Serious cross-country foot hunters should carry spare battery pairs, or consider adding a switched spare arrangement using larger capacity batteries such as NEDA #1603.

Battery life is very dependent on receiver speaker volume and use of the dial light. This is due in part to the lack of a squelch circuit on the receiver, a disadvantage according to some users. The use of an earphone can markedly extend time between battery changes if there is high ambient noise. For mobile use, the L-Per is easily connected to a car or aircraft power system of up to +30 volts. A special cord is available, or can be home brewed.

L-Tronics has set out to capture the search and rescue market by building a unit capable of operation in the extremes of weather. The receiver case is die cast alu-

Fig. 8-4. Marc Wiz, WA6HBR, using the L-Per for sniffing.

minum and water resistant. The portable antenna is not quite so rugged. It's certainly not designed for the rigors of vehicle mounting. Nevertheless, many a ham has gotten started in T-hunting using an L-Per held out the car window.

It is better to save the wooden antenna unit for foot use and fix up something more appropriate for a vehicle. The PVC-pipe/broomhandle home brew method can produce a rugged and inexpensive mobile setup, or a mobile antenna can be bought from the manufacturer. The L-Tronics literature cautions that there may be unwanted errors introduced by other antennas on the vehicle, so it's best to remove them if possible during L-Per use.

Be careful using the wooden antenna unit in heavy brush. It may be best to carry it folded and stop occasionally to unfold it and get bearings. Be sure to hold it up high for best signal strength, as shown in Fig. 8-4.

Other L-Per Configurations

A rotating antenna is not required on an airplane. A pair of fixed quarter wave antennas is recommended, and can be bought from L-Tronics. The manual gives detailed information on antenna mounts for aircraft, and pointers on airborne hunting. L-Tronics makes other DF units intended for aircraft mounting, and aircraft antennas. Some units have alarms which sound when an ELT signal is heard.

The left/right front/back technique is a method of systematic hunting that can be used in road vehicles with a rotary mount for the L-Per antennas. With the antenna aimed to the left of the car (270 degrees), a "left" indication shows a signal behind and a "right" indication means the signal is generally in front. A crossover as the team drives along indicates that the source is directly along-

Fig. 8-5. Two antenna pairs are used in place of a rotating pair in this installation.

Fig. 8-6. The switch box attached to the side of the L-Per is used to select which pair of antennas are used: front/back or left/right.

side (90 or 270 degrees) the vehicle. On a long city street there is often little signal from either side, but at major intersections the antenna should be oriented forward to determine left or right bearing down side streets.

Some successful mobile jammer hunters employ this method with four quarter wave whips mounted on top of the vehicle in a diamond pattern (Fig. 8-5) and selected in pairs with a switch (Fig. 8-6). This arrangement is waterproof, unobtrusive, and always ready when needed. The switch selects the left/right pair (1 & 2) or the front/rear pair (3 & 4) of antennas.

With this pair switch system, a form of the stairstep approach to hunting is used. The initial bearing is determined and plotted with a hand rotated antenna, or by driving in a circle using the left/right pair. The team takes off on the road that appears to lead most directly to the transmitter. En route, the indication for the front/back pair is watched.

When the transmitter is at right angles to the vehicle, the meter swings past zero. By switching to the left/right pair at this point, a decision can be made to turn left or right. After the turn, the hunters note with the front/back pair that they are nearly facing the transmitter, and can continue until another crossover. The process is repeated until the hidden T is reached.

As you can imagine, the stairstep approach with fixed antennas begins to lose its usefulness in areas where the roads aren't laid out at right angles, and where reflections can get into the act. It doesn't lend itself to the 45-90 distance measurement method. It has its greatest use for hunting jammers in city areas where signals are strong. Jammer hunters need to keep a low profile and always be ready to go, and they can with this setup. L-Tronics sells the sets of antennas and switch box, or they can be home built.

Improving the Little L-Per

Roger Chaffin, W5RGX, of the National Association of Search and Rescue suggests some improvements for L-Per owners. First, consider marking the sensitivity control dial with calibration marks to indicate the amount of signal reduction. Knowing how much the receiver gain has been reduced helps you estimate your distance from the transmitter.

This marking is best done with a calibrated generator such as a Hewlett-Packard model 608. The marks can either show dB of signal reduction or receiver sensitivity at the control setting. By using a waterproof pen on a plastic card fastened under the control nut, the receiver is not permanently marked, and the card can be replaced if the radio is returned or modified.

W5RGX's other modification uses the external power input regulator to regulate the battery voltage. The stock L-Per, as stated earlier, uses two nine-volt batteries in parallel. After about five hours battery voltage drops enough to reduce sensitivity by about 6 dB, and sensitivity remains at this lower level until the batteries fail at about twenty hours of use.

The modification consists of rewiring the batteries in series, connecting them to the input of the regulator at the power switch, and wiring the regulator output directly to the receiver circuits. Though battery current is higher, sensitivity remains at maximum for the full battery life, which is still about twenty hours.

At end of life the modified unit drops off in sensitivity very rapidly, while the sensitivity dropoff for an unmodified unit is slower. Hunters should be sure to always

carry extra batteries. Further information on this modification is available from NASAR at the address in the back of this book.

In summary, the Little L-Per is not the unit to buy for experimentation and cruising the bands. Its simplicity of operation may be deceiving due to bearing errors in strong multipath areas. But if you only hunt on a limited number of frequencies, and want a complete unit which is portable and always ready to go, you'll want to give it serious consideration.

THE HAPPY FLYERS DF

The Happy Flyers, whose purpose and programs are explained in Chapter 10, is a group interested in effective and inexpensive DF units, particularly for use on private aircraft for search and rescue. During the mid 1970's Hart Postlethwaite (WB6CQW), Jim Williams (K6HIO), Bob Broadway (WA6CZJ), and others in the organization developed an add-on phase comparison DF unit with sharp left/right indications that could be built very inexpensively. The heart of the unit is a single integrated circuit, originally developed for RTTY and other FSK applications, which contains the oscillator, detector, and indicator driver.

The operation of the Happy Flyers DF system is similar to the other dual antenna systems. The panel meter indicates the direction to turn the antenna. The dual LEDs aid DFing in the dark, and give a more rapid response than the meter. They also help distinguish direct from reflected signals. When the LEDs are flickering and pulsing rapidly at crossover, suspect multipath. If you get apparent forward crossovers in two directions, one flickering and one steady, trust the steady one.

The basic circuit (Fig. 8-7) works only with AM receivers, standard in the aircraft band. Several circuit

Table 8-2. Performance of the Happy Flyer DF in the Presence of Controlled Reflections in NASAR Test.

REFLECTION ANGLE (θ) DEGREES	REFLECTION AMPLITUDE					
	−20 dB		−10 dB		−6 dB	
	−	+	−	+	−	+
10	0	2	4	4	10	4
30	2	4	10	8	20	10
50	4	4	14	12	26	18
70	4	6	16	14	28	22
90	4	6	16	16	24	26

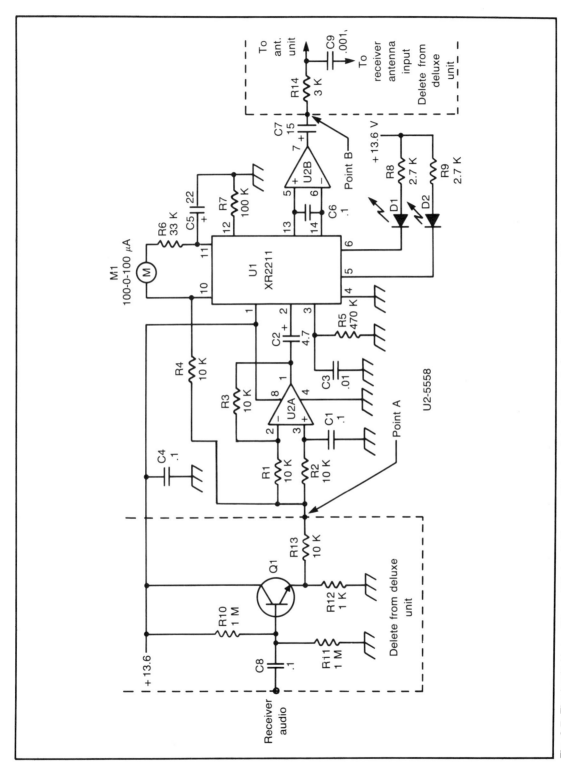

Fig. 8-7. The basic Happy Flyers DF circuit diagram. This circuit is for AM-only receivers that have AGC.

107

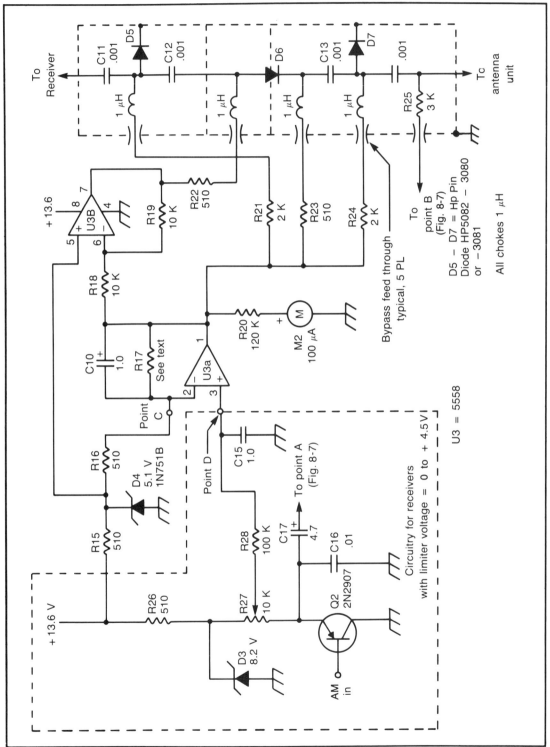

Fig. 8-8. Added circuit for the deluxe Happy Flyers unit for FM receivers. This input circuit assumes that AM input voltage increases with increasing signal.

configurations are possible to allow operation with almost any AM set. The receiver AGC and dynamic range of the DF circuit permit DFing with no sensitivity adjustment until the signal is so strong that the receiver is overpowered by leakage into the coax or case. Metal cased radios and double-shielded coax allow closer-in hunting.

FM receivers are usable if a point in the i-f chain can be found where a high level AM-detected signal can be tapped off. Usually the S-meter circuit provides such a point, if any heavy damping components are removed. If no S-meter is present, the i-f signal can be tapped and AM-detected externally. Since FM receivers usually have no AGC, external attenuation in the rf line must be added to keep the i-f stages out of limiting.

The deluxe Happy Flyers unit add-on (Fig. 8-8) includes a voltage controlled attenuator to automatically keep the rf input to the receiver below i-f limiting. As a fringe benefit, the attenuator control voltage can be connected to a meter to provide a wide range signal strength indicator. A higher meter reading indicates more attenuation is being put into the line.

Figure 8-9 is a block diagram of the complete deluxe Happy Flyers DF. The antenna pattern is switched with PIN diodes driven from the free-running square wave clock, which also drives the synchronous detector. The received signal, attenuated as required by the closed loop automatic attenuation circuit, is processed by the external receiver. The AM-detected i-f drives both the automatic attenuator sensor and the synchronous detector. Readout of the phase comparator drives both an analog meter and left/right LEDs.

In an AM receiver, the tap-off point is just after the detector diode, before the volume control, or at the input to the control. On FM sets, use the output of the S-meter detector, such as point A in Fig. 5-1. If a large capacitor is present at the tap-off point, delete or reduce its value.

The meter resistors you use at R6 and R20 depend on the meter movement you find. Values are given for 100 microampere movements (100 microamps each direction in the case of M1). For 50 microamp movements, double the values, and for 200 microamps, make them

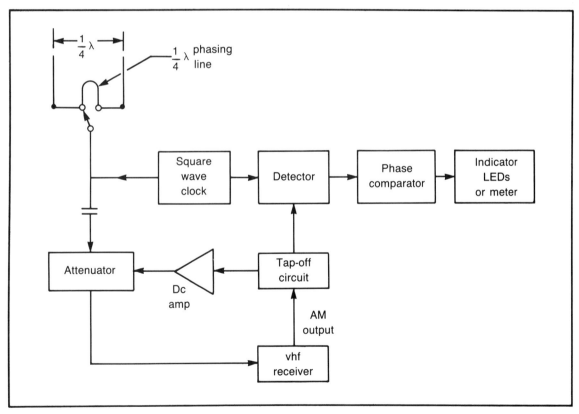

Fig. 8-9. Block diagram of the Happy Flyers vhf/DF system.

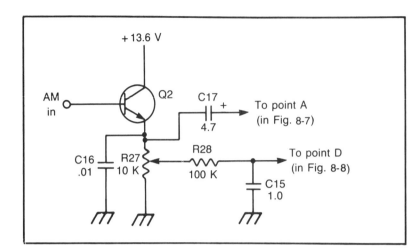

Fig. 8-10. Replacement input circuit for the deluxe unit when limiter resting voltage exceeds 4.3 V. This is for voltage increasing with greater signal strength. For opposite case, see text.

half the values. Don't use movements with more than 200 microamperes full scale. Change values accordingly if a meter pins hard or doesn't reach full scale. Change the value of C5 as necessary to suit your taste and the damping of your meter.

Whereas the basic unit uses only the audio information from the AM detector, and is capacitor coupled, the deluxe unit must use the dc information for the attenuator driver, so connection to the receiver is a bit more complex for the deluxe unit. The components in the dashed box on the left of Fig. 8-8 are for receivers which have AM output voltage levels between 0 and +4.3 volts, as do most amateur FM receivers when the S-meter output is used. Some receivers have a resting voltage above +4.3 volts at limiter test points to be used for tapoff. In such a case, use the circuit of Fig. 8-10 instead of the components in the dashed box on the left of Fig. 8-8.

Before wiring the deluxe circuit, carefully measure the voltage at your chosen tapoff point, first with no signal (V_{REST}), then with a signal that puts the S-meter at about half scale (V_{LIM}), and finally with a signal strong enough to produce maximum voltage at the tapoff point (V_{SAT}). We want the automatic attenuator to start cutting down the signal before that saturation point. The difference between saturation voltage and limiting voltage (V_{SAT} minus V_{LIM}) is the delta voltage (V_{DELTA}), used in the computation of R17.

Note the direction that the voltage goes as the signal increases. On most sets it increases (becomes more positive). If so, the wiring of Figs. 8-8 and 8-10 will be used, depending on the resting level. If the voltage becomes lower (more negative) as the signal level increases, reverse the connections at points C and D. In this negative-going case R16 goes to U3-3 and R28/C15 will go to U3-2. Note that R17/C10 goes to U3-2 in all cases. For the circuit of Fig. 8-8, choose R17 by using this formula:

$$R17 = \frac{2448 \times (7.5 - V_{LIM})}{V_{DELTA}}$$

R17 for the input circuit of Fig. 8-10 is chosen by:

$$R17 = \frac{1074 \times (V_{LIM} - 0.7)}{V_{DELTA}}$$

In the case where the voltage goes more positive with increasing signal strength, subtract 510 ohms from the calculated value of R17. If you determine when testing that the voltage at U3-1 does not get up near the supply voltage at maximum signal input, increase the value of R17.

Except for the antenna switcher and attenuator, the circuits operate at audio frequencies and construction is not critical. Wire wrap or hand wiring is fine. Troubleshooting is easiest if sockets are used for the ICs. Building the unit yourself gives you a great deal of freedom in the mechanical design. Figure 8-11 shows how large meters can be used for mobile use. For use with a handheld, a more compact package is in order. Edge-view meters are small and can be stacked to miniaturize the unit.

The attenuator is constructed of double sided circuit board material. It has three sections with partitions as shown in the photo. Construction technique is similar to the attenuators of Chapter 6. Carefully solder the seams

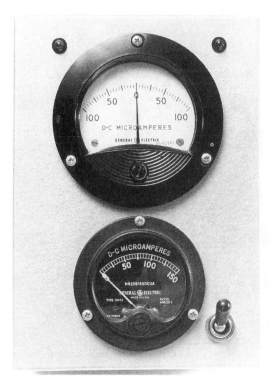

Fig. 8-11. The outside of a Happy Flyers DF. The top meter is the direction indicator and the bottom is the signal strength.

for no rf leaks. The inductors can be anything from 0.82 to 8.2 microhenries, but be sure to use ones that are as physically small as possible. A properly built attenuator should have low insertion loss on weak signals and up to 60 dB attenuation at maximum.

If that's not enough for your needs, two attenuators can be put in series, or an ordinary step attenuator used for additional attenuation. If two attenuators are placed in series, the dc leads to each go to the same driver nodes, but the values of R21-24 are doubled. Don't expect to achieve much more than 90 or 100 dB effective attenuation maximum, no matter what you do. Remember that rf will find its way into the receiver case and input coax.

The antenna switching unit (Fig. 8-3) can be built on a piece of circuit board with the various component connections routed out with a motor tool, or can be built on an old fashioned terminal strip. It should be put in a small metal box to prevent direct signal pickup. The delay line coax, and coax to the antenna, can be RG-58 or RG-174. Do not use foam coax. The delay line is 0.66 times the antenna spacing distance. That spacing distance should

be about a quarter wavelength center to center, but it can be less, down to about an eighth wavelength. Wrap it around or bunch it up in the box.

Quarter wave whip antennas can be used on cars and planes instead of the dipoles shown here. They should be clear of obstructions and other antennas. Dipoles are better for foot use, and are mounted on a "T" handle. Do not use flimsy wire for antennas. When it flops around, the meter indications are very erratic. Mark the left and right antenna and coaxes so you won't accidentally use the unit backwards.

The unit performs best if the various coax shields and circuit return connections in the antenna switcher are not connected to the enclosure, but just tied together as shown in Fig. 8-3. If the unit is installed in a car or plane, be careful how it is grounded. The receiver case should be solidly grounded. Don't depend on the antenna coaxes for the ground return for the receiver.

Figure 8-12 is a deluxe unit circuit board with components in place. The covers have not been soldered on the attenuator. When soldered together, the automatic attenuator is leakproof. Rf can still get into the output coax and receiver case, however. You may find that setup of your Happy Flyers unit is difficult when you use the tried-and-true method of having a friend with a handie-talkie nearby. The problem is likely leakage and overload into the receiver.

If values are chosen properly for the limiter voltage and meter scaling, there is only minor tune up required to put the deluxe system in operation. With all ICs out of their sockets, apply power and check for shorts and proper voltages on the IC pins. If everything looks OK, install U3 and reapply power. If you build the unit with the input circuit of Fig. 8-8, connect Q2 base to ground for checkout. If instead you used the input circuit of Fig. 8-10, connect Q2 base to the positive supply.

With a voltmeter at point D, a voltage swing both above and below +5 volts should be obtained when R27 is rotated through its range. Rotating R27 should also cause a HIGH/LOW transition at U3-1 when the input crosses about +5 volts. There should be a similar transition at U3-7, except opposite polarity. At this point it should be possible to control a vhf signal through the attenuator with R27, though the response will be very sharp.

U1 and U2 can be installed now. The switching square wave should be visible on a scope connected to Point B. If not, look for small ramps at U1-13 and U1-14 to see if the problem is in U1 or U2. With everything working properly, it's time to disconnect the base of Q2 from supply or ground and hook it to the receiver.

Fig. 8-12. An original Happy Flyers deluxe circuit board with the attenuator chambers open for clarity.

Make first checks with a high distant signal, such as a repeater. You should be outside in the clear, not inside a building. Front and back is resolved as described previously in the L-Per description. With no signal being received, M1 should be at zero center. A buzzing should be audible on received signals, and the meter should indicate proper direction to turn.

R27 should be adjusted so that the automatic attenuator starts to reduce signal input when the S-meter gets to about half scale. This is easiest to do with a signal generator. If you don't have one, use a local repeater and adjust R27 to get the S-meter reading down to about half scale. If the direction indications are backwards, reverse meter polarity or antenna orientation.

If the square wave at B is present, but proper DF indications are not present, try hooking an audio signal generator to point A through a 0.1 microfarad capacitor.

As the frequency of the generator is swept slowly through the switching frequency of the antenna unit, M1 should swing from side to side. This verifies the operation of the main circuit. If there is still trouble, check the antenna and switcher.

Only about 10 millivolts of signal is necessary at pin 2 of U1. R13 can be changed as necessary, or a similar resistor added to the deluxe unit, to optimize performance. If there is electrical noise or poor DF results with a car or aircraft installation, experiment with the grounding technique to see if that helps. Watch for ground loops (multiple connections to ground at different points in the vehicle).

Some users have reported that making the equal-length coax lines from the switcher board to the antenna exactly one electrical quarter wavelength (the same as the phasing line) instead of 10 inches or so gives more

DF sensitivity. If the antennas were a perfect match to the receiver this wouldn't matter, but in the case of ground planes this won't be true. So there is probably a matching effect in the quarter-wave lines under these circumstances. Experiment and see what works best in your installation.

The Happy Flyers design has no provision for a non-switched single cardioid pattern mode such as incorporated in the L-Per. If you lose the signal and have a beam, disconnect the DF unit, hook the beam up to the receiver directly, and use the S-meter. As with other switched antenna DFs, care must be taken to prevent transmitting through the unit, which can cause failure of the diodes in the antenna switcher or attenuator.

The ability of the unit to work with many different radios might engender thoughts of a switched setup to connect the DF to several different receivers in the same vehicle. A multiple radio selection scheme has been suggested in the organization's literature. It uses a panel of BNC feed-throughs for the radios' antenna lines and a rotary switch to select audio for the DF from the proper receiver.

It is suggested that the same DF can be used with the aircraft, CAP, 2 meter, and marine radios in the plane, so long as they are in the same part of the spectrum. Unfortunately, it isn't quite that simple. First, there are differences in antenna and phasing line lengths for the various vhf bands. A compromise has to be made if all are to be covered, and careful checking should be done to ensure that DF performance is proper on all bands.

Second, one cannot expect the AM takeoff points in each radio to have the same polarity and voltage swing. A polarity switch was suggested, but a more elegant way is to provide a conditioning stage for each radio's output to bring the polarity and level to a common value before connection to the rotary switch.

Happy Flyers organization members have put a great deal of time and effort toward encouraging amateur radio operators and flyers to build and operate these DFs for search and rescue and jammer location. At various times circuit boards have been available for the units, plus detailed literature on the operation of this and other units for airborne use. The group has been an unexcelled source of information on DF testing and airborne hunting techniques. Write to the address in Appendix A for further information on current activities.

DETERMINING BEARING INACCURACIES

Compared to a quad or other gain antenna at vhf, the homing DF's biggest disadvantage is its inability to show all signal sources and their amplitudes. The quad and S-meter show the relative strengths of the direct and reflected signal in Fig. 8-13A, though there is some distortion of the pattern depending on the phase of the reflection. But a homing DF seldom shows both correctly. The direct signal indication is "pulled off" by the reflection.

The worst part is that, in the case of (B), how far it is pulled off cannot be readily determined. It could just as well have been pulled off in the opposite direction of the reflection! The amount and direction of the error is a function of the amplitude of the reflection and its phase relationship to the direct signal, which is a function of the relative distance. Usually only a fraction of a wavelength is important in relative distances, which makes the sidestep method (to be described shortly) work. Of course when the reflected signal is stronger than the direct, both systems fail. That's why hunting from hilltops and airplanes is so important.

To assess the effect of reflections on switched cardioid DF units, controlled tests were done by the National Association of Search and Rescue (NASAR). Bearing accuracy is vital in off-road ELT searching, and NASAR wanted to know just how badly reflections can throw off the bearings.

A test signal source and reflection were set up. Data was taken on Happy Flyers and L-Per DF units under the following conditions:

☐ Angular relationship of main signal and reflection, theta (θ), was tested at 10, 30, 50, 70, and 90 degrees.

☐ Amplitude relationship of signal to reflection was tested with reflection co-efficients of .5, .316, and .1, giving reflections -6, -10, and -20 dB with respect to the direct signal, respectively.

☐ Location of the reflector was varied from 0 to 50 feet from the DF unit, to cover all possible phase relationships of the received signal and reflection.

☐ All measurements were made at two foot intervals.

The results of the 315 careful measurements on each type of DF unit are summarized in Tables 8-1 and 8-2. As an example of the results, Table 8-1 shows that when the reflection angle was 90 degrees and the reflected signal was 10 dB below the direct signal, the L-Per bearing error was between -18 and $+22$ degrees, depending on the phase relationship that was established by the distance to the reflector. The results also showed that when the reflection was -10 or -6 dB with respect to the di-

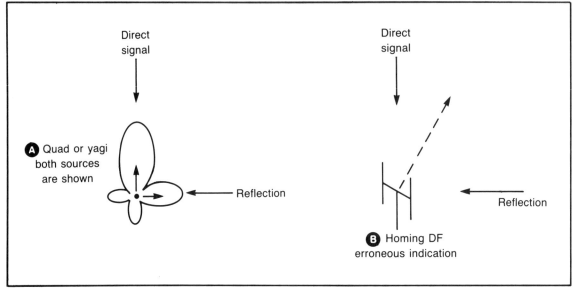

Fig. 8-13. Comparison of a quad beam versus a homing DF in the presence of direct and reflected signal.

rect signal, there were very few instances of zero-error readings, even at narrow angles.

The tables invite a comparison between the two DF units. Although the differences are not overwhelming at all, the Happy Flyers unit tested seemed to have an edge. It had less maximum absolute error in ten of the fifteen angle/amplitude situations, while they were equal in the remaining five situations. But that's not enough reason to switch units. The importance of the tests is not which unit "won;" it's the distressing fact that it's easy to get 30 degree bearing error on either type when reflections are strong.

In city hunts, it's not unusual to have an object present a reflection 6 dB below the direct signal. Luckily it's easily discovered, because when the DF unit is moved a yard (or a block), the bearing makes a large change. By staying on-the-go and averaging bearings, scientifically or by eyeball, the correct trend toward the T will be apparent.

SIDE-STEP AND BASELINE AVERAGING TECHNIQUES

When hunting on foot, try this technique to help to differentiate between direct and reflected bearings with a switched antenna DF. Hold the unit in front of you and rotate your body to get a crossover bearing with the meter or LEDs. Now keep the antenna pointed in the same direction and slowly step sideways, first left and then right. If there's no room to step, simply move the antenna back and forth in front of your body while continuing to keep it pointed to the apparent source of signal.

When receiving a direct signal in the clear, the sideways motion should make little or no difference in the centered left/right indication. A nearby reflection, on the other hand, should cause the left/right indication to fluctuate wildly with the lateral movement of the DF.

This method, attributed to Ben Bohach, K0GVS, works best in discriminating against close-in reflections. It is much less effective if the reflection is from a mountain many miles away. Try it—you may find it a big help in sniffing.

A similar technique, attributed to Gary Holcomb, WB5MVV, is the baseline averaging technique. It is used when taking bearings off-road during ELT searches. It is based on the principle that a corrugated wavefront is present when there is mild multipath, but the average value of this wavefront is normal to the source. (This concept gets more complete treatment in Chapter 10.)

A compass is used with the DF unit to take a bearing on the signal to the nearest degree. The observer then moves five feet or so in a direction perpendicular to the line of bearing and takes another reading. About eight readings are taken this way. The readings are then averaged to get a final bearing.

If the readings are widely divergent, say more than plus or minus 15 degrees from the average, the result should be discarded because the multipath is too great. A new location should be chosen for another try. If only one or two bearings are significantly different, the wild ones should be thrown out before averaging.

There is little geometrical error due to the side-stepping unless the transmitter is within a half mile or so, because the wavefront is quite planar at that distance. This technique works best when the source is a mile or more away.

In the hills and mountains, the reflection problem is very difficult. A reflection from a large distant mountain can cause consistent bearing errors for many miles of driving. In extreme cases, the reflection is actually stronger than the direct signal.

Imagine yourself in the situation of Fig. 3-8. The "wrong-way reflection" causes severe errors as long as the direct signal is blocked. If the only road is along the canyon floor, you will be unable to get a correct bearing on it from point C to point F, whether you use a homing DF or a quad. The lesson here is to get as high as you can for bearings, and to put the greatest trust in bearings taken from the highest locations, while distrusting bearing from canyons. Getting interested in flying lessons now?

THE K6BMG SUPERDF

Russ Andrews, K6BMG, is an avid T-hunter who has given many programs at clubs and conventions in the southern California area to encourage hams to get into the sport. He is also a competent engineer who has done considerable study of RDF principles and techniques. He has developed and is marketing a unique DFing system with looks and operation similar to the Happy Flyers and L-Per packages, but based on an entirely different principle.

The BMG DF unit is intended for easy use with vhf-FM mobile rigs and handhelds, which seem to be standard issue with active hams in urban areas. It does not work with AM-only or SSB-only receivers. Although the system was designed with amateur 2 meter band hunting in mind, the basic control unit works from 50 to 1000 MHz when used with appropriate antennas and FM receivers.

Only two connections to the using receiver are required. In most cases no modification is needed. The control unit is in line between the DF antenna and the receiver antenna connection. Audio from the receiver external speaker/earphone jack provides the recovered phase information. A speaker in the control unit allows monitoring of the receiver and the superimposed DF tone.

Hunting with the unit is very similar to using traditional switched pattern systems. The red LED tells the user to turn left, while the green indicator says turn right.

At the switchover point, the unit points to the bunny. Front/back switchover points are easily differentiated in the same manner as with the L-Per or Happy Flyer DFs. If no reflections are present, a straight walk or drive leads to the transmitter. Reflections will result in a "drunken man's walk," as K6BMG puts it.

In addition, a tone-pitch mode is provided. The LEDs are turned off in this mode, reducing battery drain from 30 mA to about 10 mA. The pitch of the tone of the DF oscillator is changed by the comparator such that low pitch tells the user to turn left, and high pitch indicates a right turn. The tone mode is very useful for mobile operation, allowing the driver to keep his eyes on the road and the navigator to keep his eyes on the map.

A switch selectable slow mode averages signal reflections and holds left-right information on a short signal burst for several seconds. If your base station is on a high hill, you'll find this method gets bearings on kerchunkers faster than using a rotating beam. The LEDs tell which way to turn the antenna, and there are no side or back lobes. Stationary bearings not taken from high points are suspect, as reflections can cause serious errors.

The limiting characteristic of FM receivers makes attenuators or AGC circuits unnecessary, and the unit provides DF information from the first point the signal can be heard (or sometimes even on undetectable signals, claims the inventor) to within inches of the transmitting antenna. An S-meter is not required on the receiver, although it would be of considerable help in judging distance to the fox. The lack of relative distance information is one of the biggest drawbacks in a system of this type.

Because a carrier is required for detection of the RDF tone, this unit does not DF pulsed noise, such as noisy power lines. The antennas are intended for use with vertically polarized signals, so the results with a distant horizontally polarized bunny may be poor if there is significant multipath present.

K6BMG considers his circuit proprietary and has sought patent protection for it. It is different from switched pattern systems described earlier in this chapter because the antenna pattern does not play a part in determination of the bearing. Only the relative phase of the two antenna signals, which is a function of their relative distance from the rf source, is important. When one antenna is closer to the transmitter than the other, the phase difference is detected by the FM detector in the receiver and appears as steps or jumps in the audio output.

These steps are processed by a synchronous detector in the SuperDF control unit and drive the direction indicators. The LED that lights is one corresponding to the closest antenna. At the switchover point, where both

Fig. 8-14. It's in there! Clarke Harris, WB6ADC, sniffing at night with the SuperDF.

antennas are equidistant, the unit is pointing to the bunny as in the photo (Fig. 8-14). Since AM receivers do not detect the phase difference, they cannot be used with the SuperDF.

Building the Kit

The SuperDF kit is packaged by Don Rose (WA6QAC) Electronics of El Monte, California. The first one we evaluated was the earliest model, which had good performance but left some things to be desired mechanically. The coax connectors and a few other parts were surplus, but the remainder were typical commercial types. The circuit board was of excellent quality, but the enclosure had a true home brew appearance. For example, there was no lettering on the panel, and the holes for the miniature slide switches were round instead of rectangular.

Comments by hams led to an improved model, the SDC2. It uses a larger box and has two circuit boards, one holding all the switches and LEDs, leaving almost no point-to-point wiring. The inside is now very clean-looking, and the outside markings are nicely silk screened.

Assembly instructions are fairly easy to figure out, although this may not be a good first kit for newcomers to electronics. The eight-page adjustment procedure is in the form of a flowchart, which makes it very easy to follow. Because of i-f delay variations between different models of receivers, the control unit must be adjusted for proper timing with the receiver to be used. It is claimed that adjustment can be done with only a voltmeter and a screwdriver, but we found that the test point voltage was very low and difficult to adjust, even with a sensitive VTVM. We were more successful using a dual trace wideband oscilloscope to view the timing of the internal window signal as we made adjustments.

Several antenna unit versions are available for the various frequency ranges. The most versatile unit has four folding removable whips which allow storage of the unit in a very small space without having to remove whips and risk them getting misplaced. The folding feature is accomplished very simply by the use of right-angle uhf type coax adapters.

A number of optional features were described in the manual for the first model, but no parts for them were supplied. Two features, a spare battery selection switch and a switch to kill the DF tone when monitoring with the receiver, are so important that they are now all incorporated in the SDC2. That's good. Can you think of anything more agonizing than to have your only battery die as you close in on the bunny?

One option that is not part of the stock SDC2 is a low battery warning LED. We'd recommend adding it for rapid battery checks before going out to hunt. If you don't want to add extra circuitry, put a test point on the box for quick checks with a digital voltmeter. For those that are used to L-Pers and Happy Flyers units and want a meter readout, a 1 mA meter can be added externally to the SDC2. When used at a fixed radio volume control setting, a better idea of number of degrees off axis can be gotten on kerchunk signals.

The supplied LEDs are small and not very bright. For DFing in daylight we replaced them with larger superbright LEDs and added black hoods to block out sunlight. Some hunters may want to add a jack for an earphone on the tone oscillator. This allows DFing in the tone mode without having to listen to the hider's tone box.

If the SuperDF is to be used with only one model of receiver, the phase selection switch can be deleted or mounted backwards so it does not go through the front panel. This prevents it accidentally getting set to the wrong position, which results in bearings in the opposite direction of the hidden T. With the switch deleted, the SDC2 can still be used with other receivers, as there's another easy way to change phase. When we used the unit with a 1-1/4 meter mobile rig that had opposite phase requirements from the usual 2 meter handheld, we merely inverted the pointer on the antenna mast.

For mobile use it would be desirable to power the unit from the car battery, but this is not easily done. The nine volts from the internal transistor radio battery is split to be plus and minus 4.5 V with respect to ground. To use mobile power, a source of regulated positive and negative 4.5 V relative to the car frame, and appropriate connectors, would be needed. Since the expected life of an alkaline battery is about 15 hours with the LEDs on, the internal battery and spare meet most needs and prevent problems when switching from mobile to on-foot use in the heat of battle.

The control box and antenna unit can be configured in a number of creative ways for various hunting needs. Figure 8-15 shows how they can be connected together for hunting on foot, with a compass attached to the antenna unit with strips of Velcro® for quick removal. The aluminum antenna boom does not distort the compass indication when offset slightly as shown.

The double male coax adapter was made from two PL-259 plugs soldered shell to shell, with stiff wire tying the center pins together. By cleaning the rear of the connectors thoroughly and positioning them carefully before soldering all around, an adapter was produced that will not twist loose like the commercial adapters we tried. It's so rigid that when the setup was bounced around in the car trunk, the cast-metal SO-239 fitting supplied on the antenna unit broke off at the mounting ears, while the adapter was undamaged. Owners expecting hard use should consider replacing these surplus SO-239's with ones having a stronger flange.

For vehicle hunting, the control box is separated from the antenna and put on the dash. A broom handle mast works well. Clips grasp the antenna boom, so that it is easily removed for rapid transition from car to foot use. Don't forget to install some sort of indicator to tell the driver which way to the bunny. The kit does not supply the mast, compass, or double-male adapter.

Hunting with the SuperDF

The setup just described worked well on our test hunts. At freeway speeds, the dual whips are far easier to turn than a four-element quad. The tone change mode allows DFing while watching the road. We found it almost mandatory to dedicate a handheld rig for use only with the SuperDF, and have another mobile rig in the car hooked to a separate whip antenna through an attenuator. For long hunts, the regular/charger described in Chapter 14 can be used to keep the handheld's batteries up.

The separate mobile rig is useful to get S-meter readings, to listen on the repeater output for the bunny on repeater hunts, and for transmitting (attenuator out) when required. Transmitting through the SuperDF antenna at more than a watt or so will destroy the diodes. Doing all transmitting on a separate rig prevents any possible accidents. We noted that the SuperDF buzz can be heard in the separate mobile receiver. This is caused by re-radiation of the bunny's signal from the SuperDF antennas.

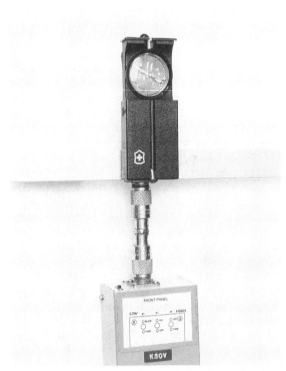

Fig. 8-15. The SuperDF configured for hunting on foot. The control box is connected to the antenna with a double-male connector. The compass lets you take direct bearing of the signal.

The important things to remember when hunting with the SuperDF are to keep moving and place the greatest trust in bearings taken at high points. The SuperDF is very prone to inaccuracies caused by multipath, and continuous movement results in better bearings by averaging the reflections out. When multipath propagation or re-radiation is not present, bearing accuracy is excellent, and this accuracy can be maintained over a very wide frequency range.

To determine the unit's accuracy, a test was performed for an FCC field office using the unit mounted atop a van in a clear area. The lower tips of the antenna were about 3 feet above the van roof, and a non-metallic mast was used. The bearing accuracy on a signal source 200 feet away was checked at 5 MHz intervals over the range from 110 to 260 MHz. Mean bearing inaccuracy was + 0.137 degrees, with a standard deviation of 1.2 degrees. The worst inaccuracy, at two frequencies, was + 3.5 degrees.

How can a hider fool the SuperDF? Perhaps the easiest way is to be sure there are lots of signal reflections

near the hidden T. The first time we tried the unit in competition, the hider's very low rig and J antenna were lying on the ground under a railroad overpass. The chain link fences on both sides of the road and along the tracks were so well "lit up" that it seemed like no matter where the hunters stood, the signal seemed to be coming straight from the nearest point on the fence.

Horizontal polarization for the transmit antenna accentuates the effect of reflections on the SuperDF. When Walt Le Blanc, WB6RQT, and Vince Stagnaro, WA6DLQ, hid a low powered transmitter with a horizontal dipole antenna in the succulents on the bank of a freeway overcrossing, three hunters with SuperDFs found they gave very inaccurate and inconsistent readings. Tilting the antenna unit at a 45 degree angle helped somewhat, but was not a total cure.

On these particular hunts, the sniffer described in Chapter 12 worked much better than the BMG for closing in. The SuperDF may also give inaccurate bearings on off-channel signals, pulsed signals, and signals which deviate out of the receiver passband. Very strong adjacent channels or even out-of-band signals can also cause DF errors.

Using the SuperDF with a handheld rig can pose a problem in areas with high levels of vhf signals. A metal cased receiver may be needed to cut down on such interference. Most of the time, though, the setup works surprisingly well. On one Saturday night WB6RQT had to hide transmitters for two hunting groups at the same time. He used two 10 watt 2 meter rigs with their antennas 15 feet apart. The hunt frequencies were about 1 megahertz apart. Hunters had to identify which transmitter was the one they were hunting. This was a difficult trick for ordinary sniffers with single tuned circuits, but the BMG unit with a handie-talkie DFed them separately, even when the hunter stood right between them.

In summary, the SuperDF is a compact unit that can turn any vhf or uhf FM rig into an instant T-hunting setup. It can be kept in the car trunk to be ready for emergency use whenever needed. It is not the most sophisticated setup by far, but can find jammers, provide lots of fun, and serve as a basic unit for experimentation and improvement.

TESTING HOMING DFs

When a homing DF set-up doesn't work, is the problem in the DF, the antenna, or the system installation? It's possible to check out the DF independently with just a stable signal generator, a couple of fixed attenuators, and two equal lengths of coax. Figure 8-16 shows how

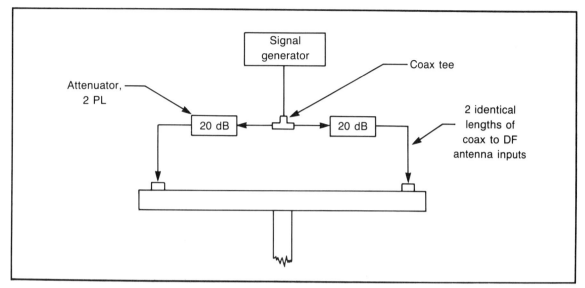

Fig. 8-16. A homing DF test setup. Coax lengths must be exactly matched.

to hook them up. Be sure that all components, including the source and splitter, are well shielded, to prevent direct rf pickup through the receiver case. The attenuator pads must be identical to avoid phase length differences. (Chapters 6 and 14 give details on attenuator pads.)

The pads serve to decouple the DF antenna inputs from each other and the source. With an on-frequency signal from the generator, the indication on the DF should be centered at all signal levels. The DF system sensitivity can be measured with this setup and a calibrated output generator. Just add 20 dB to the generator output indication to account for the pads.

Use this setup to check for effects of ignition or other noise sources in the vehicle, which should not affect the DF indication if everything is working properly. Add a six inch length to one of the coaxes, and the indication should go to one side. Take the coax from one side and put it into the other line and watch the indication show the opposite side. The deflection in each direction should be the same.

Chapter 9

Doppler DF Units

Manually rotated antennas are at their worst when the target signal is intermittent or in motion. Trying to find someone who enjoys driving around town while briefly kerchunking the local repeater can be maddening. Homing units such as the Super DF indicate which way the antenna must be rotated to achieve a null or peak, even with very short transmissions. But with a rapidly moving transmitter and/or hunting vehicle, particularly on winding hilly roads, getting a bearing is like trying to thread a moving needle. Wouldn't it be nice, in such a case, to have a DF system that gives actual bearings relative to the car in real time? In essence, it would be a moving finger that says, "The signal is that way."

One approach to this goal is the Doppler scanning direction finder. It has no moving parts, nearly instantaneous response time, and a low profile on the vehicle. The Doppler is excellent for use in a moving car on a one-man hunt because it's easy for the driver to read. In this chapter we'll explain what makes a Doppler DF tick, how you can build one, and what to expect from it and from commercial units.

HOW DOPPLER DFs WORK

The principle behind the Doppler DF was first dis-

covered by Austrian physicist Christian Doppler in the nineteenth century. When the distance between a transmitter and a receiver is changing, the received frequency of the transmitter is shifted—lower if the two are moving apart, higher if moving closer. The Doppler effect is readily observed in sound waves, as you have surely noticed when a train passes you with the horn sounding. The pitch suddenly drops as the engine passes you, at the transition of approaching to receding. The effect is not as readily observable with light or radio signals. The speed of those waves is so much greater than sound waves that the velocity must also be much greater for the effect to be noticeable.

So how is the Doppler effect used in direction finding? Imagine a movable vertical antenna as shown in Fig. 9-1. As the antenna is rotated at a constant rate around the center axis, it moves alternately closer to and farther away from the signal source. The receiver connected to this antenna perceives a Doppler frequency shift in every received signal that varies sinusoidally with the antenna rotation. Figure 9-2 shows this apparent Doppler shift at each point around the circle. At the nearest and farthest points, the shift is zero. Shift is greatest 90 degrees from these points.

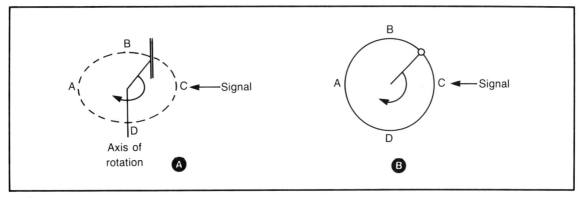

Fig. 9-1. A rotating antenna.

If the rotating antenna is connected to a radio receiver with FM detection, the varying Doppler shift appears as superimposed frequency modulation on the signal. The FM detector output includes a sine wave at the frequency of the antenna rotation in cycles per second, in addition to any other modulation on the signal.

By comparing the phase of this demodulated Doppler signal with a reference signal acquired from the antenna rotating circuitry, the direction of the incoming signal can be determined. The first description of DFing using the Doppler modulation induced by a rotating antenna was by H. T. Budenbom in 1947. Today there are over 300 patents on Doppler type DF schemes for the various frequency ranges in the spectrum.

Now let's look at a typical case at vhf. The formula for frequency shift is:

$$S = \frac{R \ W \ Fc}{C}$$

where S = Zero-to-peak Doppler shift in Hz (same as peak FM deviation)

R = Radius of antenna rotation in meters

W = Angular velocity of rotation in radians per second

C = Velocity of light in meters per second

Fc = Carrier frequency of received signal in Hz

Converting to the British system:

$$S = (R \ Fr \ Fc) / 1880$$

where R = Radius of antenna rotation in inches

Fr = Frequency of rotation in Hz

Fc = Carrier frequency in MHz

For example, assume +/− 0.5 kHz deviation of the Doppler signal is desired at 10,000 MHz (though that's not very likely). If the antenna radius of rotation is ten inches, the antenna rotation rate must be 560 rpm. That's plenty high, but for the same deviation at 2 meters, the rotation speed must be more than 38,600 rpm. So much for mechanically rotating the antenna!

But it isn't always necessary to mechanically rotate the antenna. We can scan a circular array of antennas electronically, connecting one antenna at a time to the receiver. As the antennas are sequentially selected the radio perceives this as if it were a single rotating antenna. The same superimposed Doppler shift is created, though it is a piecewise approximation of the sine wave of Fig.

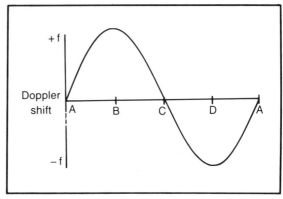

Fig. 9-2. The perceived Doppler shift at each point around the circle.

9-2. The most correct names for this electronic rotation scheme are pseudo-doppler and sequential phase, but the industry tends to just call any scanned antenna system a Doppler DF.

Dopper DFs are practical from the hf bands on up. Design of a particular system must take into account many factors to determine the ideal number of elements, rate of rotation, and antenna system aperture. The design of high performance Doppler DFs has been covered at great length in the scientific literature.

DOPPLER DFs FOR RADIO AMATEURS

Doppler DFing for Amateur Radio first began to take off in 1978, when Terrence Rogers, WA4BVY, described a practical unit for add-on use with 2 meter mobile rigs in a *QST* article. It used eight antennas and 16 LEDs. Phrases such as "key to success" and "secret weapon"

in the article caught the attention of hams looking for quick ways to counter the growing jamming problems.

The DoppleScAnt, as WA4BVY calls it, works well and was used as a basis for many hams to experiment and upgrade the design. Some found ways to use rings of 32 or more LEDs, or digital displays. Others didn't like the mix of CMOS and TTL logic, and redesigned the system to use CMOS logic only, for less power consumption. W6AOP sold kits of his CMOS version to many hams.

In 1981 several commercial firms began advertising add-on Doppler DFs for the Amateur Radio market. Some of these companies never actually delivered any units, or soon stopped making them. But Doppler Systems of Phoenix, Arizona, survived the shakeout and still continues to upgrade its line. Datong Electronics of Leeds, England, also markets a mobile Doppler DF for hams. It may be available from some mail order houses in the

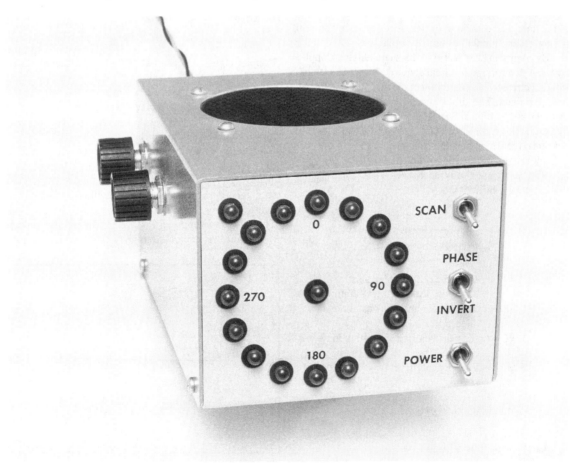

Fig. 9-3. The Roanoke Doppler display unit with the 16 display LEDs but without the low level LED.

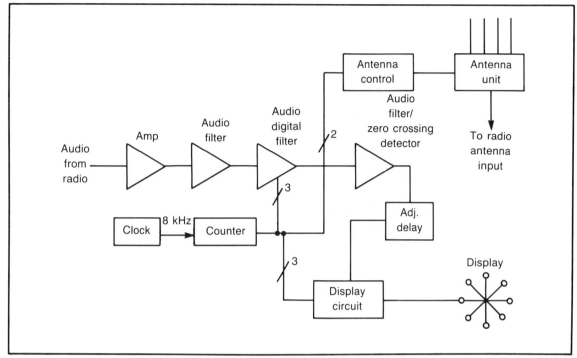

Fig. 9-4. Block diagram of the Roanoke Doppler.

USA. More recently, Dick Smith Electronics of Australia has marketed a very inexpensive DF, both "down under" and in the USA.

Doppler Systems reports that two thirds of its sales are to non-ham entities such as cable TV companies, government, and commercial firms. This is no doubt because the prices for Amateur Radio market DF units, though considered high by many hams, are very low compared to DF equipment targeted for the commercial/government market from companies such as OAR or Watkins-Johnson.

THE ROANOKE DOPPLER FOR VHF

Because of the low cost and availability of digital ICs, a Doppler DF unit for vhf can be built cheaply and easily. The unit described here, shown in Fig. 9-3, was designed by Chuck Tavaris, N4FQ. Truly a bare-bones design, it was intended to be as simple as possible and inexpensive to reproduce. Several have been successfully built in the Roanoke, Virginia area. If you hunt around there, chances are you'll be competing against at least one.

Figure 9-4 is a block diagram of the Roanoke Doppler DF. The 8 kHz oscillator increments a counter to drive the display, sequence the antenna assembly, and clock the audio digital filter. As the four antennas are sequentially selected, the LED indicators are scanned in step such that each indicator corresponds to a particular point in the electrical rotation of the antenna system.

The rotation of the antennas causes the 500 Hz audio tone to be superimposed on the receiver audio. An audio filter with active analog stages and a digital filter isolate this tone from the rest of the audio. After the last audio stage, a comparator looks for the ac zero crossing of the demodulated Doppler sine wave. This zero crossing, going through an adjustable delay, is used to update the LED display.

Detailed Circuit Description

Refer to Fig. 9-5 for the schematic and Fig. 9-6 of the parts list for the Roanoke Doppler. Master timing of the unit is controlled by oscillator U8. The 8 kHz output is divided by counter U9. The Q3 and Q4 outputs of U9 are used, through U10, to drive the antenna unit. U10 is a BCD-to-decimal decoder. As Q3 and Q4 count from 0 to 3, the antennas are sequentially selected by the low output states of U10. As these outputs (0, 1, 2, 3) are selected in turn, the antenna is electronically spun. The

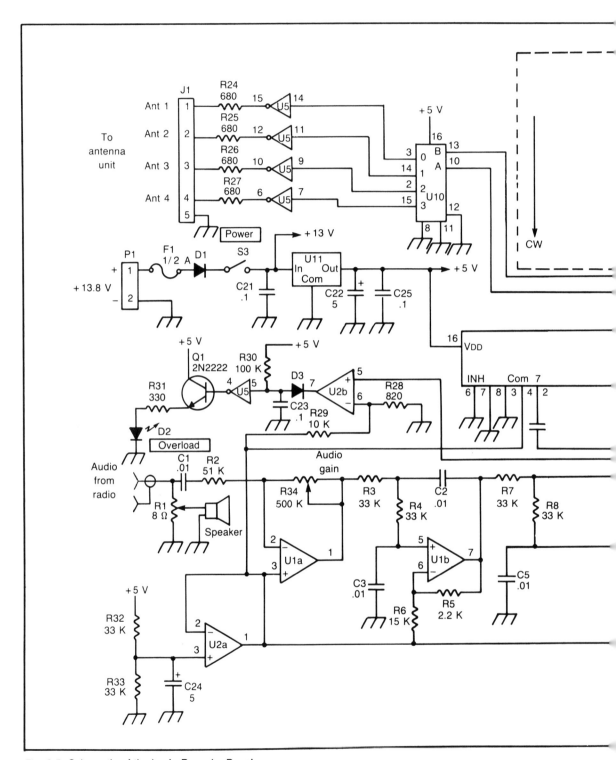

Fig. 9-5. Schematic of the basic Roanoke Doppler.

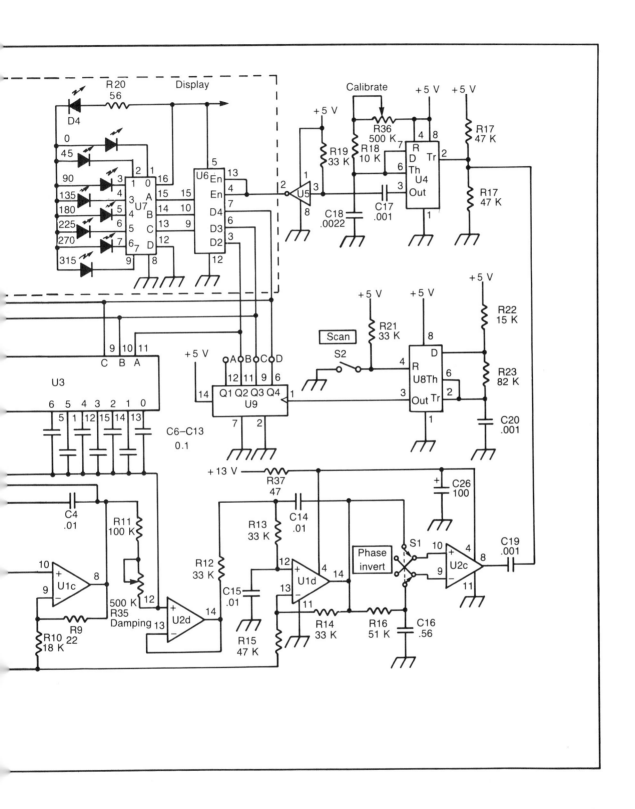

125

```
Parts List for Roanoke Doppler

U1, U2 = LM324
U3 = CD4051
U4, U8 = 555
U5 = CD4049
U6 = 74LS75
U7 = 7445
U9 = CD4024
U10 = CD4028
U11 = LM7805
D1 = 1N4003
D2, D4 = LED
D3 = 1N4148
```

Fig. 9-6. Parts list for the basic Roanoke Doppler.

rotating antenna pattern causes an apparent Doppler modulation of the received FM signal.

Receiver audio with this added modulation is patched into the DF unit from the radio's external speaker jack. Operational amplifier U1A buffers the input level to the proper setting with the radio's volume control at its normal position. C1 and R2 form a high pass filter, rejecting subaudible tones and other audio information below 300 Hz.

U1B and U1C are active low pass filters to sharply cut high frequency components of the audio signal. This enhances the incoming Doppler frequency component by removing noise and harmonics. U3 is a CMOS analog multiplexer chip connected as a switched capacitor filter. This filter is extremely narrow, with bandwidth depending on the setting of R35. The filter is switched by the same clock that scans the antennas, ensuring that the center frequency of the filter is always exactly the same as the Doppler tone coming from the receiver.

The filter output is buffered by U2D and post-filtered by active low pass filter U1D to get rid of the steps in the switched filter output. The signal at this point (U1-14) is a clean sine wave. The output of U1D is fed to U2C, a comparator with its trip level set to the average U1D output by R16 and C16. As the sine wave from U1D-14 passes through this average level, the output at U2-8 switches state.

U6 is a 4-bit latch. As the antennas are sequentially selected by counter U9, these addresses are also sent to

U6. On every negative transition of U2-8, a 555 timer connected as a one-shot (U4) fires. The falling edge of the pulse from U4, coupled by C17 and inverted by U5, momentarily enables U6, storing the address from U9. This address is decoded by U7, a one-of-ten decoder driver. One of the outputs of U7 will go low, turning on one of the LED indicators. By adjusting the pulse width of the U4 output with CALIBRATE control R36, the LED that comes on, relative to the phase of the zero crossing point, is changed.

When the system is up and running, the display direction is calibrated by adjusting R36. It should have sufficient range to move the display 180 degrees, if necessary, to accommodate any convenient antenna mounting arrangement. Phase reversal switch S1 makes a 180 degree change in the indicated output. It is needed to allow rapid substitution of receivers. Different receivers may have opposite audio output polarity due to different numbers of inverting audio stages. Once calibrated, mark the switch position on the panel for each radio you'll be using, to avoid accidental errors. And be sure to always mount the antenna assembly with same orientation for each hunt!

If the audio input to the DF unit is too high, or the gain of the first stage is set too high, the filters will severely clip the audio. U2B is used as a comparator to detect excessive audio signal levels. If the peak audio at U1-8 approaches clipping level, overload LED D2 is turned on by Q1. C23 and R30 hold D2 on for several milliseconds each time an overload is sensed, to ensure visibility of the peaks.

Construction

Parts for the Roanoke Doppler are all readily available. If you can't find them locally, try one of the many mail order suppliers. The display unit logic (Fig. 9-7) can be built with wire wrap techniques on perforated board. The box is 3 × 4 × 6 inches.

Since most of the ICs are CMOS types, the LED display constitutes a large portion of the power drain. The 5 volt regulator (U11) is in a TO-220 package (RS 276-1770) and is bolted to the box. There are no other heat dissipation problems. D4 is the LED in the center of the display. Don't omit it—otherwise, you'll have a difficult time reading the direction in the dark.

A speaker should be mounted in the box, or a jack provided for your present external speaker. On most sets the internal speaker is cut off when this unit is plugged in to the external speaker or earphone jack. This is good because the radio's volume control is used to set the level

Fig. 9-7. Photo of the display unit with the cover off. The 16 LED display modification is built on the piece of perforated board next to the speaker in the cover.

into the DF, then left alone and the external speaker control (R1) is used to set listening level. It is important to hear the signal when hunting to help determine when multipath is present. Some receivers have a resistor in series with the earphone jack. If you plug in to this jack and notice very low speaker level in the box with R1 set for maximum level, check for this resistor and short it out if necessary.

The calibration and damping controls should be accessible from the outside. A screwdriver adjust pot is best for the calibration control to prevent accidental misadjustment. The gain control and phase switch can be mounted inside. For ease of servicing and adjustment, make test points for important nodes such as the output of each op-amp. They can be just little pieces of stiff wire

sticking up, to which the scope probe can be clipped.

The antenna unit schematic is shown in Fig. 9-8 with mechanical construction details for a 2 meter assembly given in Fig. 9-9. It is constructed from a 19 × 19 inch piece of double-sided copper-clad board screwed down on a frame of half-inch plywood. Capacitors, diodes, and chokes are mounted on standoffs or terminal strips. Copper-clad board makes the job easier since the leads and coax shields can be soldered directly to it for a good ground plane. If it's not available, use sheet copper. Aluminum can be used as a last resort, but connections to it must be made with lugs instead of solder.

The spike antennas are attached using BNC connectors mounted on the copper-clad board. They are made from 3/32 inch bronze welding rod, available at welding

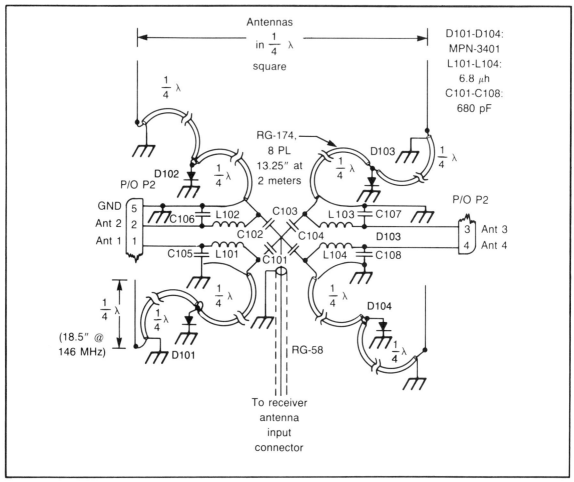

Fig. 9-8. Schematic of the antenna unit.

supply stores. Ends of the rods are filed down and soldered to the center contact pin of a male BNC fitting, as illustrated in the detail of Fig. 9-9. The rod, with center pin attached, is installed in the connector shell and the inside of the shell is filled with epoxy from the top.

The plywood base is cut to match the size of the sheet. Large holes (greater than 3/4 inch) are drilled in the plywood beneath each antenna and a large rectangular hole (about 4 × 6 inches) is cut beneath the center of the antenna array to clear all of the switching components (Fig. 9-10). The control cable and individual pieces of coax can be run between the copper-clad board and the plywood sheet if slots are routed on top of the plywood.

Antenna switching is accomplished by diodes D101 through D104. Three diodes are on at all times, the fourth being off to allow its associated whip to be active. Diodes

should be identical PIN types, such as Motorola MPN-3401. An equivalent part to the MPN-3401 is commonly available in the ECG and NTE replacement semiconductor lines as ECG-555 and NTE-555.

Though rf PIN diodes are highly recommended, ordinary silicon switching diodes, such as 1N4148, can be substituted in a pinch. With ordinary diodes, you'll have much higher levels of switching noise in the received signal and a greater chance of cross-modulation products from nearby high-powered stations. But when the hunted signal is strong, they'll work. So use them to get started if you can't find the PIN diodes right away.

Lay out the inductors and capacitors of the antenna unit in a symmetrical pattern, keeping all component leads as short as possible. The four feed coaxes must be exactly the same length. RG-174 is easiest to use for the

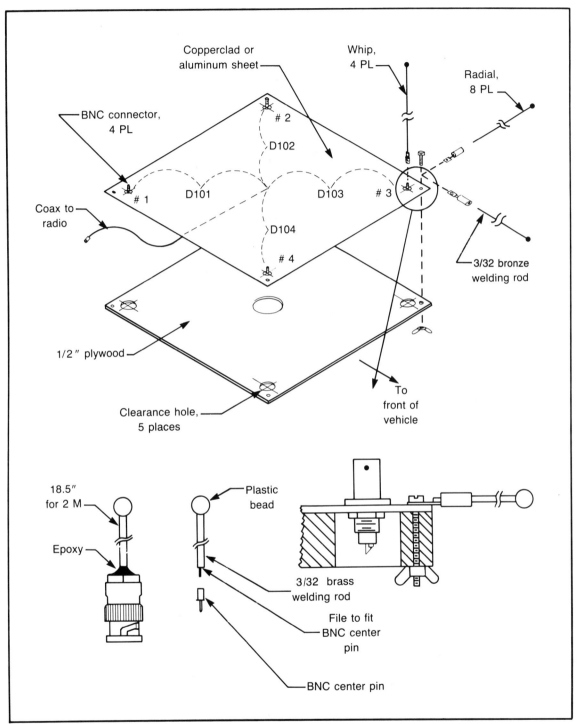

Fig. 9-9. Construction and assembly details of the antenna, with dimensions for two meters. Don't omit the radials—performance will be poor without them.

Fig. 9-10. Closeup of the antenna feed point. Note the short lead lengths on the components.

four feeders, but RG-58 can be substituted in a pinch. Each of the eight coax lines, two between each whip and the common point, must be one electrical quarter wavelength (or an odd multiple) long. This is necessary so that when the PIN diodes are on, the rf short is transformed into an apparent open circuit at both the antenna base and the common point. Open circuits are reflected to the common point to prevent loading. Open circuits are also required at the whip bases when off to "float" the switched-off whips. Grounded whips are undesirable because they act like parasitic elements and re-radiate the incoming signal, causing erratic and erroneous readings.

Suction cups are mounted on the corners of the plywood to protect the car and raise the antenna slightly. Luggage hold-down straps are used to hold the antenna unit securely on the roof. Both the suction cups and straps are available from automotive and hardware stores. Figure 9-11 is a photo of the completed antenna unit on top of the car, ready to hunt.

The radials are made from the same bronze rod as the antennas. They attach to the copper sheet with lugs, screws, and wing nuts so they can be folded up for removal. Don't omit the radials—performance will be poor without them. A Cannon DP-9 plug/jack combination on the control box (P2/J1) allows quick disconnect of the antenna control wires. If you substitute these connectors, make sure to use polarized ones so that they are always connected properly.

For use on another band, make the antenna elements and radials one quarter wavelength long, and space the verticals in a square which is one quarter wavelength on a side, or a little less. The four feed coaxes are each an electrical quarter wavelength long, which is 0.66 times the length of each vertical whip for ordinary coax (not foam dielectric).

Checkout

If you have or can borrow a good dc oscilloscope with response to 15 MHz or higher, and an audio frequency generator, they will be very useful for checking out your Roanoke Doppler unit. They'll also help you understand just how it works. You'll also need a good dc voltmeter. The first step is to carefully go over your wiring and look for errors. Before installing U1, put the ohmmeter probes across trim pot R34 and adjust it for 10 k ohms. Adjust R35 for minimum resistance. S2 should be open.

The first power application smoke test should be done before the ICs are put into their sockets. Leave the an-tenna unit disconnected for the time being. Check power drain, which should be very low. Verify supply voltages and ground at the appropriate pins of the IC sockets. All digital ICs are powered from the 5 V source and the LM324s get power from S3 (+13 V).

Now remove power, install all the ICs, and reapply power. The voltage at the output of regulator U11 should be 5 Vdc. The bias source at U2-1 should be 2.5 Vdc, and so should the audio stage outputs at U1-1, U1-7, U1-8, U2-14 and U1-14. The voltage at U2-6 should be 0.5 Vdc.

Connect the oscilloscope to view the waveform at U8-3. You should see the main clock signal as in Fig. 9-12. It is a rectangular waveform of slightly more than 50 per-cent duty at about 8 kHz. Closing S2 inhibits this signal. Next check the outputs of U9 (with S2 open, of course). They should look like the illustration. Now look at the

Fig. 9-11. The antenna unit mounted on the car ready for hunting.

antenna drive outputs at U5-15, -12, -10, and -6, comparing them with the figure.

Temporarily disconnect the top lead of the 8 ohm speaker control (R1) to prevent it from loading the audio generator. Set the audio generator for a one volt peak to peak (P-P) sine wave at 350 hertz and connect it to the audio input at C1. Verify that the audio signal is present at U1-1, with amplitude of one volt P-P. Then check the signal at the filter outputs, U1-7 (0.7 V P-P) and U1-8

(2.6 V P-P). Reset the generator to 600 hertz, verify that the level into C1 is still one volt P-P, and look again at the filter outputs, this time for 0.45 V P-P at U1-7 and 0.8 V P-P at U1-8. This checks operation of the low pass filter.

Set the generator for 500 hertz. When the input signal level is increased, the sine wave output of the last filter op-amp (U1-8) eventually starts clipping at the negative peak. The overload LED (D2) should light at just

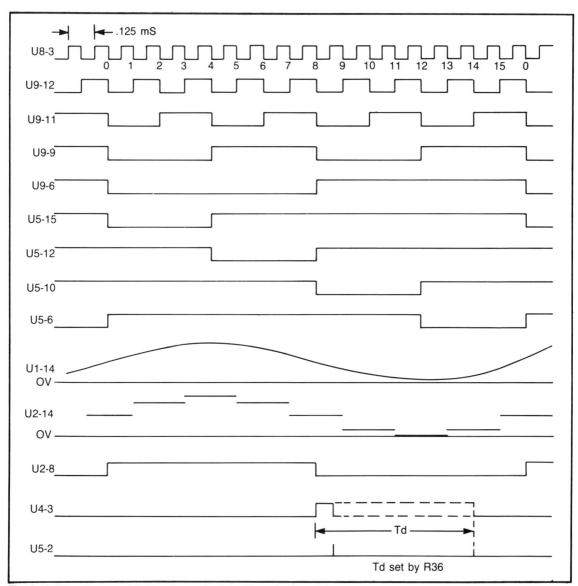

Fig. 9-12. Waveforms and timing diagrams of the Roanoke Doppler.

about the point where clipping at U1-8 begins. Reset the input level at C1 to one volt P-P after this test.

Connect the oscilloscope to U1-14, the output of the digital filter. Vary the audio oscillator frequency around the 500 Hz region and look for a peak in the waveform amplitude. You may have trouble finding the peak in the audio. This filter is extremely sharp, with a bandwidth of less than 2 hertz with R35 at minimum. When R35 is increased, the filter is even sharper. With the generator input centered in the filter, move the probe to U2-14 and notice the chopped sine wave effect. This is normal. The U1D filter stage cleans up the switched capacitor filter output, as Fig. 9-12 illustrates. The phase relationship of U2-14 and U1-14 versus the other points varies.

When the audio oscillator frequency exactly equals the antenna rotation frequency, only one LED indicator in the circular display is on and the display is stationary. By varying the input frequency just a slight amount, the display can be made to rotate—clockwise for lower frequency, and counterclockwise for higher frequency. The greater the frequency difference from filter center, the faster the rotation.

Now look at the waveform at U2-8. Use the second trace if you have a dual trace scope. The comparator output should be a nice square wave (rise and fall times about 40 microseconds) of near supply amplitude, switching at the midpoint of the sine wave of U1-14. Move the first trace probe to U4-3. At each negative transition of U2-8 the 555 should trigger. The pulse width should change when R36 is adjusted. Set R36 for minimum pulse width.

Connect the antenna assembly to the DF unit. Connect the oscilloscope probe to each vertical antenna in succession. The signals should appear as Fig. 9-12 (U5-6, -10, -12, -15), except that the amplitude will be only one diode drop (about 0.6 V). Disconnect the audio oscillator and reconnect R1 to the input circuit.

Make a quick bench check of the complete system before mounting it in the vehicle. Connect the antenna and audio leads to the radio. Tune it to a full-quieting station, such as an active local repeater. Set the speaker control (R1) full clockwise, and adjust the radio volume control for the loudest audio you'll ever want during a hunt. Back off on the radio volume control if you note distortion in the receiver audio amplifier. Note this receiver volume control position. Reduce R1 for a comfortable listening level. You should now be hearing the Doppler tone superimposed on the receiver audio. Adjust R34, the internal audio gain trimming pot, for occasional flashing of the overload indicator on voice peaks.

If the unit is working properly, the direction indication can be changed by rotating the antenna unit. Proximity effects will cause changes in the apparent direction when working indoors, so don't expect to get correct readings there. When everything seems to be working properly, mount the unit on the car and hook everything together.

Calibration

The method of calibrating and checking a vhf Doppler seems obvious: Just walk around the vehicle with a transmitting handheld and adjust the calibration control on the DF for correct bearings. But take it from the many hams who have tried it—that method won't give an accurate calibration. It is OK only for a rough check. Nearby reflections and the near-field characteristics of the signal give you inconsistent indications. You'll be convinced the unit isn't working properly if you try a final calibration with this method.

A repeater or strong base station that is a mile or more away makes a much better first check. The signal should be strong, the path should be unobstructed, and the vehicle should be in a relatively clear area, such as a large empty parking lot. Drive the car around in a circle and verify that the bearing is reasonably consistent. Again, don't expect super accuracy on this check, particularly if the repeater is many miles away. As described in Chapter 18, propagation can cause errors.

The best final calibration is done with the vehicle moving. N4FQ suggests calibrating the unit while moving slowly down a long stretch of straight (and vacant) road with a friend a quarter mile or so ahead. With the other vehicle keeping pace ahead of you (and preferably with someone else driving!), adjust the calibration pot, R36, until the top LED is on. Now pass the signal source—the bottom LED should now be on. Doing the calibration while in motion helps average out the local reflections which can throw off stationary bearings.

The receive frequency used for calibration can be anywhere in the band of interest, and it isn't necessary to recalibrate when you QSY. Be sure to run a check on signals to the right and left of the vehicle. This ensures that an error in wiring has not caused a mirror image display.

The level of the radio's volume changes the apparent direction of the signal. You can see this by tuning in a signal and varying the volume up and down. Make sure not to take bearings with too low an audio input level. Don't trust the display to remember a bearing after the signal disappears—take the reading while the signal is still there.

As you drive along note the amount of flutter on the

display caused by multipath and reflections. The amount of flutter can be controlled by the setting of the damping control pot, R35. You'll probably want the damping to be at the maximum setting, unless the transmissions are extremely short.

MODIFICATIONS AND IMPROVEMENTS

The chief advantage of this super-simple Doppler DF is its very low cost. You should be able to duplicate the basic unit at home for about 50 dollars, even if you buy everything. If you have a well-stocked junk box, it might cost half that. Once you understand how it works, you may think of some ways to improve it or customize it for your own applications. Here are two improvements developed by the authors.

A 16 LED Display

When using 8 LEDs the best bearing resolution is 45 degrees. We modified the Doppler unit to have a 16 LED display, improving the resolution to 22.5 degrees. It's much easier to hunt with 16 LED display than with only 8. Since the modification is so simple, we highly recommend it.

The circuit operation is identical except that the display circuitry enclosed by the dashed lines in Fig. 9-5 is replaced by the circuit in Fig. 9-13. The CMOS decoder/latch chip (U201) replaces both the old U6 and U7 and works in exactly the same manner. The only differences are the 16 outputs and the fact that this chip will not drive the LEDs directly. U202, U203, and U204 are hex MOS-to-LED drivers which provide the drive current requirements of the LED display. The 75492 ICs are somewhat uncommon, but are made by several manufacturers and are available from Digi-Key for about a dollar each.

As a fringe benefit, this IC substitution eliminates all the TTL-type ICs from the display unit. This lowers the total current drain of the system to 36 milliamperes (48 mA with the overload LED on). That's low enough for

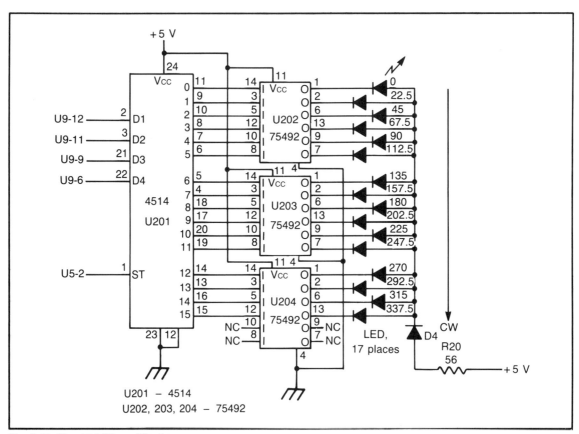

Fig. 9-13. The 16 LED modification.

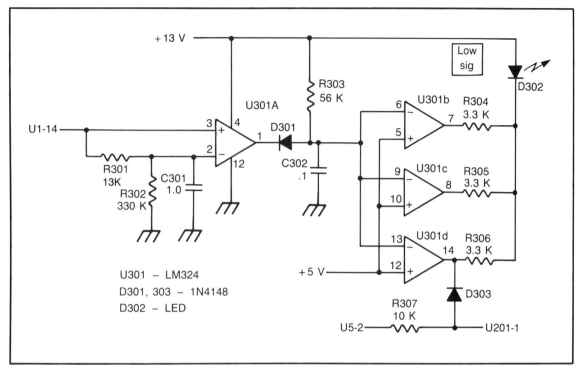

Fig. 9-14. The low signal lockout add-on circuit. The 16 LED modification must be installed for this circuit to work.

dry cell or nickel-cadmium battery operation. Does this give you some ideas for a fully portable Doppler for use with a handie-talkie? Try it! You'll find that the unit will work with supply voltages as low as 7.5 volts.

A Low Signal Level Lockout

The zero-crossing detector comparator (U2C) does not produce a good square wave when the Doppler tone level is low. That circuit is also sensitive to noise from the clock and antenna rotation logic. The result is display error at very low audio levels and a tendency for the display to return to the same compass position when there is no input signal. The setting of the receiver volume control may make 45 or more degrees of difference in the bearing indication on the basic unit.

The lockout circuit addition of Fig. 9-14 prevents the display error and automatically holds the last direction indication when the input signal stops. No parts in the basic display unit are deleted, but the lead from U5-2 to U201-1 is broken for the insertion of R307. The lockout circuit as shown works only with the 16 LED CMOS display of Fig. 9-13. To use it with the 8 LED TTL display, R307 and D303 must be replaced with logic gates.

The reference for comparator U301a is set to 2.4 volts by divider R301/302. When the filtered Doppler tone level is greater than 0.1 volts peak (2.5 V operating point minus 2.4 V), the negative peaks cause U301-1 to go LOW. R303 and C302 hold these peaks. This turns off low signal LED D302 and enables the display through D303. Paralleling the three sections of U301 is necessary to supply the drive current for D302.

To check out this modification, first put a clip lead across C302 and verify that the unit works as before. If you build your unit with this mod in place, short C302 while performing the initial checkout procedure outlined earlier in this chapter. When everything is working properly, remove the short and verify that when the receiver audio is not present or at a very low level, the display halts and D302 is on. The displayed bearing of a strong signal in the clear should not change as the volume control is varied from the low signal to overload points.

As you drive along with this feature in operation, you'll note that when the signal gets noisy, the low level indicator flickers and the display hesitates. It may also occur in some areas of multipath. This is good because the circuit is suppressing readings that have a high probability of being inaccurate. On very noisy signals, you may also find that turning the gain up to get rid of the low

135

level indication may result in an overload indication. Try to find a compromise setting of the receiver's volume control. Don't worry about occasional or regular flashing of either LED, but don't allow either to be on almost continuously.

HUNTING WITH DOPPLER DFs

As with any DF system, it's important to get to know the gear before taking it out for an actual hunt. The instantaneously updating display of a Doppler is easy to read, but there may be some subleties to its interpretation. The display will dance around on multipath signals and seem to be useless. Remember that it can indicate only one direction for each rotation of the antenna, and if there are nearly equal strength signals coming from two or more directions, the resultant indication will be wrong. The secret is to take the time to learn to read the display (and listen to the audio out of the speaker) in different types of terrain and with different transmitting sites, power, and antennas.

A great way to learn is to keep the DF on the car as you drive around town on your daily errands. Keep the radio tuned to a simplex frequency or to the input of the local repeater. Watch what the display does with known location transmitters. You will find that you will get better bearings when the vehicle is moving because the minor peaks, nulls, and reflections found almost everywhere are averaged out.

As you become familiar with the unit, try to hunt local signals for practice. You may want to put your attenuator in line with the antenna coax to the receiver. The S-meter and attenuator are of great additional help when you get to know how to read them. After practice, you will know when you are close to the transmitter because of the amount of attenuation needed to keep the S-meter on scale.

Adding attenuation moves the signal level toward the noisy region where bearings are less accurate, so use the attenuator only for quick strength checks, and leave it at zero the rest of the time. Another way to tell when you're very close is to check to see if the signal can be heard with the antenna disconnected from the receiver. This test can also be done with your separate handie-talkie.

The biggest problem in interpreting the indication is multipath. On an elevated freeway in the clear, the bearings may be steady and highly accurate, only to become jumpy and inconsistent upon exiting to surface streets. Some users report that bearings can go haywire in wooded areas, too. The reason is simple: When reflected signals

approach the strength of the direct signal, the result in a Doppler DF is a bearing indication that is incorrect for both the direct and reflected sources. The best strategy is to keep moving and watch the general trend of the indications. By moving along, the effects of close-in reflections are averaged out.

As you drive around you will note that the DF tone in the receiver audio changes quality from smooth to raspy. Raspiness of the tone generally indicates multipath is present. This is caused by multiple Doppler signals summing together in a random fashion, giving a high harmonic content to the resulting DF tone. At worst the tone seems to jump in pitch by exactly one octave, and may stay high almost continuously. Trust most the bearings indicated when the tone isn't raspy or an octave high. This advice is true for all Doppler DFs, not just the Roanoke Doppler.

Vertical whips on the front and rear deck of the car probably won't affect the Doppler DF operation. A quad or beam, on the other hand, may cause trouble. With the car stopped and a bearing on a fixed station displayed on the Doppler DF, rotate the quad. Does it cause the Doppler display to move around? Try this again at highway speed. You may have to relocate the quad, or perhaps give up trying to use two DF systems in one vehicle.

Make sure that the antenna unit is mounted squarely on the car roof to avoid bearing errors from any rotation with respect to the car from hunt to hunt. A one-piece antenna unit as described here is better than using four magnetic mount antennas in that regard, as the antennas in the fixed unit are always in the same relative position.

The Watkins-Johnson Company, a manufacturer of portable DF units for tactical military use, has done extensive studies of the factors causing error in Doppler DF readings. The theoretical error in a four antenna unit such as this is worst at 0, 90, 180, and 270 degrees in one direction, and at 45, 135, 225, and 315 degrees in the other direction. This error is caused by parasitic re-radiation among the antennas. So with typical vehicle mounting as in the photo, best accuracy occurs around the 30, 60, 120, 150, 210, 240, 300, and 330 degree regions. But the error amounts are so small that it's not worth worrying about it or trying to compensate for it in mobile use.

The hard-switching antenna rotation system of the Roanoke DF may make it subject to QRM problems in some high-rf locations. The commutation of the antenna at an audio rate causes sidebands at the commutation frequency (about 2000 Hz) and its harmonics to be superimposed on all incoming signals. For example, the signal being hunted has added sidebands at plus and minus 2,

4, 6, and 8 kilohertz, and so on. The steep switching transients result in significant amplitudes of the high order harmonics. This is no problem on the signal being DFed, but these same sidebands are also impressed on other signals entering the receiver input. If the sidebands from a strong in-band repeater or out-of-band paging system land on the bunny's frequency, they pass through.

In Fig. 9-15, sidebands of the 146.61 repeater, generated in the DF, are causing QRM to hunt on 146.55 MHz. When this happens, you'll know it. The desired signal may be blotted out, or you may hear typical cross modulation effects. The better your receiver, the less you'll notice this potential problem. Receivers with the best capture performance, most linear discriminators, and narrowest and steepest i-f filters perform the best.

In our road tests of the Roanoke DF, cross modulation has not turned out to be a problem. In its first competitive trial, the unit did a respectable job of tracking the half-watt transmitter on the back of WB6GCT's motorcycle, as he rode in a big circle around the perimeter of a local shopping mall. Gary also hid on the second trial, when the unit quickly found the closest parking place to the T. The antenna that time was a stubby rubber duckie mounted upside down underneath a rail on an abandoned railroad track (Fig. 9-16).

Hiders who want to make it tough on vhf Doppler DFers should use horizontal antenna polarization for the fox. This accentuates the effect of reflections. The more reflective surfaces around the hidden transmitter, the tougher time the hunters will have. If the rules allow two or more transmitters, make them all about equal power and equal antenna gain, and keep them relatively close together (within a block or so). That's because it's easier for the Doppler to discriminate between two signals, or between the direct or reflected signal, if there is a significant difference in amplitude between them. The DF indicates the bearing on the much stronger signal only.

COMMERCIAL DOPPLER DFs

Doppler DF systems are available to commercial, marine, and military users for just about any frequency range. At super high frequencies, mechanically rotating antennas are possible. At hf, the radius of the antenna ring and the number of antennas can reach gigantic proportions. Some units filter the Doppler tone out of the receiver audio, or design the system so that the tone isn't in the voice frequency range. That way the user doesn't have to listen to it. But then again, he also can't get any clues about the accuracy of the bearing from the quality of the tone.

Many sophisticated methods are used to improve performance. For example, one manufacturer adds a vertical sense antenna right in the middle of the array. This antenna feeds a separate communications receiver. The i-f signal from this receiver is beaten against the i-f signal from the Doppler receiver in a mixer. This cancels out the voice or other baseband modulation on the received signal, preventing it from interfering with detection of the Doppler tone.

Doppler Systems RDF Units

Doppler Systems manufactures a line of Doppler type RDF units which can be used on any vhf-FM receiver in the 30 to 500 MHz range. The basic display unit, model DDF-4001, provides a readout with 16 light emitting diodes (LEDs), spaced at 22.5 degree intervals. Since 16 LEDs do not provide a great deal of accuracy, the DDF-4002 is more suitable for long distance bearings. It includes a three-digit readout in addition to the circular LED display, providing indication to the nearest degree (Fig. 9-17).

Two deluxe models have been added to the line. For computer analysis and data taking, the model DDF-4003 adds rear panel output of the received bearing in RS-232C

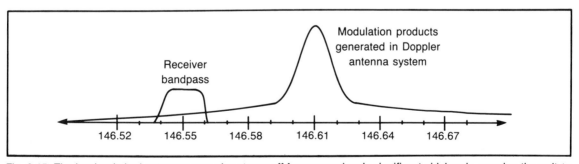

Fig. 9-15. The hard switched antennas may give strong off-frequency signals significant sidebands, causing the unit to give erroneous bearing.

Fig. 9-16. The hidden T was hidden under the railroad bed; the antenna, a rubber duckie stuck to the bottom of the rail.

format at 300 baud. Selection of 7 or 8 bits and parity is available. Model DDF-4004 has the same readouts as the DDF-4002 and includes a speech synthesizer to sound off the bearing. With its internal speaker, the DDF-4004 allows the driver to keep eyes on the road while hunting. The voice output can also be tape recorded for evidence and record keeping, or tied to a repeater audio input or phone line in an unattended fixed setup.

All of these display units are available in kit form or completely wired and tested. They operate from a nominal 12 volt dc source at less than one ampere current drain. The instruction manual gives very complete assembly instructions, schematics, a complete theory of operation section, and four pages of oscilloscope photographs to aid in checkout and troubleshooting. However, the manual gives very little information on how to use the

unit on actual hunts.

Since no ac power supply is included, the display unit is quite light and may be placed on the car dashboard for easy visibility. In response to buyer requests, a mounting bracket is now provided. The circuit boards mount vertically in the box and are connected by ribbon cables. Construction and layout is very clean. No special test equipment is required to build and calibrate the unit, unless troubleshooting is needed. In that case, a good dc oscilloscope is vital.

Don't expect accuracy to plus or minus one degree just because a three digit readout is provided. Specified accuracy is $\pm$ 5 degrees on 2 meters or vhf high band, and only $\pm$ 25 degrees at 480 MHz. These specs reflect the conservatism of the manufacturer, so your accuracy on non-reflected signals will probably be better than that if you're careful in calibrating the unit.

Four identical antennas are required to feed the display unit. Feed line lengths must be identical within one-half inch. Complete dimensions, parts list, and instructions for assembly are in the manual for a one-piece antenna assembly with quarter wave elements, radials, and a supporting frame, part number DDF-3000. It mounts to any vehicle by means of suction cups and hold-down straps.

This assembly was the original standard vehicle antenna setup for Doppler Systems DF units. It was found that quarter wave magnetic mount antennas performed well when carefully placed on a metal car roof. They are quick to set up and attract less attention. Spacing can be between 1/16 and 5/16 wavelength, with 1/4 wavelength being nominal. Doppler Systems markets a set of four matched mobile whips, part number DDF-4060, usable from 136 to 500 MHz.

For base installation, the DDF-4050 is a mast-mountable antenna for use from 120 to 300 MHz. An alu-minum X-frame holds four whips and provides their ground plane. Matched cable assemblies are also available.

The Doppler Systems units differ from other Doppler RDFs in the method of electronically rotating the array. Simple designs, such as the Roanoke Doppler, sequentially switch between a series of antennas in a circular array to simulate a rotating antenna. This hard switching can result in transients which may desensitize the receiver or cause intermodulation products, as mentioned earlier. The DDF units linearly mix the signals from the antennas in a summing box using a dual-gate MOSFET for each antenna.

The MOSFET driver waveforms are generated in a programmable read-only memory (PROM) and converted to an analog signal appropriate for each antenna. The output of the sequencer box is a weighted sum of the individual antenna voltages as programmed by these special gate driver signals. Besides reducing the problem of switching transients, this method allows more accurate

Fig. 9-17. The Doppler Systems control unit. (Photo courtesy Doppler Systems.)

DFing with only four antennas.

We tested the DDF-3002 (the DDF-4002 was not yet available) on a nighttime 2 meter sport hunt and found that, once in range, it was very effective in giving consistent bearings, even while driving down a freeway with low hills on either side. Strong reflections sometimes caused inaccuracies when the vehicle was stopped, but while in motion these multipath sources averaged out, and the bearing was generally accurate so long as there was a good direct signal.

Of course if the only signal being received is a reflection, then the bearing is consistent to the source of that reflection. Some users have complained that the DDF units are very difficult to use in downtown areas with very tall buildings giving high level reflections. Others say that careful eyeball averaging, and listening for the DF tone raspiness that indicates multipath, allows successful hunting under these circumstances.

In our test, the hidden transmitter signal was weak at the starting point, and it was fortunate that we brought a quad along on the hunt. We were more than halfway to the transmitter before the signal was strong enough to use the DDF-3002. In fairness, it should be pointed out that the signal was so weak that it would have been difficult to hear with just the receiver and a quarter-wave antenna at the start. Experienced hunters in our area expect weak signals, so most have a gain antenna with them.

The manufacturer states that the sensitivity of 4000 series units has been improved over the earlier 3000 series by about 3 dB. Broad band rf amplification has been added to the antenna sequencer. Gain through the sequencer is specified to be 8 dB at 30 MHz, tapering down to 0.5 dB at 480 MHz.

By the way, if you're thinking of adding rf gain like this on your home brew DF unit, be aware that the use of amplification in the antenna unit must be done with great caution. The phase relationship of signals from each antenna must be preserved exactly for proper DF accuracy over the entire frequency range. High-Q tuned circuits won't do! The DF4000 series units use very broadband preamps to preserve relative phase.

Antenna unit gain in the 2 meter ham band is 2 dB. Noise figure is stated to be 5 dB typical (frequency not specified). This compares to 1.5 dB or better for a typical 2 meter receiver and preamp. Because of the rf amplifiers in the sequencer, it is very important not to key the transmitter into the DF unit.

So even with the improvement, the user is at a sensitivity disadvantage compared to a quad hunter because of the gain of the quad and the noise figure degradation

of the sequencer. If the disadvantage is 3 dB, the DFing range is about 30 percent less, depending on the terrain. To maximize Doppler sensitivity, the antennas should be cut to resonance. Longer resonant antennas for some gain are a possibility, but beware of the effect of long antennas flexing and changing position relative to each other as the car moves. Of course if the unit is put on top of a hill for remote DF, sensitivity won't be a problem.

On the other side of the coin, it would have been nice to have an attenuator in line between the sequencer circuit in the DDF-3002 and the receiver input. Though the unit gave good directional indications up to the ending point a few yards from the one-watt transmitter, the S-meter on the receiver pinned long before that. The ability to use an attenuator's calibration to estimate the distance to the transmitter is an asset in any system.

The readout updates about twice per second. This is a compromise as longer time would be useful to hold kerchunker signals, but would be a disadvantage in situations with severe multipath. A shorter time constant would result in modulation affecting the bearing indication. No external adjustment of the update time is provided. Doppler Systems specifications claim DFing is possible with input pulses of 150 milliseconds or longer.

To hold the display between incoming transmissions, the user can select an audio squelch, which halts the display when the 300 Hz DF tone is not being received. There is also an external hold input. Of course neither of these methods automatically holds a bearing when the squelch stays open, as it does when a jammer momentarily overrides a user signal.

If you want more information on the Doppler Systems units, read the detailed article by designer David Cunningham, W7BEP, in the June 1981 issue of *73 Magazine* (see bibliography). It includes schematics and construction details of the 3000 series DFs and the DDF-3000 mobile antenna assembly mentioned earlier. Of particular interest is the description of the antenna sequencing system.

The Datong Electronics Mobile DF

Datong Electronics, Limited, is a British company that makes antennas, transverters, amplifiers, and accessories for Amateur Radio operators. The RDF I model Doppler DF has two features not found on others in this price range. An internal audio notch filter eliminates most of the DF whine, making the signal modulation more understandable. This is generally a good idea, but the ability to judge effects of multipath by the quality of the DF tone is lost. There is no speaker in the DF unit. An external speaker must be added to take advantage of the filter.

The other special feature of the Datong DF is an rf-triggered relay, which switches the transceiver over to the regular mobile whip antenna when the transmitter is keyed. To use this feature, the power into the DF must be 20 watts or less. If a mobile rf transmit power amp is part of your installation, it must be between the DF and the mobile whip.

The system consists of a display unit and a separate antenna sequencer, which has a mounting magnet and is intended to be placed on top of the vehicle with four short coax lines to the antenna of the user's choosing. To prevent water from getting into the sequencer box, water-proof coax feed-throughs are used for the four antenna coax lines, with no connectors. This makes it necessary to carry, store, and set up the sequencer and antennas as a group, and the buyer must put connectors on the coax if the unit is to be used with different antenna assemblies to cover more than one band.

The sequencer can be used (with appropriate antennas) from 20 to 200 MHz. Smoothed switching between the antennas is employed to minimize cross-modulation problems. Antennas can be bought from the manufacturer as optional extras.

Claimed accuracy of the system is ± 5 degrees at an unspecified frequency. The calibration control knob is on the front panel of the display, right in the center of the LED ring. That's convenient—perhaps too convenient. It's easy to misadjust it by accident. A response control, similar to the damping control on the Roanoke DF, is also on the panel. So are high and low audio indicators and an audio polarity switch.

The unit is not available in kit form, and no mounting bracket is supplied. The manual gives a lot of information on calibration and theory, but does not include a schematic or other servicing data. That must be requested separately from the factory.

In summary, the commercial Doppler units are good choices for someone eager to get a quick and easy start into hunting, as they can be purchased wired and ready to install. Prices vary greatly, but all of them cost more than the parts for the Roanoke Doppler. The expensive ones cost as much as you'd pay for a new full-featured 2 meter mobile rig. Antennas will add to the total cost. They are fine for rapidly hunting strong jamming signals coming from fixed stations in suburban areas. On the other hand, long distance hunters expecting weak signals would be well advised to carry a gain antenna along.

OTHER DOPPLER DF APPLICATIONS

Vhf Doppler DFs have many uses outside the amateur bands. As stated earlier, the majority of commercial units are bought for government or other non-ham use. Stuck transmitters in the business or public safety bands are common and must be found rapidly, so a number of police and fire departments have bought Doppler DFs. Some cable TV system operators use them to search for leakage. About the only hunting they are not used for is pulsed noise locating (narrow pulses can't be DFed with them) and aircraft/ELT searches. Remember that AM is used in the aircraft band, and FM detectors must be used with a Doppler DF. With a special FM receiver, ELT searches could be done with Dopplers, but their size limits their use to vehicles rather than use by operators on foot.

Chapter 10

Search and Rescue Hunting

Some transmitter hunters save lives. To the boater or aviator in distress, the ability to send out a radio signal to guide searchers means less time waiting for help. It's important work, and it's being done by both amateur volunteers and by professionals. This chapter explains what emergency transmitters are all about, who does the searching on land and sea, and some specialized techniques for search and rescue (SAR)—airborne DF and interferometers.

Our coverage of various search and rescue agencies is by no means all-encompassing: instead, it gives you a good picture of the kind of organizations that have proven their abilities in saving lives. All of them need the support and interest of amateur DF enthusiasts.

ELTs AND EPIRBs

Most amateur radio operators have heard of Emergency Locator Transmitters (ELTs) and what they're for, but most hams know little else about them unless they are also into flying. FAA regulations require ELTs on all aircraft, with a very few exceptions such as local trainers and flight test aircraft. An estimated 200,000 are in use worldwide.

ELTs are made by several manufacturers, and are usually rigidly installed in the tail of the aircraft. Figure 10-1 shows what one looks like. They are kept in an armed mode that causes the transmitter to come on when subjected to an impact of 5 g or more. The ELT makes simultaneous transmissions on 121.5 MHz, the civilian international distress frequency, and 243.0 MHz, the military distress frequency. Output power is usually less than 100 milliwatts. The audio modulation is a 3000 to 1600 Hz sweeping tone that sounds somewhat like a high-pitched siren.

ELTs have their own internal batteries capable of powering the unit for 48 hours, if they're fresh. This is none too long. Statistics for 1977 show an average of 38.3 hours per ELT search. With improving techniques, this time is decreasing.

The ELTs used on small aircraft can be turned on manually as well as by impact. This may be necessary when a pilot makes a "soft" forced landing in an inaccessible spot and needs help. The manual mode also allows tests of the unit. FAA regulations limit ELT testing to only the first five minutes of each hour and only for three audio sweeps maximum.

Since ELT transmissions are on vhf, the signal is propagated primarily in a line of sight path. Many air-

Fig. 10-1. An Emergency Locator Transmitter (ELT) and antenna.

craft accidents occur in remote areas, often in deep canyons, where there is no radio path to nearby airports. Usually the ELT is heard first by an aircraft. Pilots are encouraged to monitor 121.5 MHz whenever possible for this reason. ELT-to-aircraft range is dependent on many factors, such as antenna configuration and the plane's altitude, but detection range seldom exceeds 200 miles.

We have been discussing only one type of ELT, the Automatic Fixed type. There are three others:

☐ The Automatic Deployable ELT becomes detached from the airframe during a crash and commences transmitting.

☐ The Automatic Portable unit is intended to be easily removed from the plane and used by survivors.

☐ Personnel (Survival) ELTs have only an On/Off switch, and must be carried and turned on by hand.

ELTs have been required on general aviation aircraft since 1970. The success of this program caused the National Transportation Safety Board to recommend in 1972 that the Coast Guard require oceangoing vessels to carry a similar device, called an Emergency Position-Indicating Radio Beacon (EPIRB). International Civil Aviation Organization regulations require that aircraft making long overwater flights monitor for EPIRBs. It is interesting to note that there is no similar requirement for over-land monitoring for ELTs.

There are three classes of EPIRBs:

☐ Class A types are activated when they float free of the vessel after an accident. They are required on all commercial inspected vessels.

☐ Class B EPIRBs are stored on the ship and taken aboard the lifeboat for manual activation.

☐ Class C units are for use on vessels that stay within 20 miles of the coast. They transmit on Channels 15 and 16 in the vhf FM marine band.

EPIRBs from various manufacturers are available from marine supply outlets. Prices range from as low as $120 for manual Class C types to $500 for automatic Class A units.

Despite its problems, the ELT/EPIRB system is well entrenched. Over a quarter of a million aircraft and watercraft have ELTs or EPIRBs, but a better system is in the wings, as we will see in the section on T-hunting via satellite in Chapter 23.

THE CIVIL AIR PATROL

The Civil Air Patrol (CAP) is a civilian auxiliary of the U.S. Air Force, which dates back to World War II. Its purposes are emergency service, aerospace education, and cadet training. It is one of the most active groups in the country providing emergency and SAR services. Radio direction finding plays a vital part in its operations.

More than a thousand times a year, CAP crews take part in emergency service missions. Search and rescue missions in the continental United States are directed by the Air Force Rescue and Coordination Center at Scott Air Force Base, Illinois. In the 49th and 50th states, direction comes from the Alaskan Air Command Rescue Co-ordination Center and the Pacific Air Forces Joint Rescue Co-ordination Center. CAP members fly about 75% of these missions.

Though practices vary in different parts of the country, commercial homing DF units such as the L-Per are predominant. With their integral fix-tuned receivers, they are so simple to use that only a few minutes of training is required. They are used both on aircraft and on four wheel drive vehicles, with appropriate antennas for each. Though the CAP owns some of the units, most are owned by member DFers. Military DF equipment is considered too bulky and too difficult to use.

When an ELT signal is detected, CAP members may be called out immediately. When a small airplane is overdue without a detected ELT, the CAP is not called out until ground checks are made. When a pilot is one half hour overdue on his flight plan, a notification is sent out to all airports along the flight path, requiring them to check on the ground for the aircraft in question. At one hour overdue, the ground check is expanded to airports within 100 miles of the flight plan route.

At the two hour overdue point, the Rescue and Coordination Center is notified, a mission coordinator is selected, and a search communications base station is set up. Members throughout an entire wing are notified, usually by a pocket pager system, of the details of the incident. More than one wing may be alerted if the flight plan called for travel over more than one wing's territory. Some members call in their response, but many just go directly to the search base station.

When a crash occurs in rural areas, it is first located from the air (about 90 percent of the time), using a combination of DFing and visual sighting. If the site is accessible by land, ground units are dispatched through the communications system, working with the sheriff's department. The rest of the time, ground searchers first locate the target.

In urban areas, transmitting ELTs are located from the ground about half the time, and from the air the other half. A very high percentage of ELTs found in urban areas are the result of accidental transmissions from airports.

The CAP supports other emergency services besides search and rescue, such as airlifting supplies to disaster sites, air evacuation, and air surveillance. Communications is vital in making all this possible, both for coordinating operations on the ground and in the air, and for ensuring the safety of all participants. The organization has a network of hf and vhf stations for backup communications for all government levels, and is a vital part of many state disaster plans. It is not unusual for a CAP group to support a county sheriff, the state Office of Emergency Services, and NORAD all in one week.

The CAP claims to have the largest continuously operating radio communications system in the USA, and is proud of the fact that it is moving forward with new technologies. Packet radio and computers are now part of the communications system. Satellites are being used to help DF downed aircraft (see Chapter 23).

There are eight geographical regions in the organization, each having five to nine wings, subdivided into groups, squadrons, and flights. About 66,000 persons are now volunteering their services to the CAP. Twenty-five thousand are teenage cadets, who begin as early as age 13. The largest wing, in terms of members, covers the entire state of California. It has 4800 volunteers and also has the largest number of search callouts.

The CAP owns only 600 aircraft, but 9000 more private craft are available from the members. One sixth of the members are licensed pilots. Members flying their own planes on actual CAP missions are reimbursed for fuel and some other expenses. They are not paid for their time or for non-mission flying—in fact, they pay dues to the organization and their squadron for the privilege of belonging. The CAP provides a $100 million fleet available to the government for public service at low cost to the taxpayers.

It seeks volunteers, both flyers and non-flyers, for

search and communications. Expenses for equipment, supplies, uniforms and the like, when used for CAP service, are tax deductible. More information on the organization and its programs can be obtained from a local office or the national headquarters (The address is listed in the back of this book.)

THE US COAST GUARD AUXILIARY

As a civilian adjunct to the U.S. Coast Guard, the Coast Guard Auxiliary (USCGA) supplies vital equipment and manpower needs at minimal cost to taxpayers. There are 18 districts in the Coast Guard Auxiliary, subdivided into divisions and then to the smallest units of organization, the flotillas. Each flotilla has at least ten qualified members providing equipment and service to the organization. There are many activities, but the one of most interest to T-hunting enthusiasts is the operation of the DF monitoring stations.

The Coast Guard has found that over 50 percent of all boats in distress cannot give their position correctly to within 5 to 30 miles. This is the reason for the establishment of DF stations in important shore areas, run by the USCGA. By giving rapid fixes of the location of boaters in distress, these stations have saved uncounted amounts of fuel, time, and man-hours for the U.S. Coast Guard.

The southern California DF stations are called over 100 times a year to assist in locating the source of a distress signal. Usually the distress call comes by vhf marine radio, but their lifesaving work is not limited to boats. In one case, when a private plane went down at sea, a nearby boater's radio transmission was DFed and used to locate the wreckage, from which two people were saved.

The DF station operators such as the one pictured here (Fig. 10-2) report bearings when requested by the

Fig. 10-2. Jim Grove, N6AXN, using a Coast Guard DF unit.

Rescue Command Center, and may also communicate with aircraft which are sent up on weekends for patrol, spotting distressed craft, and guiding the USCG vessels. Occasionally stuck radios or jamming in the harbor also must be triangulated.

The USCGA needs good people. It also needs equipment, such as boats and planes. More specifically, the Auxiliary needs volunteers to use their boats and planes for SAR work. Presently 82 percent of the search and rescue man-hours for the Coast Guard is done by the Auxiliary. Any US citizen over 17 years of age who owns equipment of use to the USCGA can become a member. There are also provisions for membership of persons not having equipment. In all cases, a basic qualification training program is required. If you live in an inland area, you may still be of service, because the USCGA has jurisdiction on lakes and rivers when they are on a state border; however, there is much less RDF work done on inland waters.

THE HAPPY FLYERS

Started in the 1970's in the San Francisco Bay Area, Happy Flyers stands for "Hams And Pilots, Piloting and Yakking." Membership is not limited to either pilots or hams—anyone interested in flying is welcome. The group is loosely organized into squadrons; a squadron forms whenever there are enough interested pilots. There are no dues, and most other formalities are dispensed with.

Happy Flyers is primarily a social organization, with fly-ins, group trips, and flying poker parties, in which costs are shared among the pilots and passengers. Another important aspect of the group is public service work in speeding up the location of downed aircraft. These activities have included testing and development of equipment, DF seminars, flight checks of DF gear, and actual search and rescue work.

Members of Happy Flyers, through their own independent efforts, have amassed a large body of knowledge of airborne DF techniques, many of which are also applicable to ground hunting. To make this information available to as many persons as possible, much of it has been put into a slide show (with many graphic illustrations by Paul Hower, WA6GDC), which is available through the Happy Flyers organization.

AMATEUR DETECTION OF ELT ALARMS

One major effort of the Happy Flyers founders is to have amateur radio repeaters on high mountains in desolate areas be equipped with ELT receivers. Many amateur repeaters are located in remote, high mountain peaks, which make excellent monitoring points for ELTs. If the accident occurs within earshot of a repeater equipped with an ELT monitor, the rescue effort can begin immediately, instead of some time later when an aircraft happens upon the signal.

The method is simple. An additional receiver is put into service at the repeater, tuned to 121.5 MHz AM. When a signal is detected, a special low level tone is sent out over the repeater transmitter, alerting users that the ELT is being heard. The tone is used because:

□ Direct rebroadcast of the ELT audio is against ham regulations.
□ The tone doesn't disrupt the repeater's voice traffic.
□ The tone can activate modified pager receivers to alert search and rescue team members at any time of the day without their having to constantly monitor the repeater.

To prevent activation of the repeater alert system during tests, a delay circuit is added. The ELT must be on for ten minutes before the repeater sends the alert tone.

One of the biggest problems with the present ELT system is false alarms. They have accounted for 96 percent of all ELT transmissions in recent years. Sometimes they are caused by faulty g-force switches or corroded batteries, but more often it is a bumpy landing or operator carelessness. Pilots are asked to check their ELT by listening to 121.5 megahertz before leaving their plane, but many don't. Searching out the culprit in a row of 100 parked and locked planes at an airport is a DFer's nightmare, but it is frequently necessary.

For this reason, repeaters near airports are not important in the ELT monitor program, but remote high sites are vital. They have good coverage, and are far from the airports where most falses occur.

AIRBORNE HUNTING

Since downed aircraft are often in inaccessible terrain the best way to locate them quickly is by using DF-equipped aircraft. Hunting from an aircraft is quite different from hunting on the ground. Altitude gives a significant advantage in range. Airborne hunting is freer from the multipath problems found in ground DF. However, since many, if not most, aircraft accidents happen in rugged terrain, the chances of having some amount of multipath in actual ELT hunting are probably as great as on any ground hunt.

The equipment used in airborne DFing is somewhat simpler than in ground hunting since there is no need to rotate the antenna. There are no roads, so it's easy to turn the aircraft until the signal is directly ahead. Instrumentation is vastly superior, with a calibrated compass to give a true, accurate, bearing to the signal and VHF OmniRange (VOR) to give the exact location of the aircraft.

The majority of airborne SAR hunters use a switched antenna DF system such as the Happy Flyers or L-Per. Some of these units include a complete receiver while others use the aircraft radio. The accuracy of these units in airborne use with no multipath or reflections can be as good as 0.5 to 2 degrees. One user reports that a 2 degree kick of the rudder pedal gives a one third scale deflection on the meter. Experienced pilots can tell which of two cars parked together has the transmitter.

Installation can be with inside antennas. Some pilots mount small jacks on the windshield trim and mount all the cabling inside the panel. Small metal rods are used as antennas. Another method is to use small coax (RG-174) and strip back the shielding for the length needed at the desired frequency. The coax can be taped to the windshield for a temporary installation.

There is plenty of information in print on installation of aircraft DF equipment. The L-Per manuals cover it in detail. The Happy Flyer group has a complete manual on DFing from the air, including problems and specialized techniques. As we've said with every other aspect of RDF, there is no substitute for practice hunts, with known targets, for checking out the equipment and developing skill and confidence of the user. During a true ELT emergency there is no time to be learning how to use the set, or worse yet, finding out the hard way that it doesn't work properly.

Two typical problems that need to be looked for in every installation are interference from the aircraft electrical system and from other radios in the plane. It's not unheard of to have 400 Hz noise riding on aircraft dc lines be the cause of totally erroneous RDF readings. Sometimes the DF goes wild when another transmitter is keyed. One way to check for the former is to run a check with the RDF and/or associated receiver powered by a separate battery. The DF test system discussed in the homing DF chapter is also useful.

While certainly not the most desirable situation, it is possible to search for an ELT from the air with only a receiver and its whip antenna. One way is to just use signal strength, flying back and forth and watching the signal get stronger and weaker. You can probably think of several reasons why this isn't a very reliable method.

The other method for DFing without a DF unit is to use the wing fade technique. Bank the aircraft first one way and then the other until the signal fades due to blockage of the antenna by the wing. For this to work the antenna must be in line with the wings, either above or below the fuselage, and the wings must be metal. If your Cessna has its antenna in the tail, forget it. It's necessary to fly a 360 to get a bearing, and when the signal disappears, the ELT is at 90 or 270 degrees to the plane, depending on antenna position and direction of the bank.

ADVANCED INTERFEROMETER TECHNIQUES

Far too often, an aircraft goes down and an ELT search must be made in bad weather, when it is not possible to send out search aircraft. In these cases, the best long range accurate-reading DF systems are needed for cross country work. Triangulation from highly accurate field bearings can save valuable search time. Loops and switched antenna units have the advantage of a very sharp null, readable to a degree or so, but they often have insufficient gain to get good bearings on distant transmitters. Quads and beams have sensitivity, but their directional indications are, by comparison, somewhat broad and difficult to resolve.

Wouldn't it be nice if there was a way to get higher gain and a sharp null? There is, and it's just right for certain applications such as this. These interferometer methods can also be used to overcome some of the bearing error problems of switched antenna units.

Interferometer Theory

Yagi and quad arrays usually have only one driven element. Other elements are undriven, and are thus called parasitic. Other antenna systems, such as phased arrays and ZL Specials, use multiple driven elements with control of the relative phase relationship of the rf drive current in each element.

Suppose an antenna system is made up of two vertical elements (monopoles or dipoles) spaced a half wavelength apart. The feed line to the receiver is split with proper impedance matching to accept signals from both antennas, but the polarity of signal from one antenna is opposite that of the other.

The resultant pattern in azimuth is the familiar figure eight. Along the line defined by the two antennas, the combination results in a peak in the response on either side. Perpendicular to the line at the midpoint, there is a null on each side. This pattern is the result only when spacing is a half wavelength. As spacing increases, the number of peaks and nulls increases. For example, with spacing of two wavelengths, the pattern resembles a flower with eight petals.

What we have now is a special phased array called an interferometer. It always has a null perpendicular to the straight line joining the antenna elements. Now let's try this trick with two long beams. Figure 10-3 shows that the patterns of the individual beams will combine to form a system having a high gain lobe with a sharp null right in the middle.

How close the system comes to this perfect pattern, and the accuracy of the null, depends primarily on how good the directional antennas are by themselves. The pattern of the figure assumes low side and back lobes on the beams. Obtaining opposite phase in the output of the two beams will be covered shortly.

Increasing the spacing between the beams increases the system gain, up to about 1.3 wavelengths. Beyond that, additional lobes and nulls in the forward direction begin to appear. At 4.5 wavelength spacing (about 15 feet at two meters), for instance, there are eight lobes, four on each side of the center null, about ten degrees apart.

Wide Aperture Interferometers

For even greater accuracy and rejection of multipath, the antennas should have much greater separation and

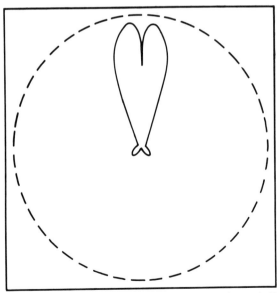

Fig. 10-3. Pattern of two long beams, spaced one half wavelength and connected as an interferometer.

be used with a special technique that gives both bearing and an idea of range. The wide aperture (also called wide baseline) interferometer has been studied and perfected by Rick Goodman, W5ALR, and Roger Chaffin, W5RGX, of the Amateur Radio Emergency Service in Albuquerque, New Mexico. The two antennas are spaced 30 to 150 feet apart. Long beams can be used, but two-element yagis do the job and are easier to carry and set up.

The system is based on the plane nature of the signal wavefront, as illustrated in Fig. 10-4. The diameter of the wavefront for a source a mile or more away is essentially planar over the short portion represented by distance d between the antennas. The front has become somewhat irregularly corrugated by wave interference from local reflections, but the average is still planar.

Wavefront lines A through H represent not only equal range, but equal phase of the signal. Visualize these circles as propagating outward at the speed of light, like waves in a pond. Though only a few are shown, there is actually a line for every wavelength along the path.

Each line represents a crest, so if the observer moves from one line to another, he receives a signal of equal phase but different time delay. Along any one line, such as H, the time and phase is the same. Now if we have antennas at positions #1 and #2, and if we can show that both the phase and time delay are identical, then they must be indeed on the same wavefront, and thus the line between the antennas is exactly perpendicular to the line of bearing to the transmitting antenna.

Looking at the phase over many feet along the wavefront this way is potentially far more accurate than looking at only about two feet of it, as a homing DF does. Assuming there are 20 degrees peak corrugations along the wavefront, a homing DF is about 12.5 degrees in error. With these same corrugations, a wide aperture system with 82 foot spacing gives an 0.3 degree theoretical error. Actually spacings greater than 30 feet are seldom necessary to obtain azimuth, because the limitations of compass-reading prevent the accuracy from being greater than can be gotten with the 30 foot spacing. It's hard to find even that much clear space in the field, anyway.

Determining that the antennas are on the same phase front is done by having two low gain beams with outputs combined out of phase as with the narrow aperture system. This could be done by using coax cable lengths differing by one half wavelength, but the resultant system then has correct phase only in one frequency band. It's easier to use a combiner with equal coax lengths, and use one of the beams upside down from the other. For example, with vertical polarization, one is used with the

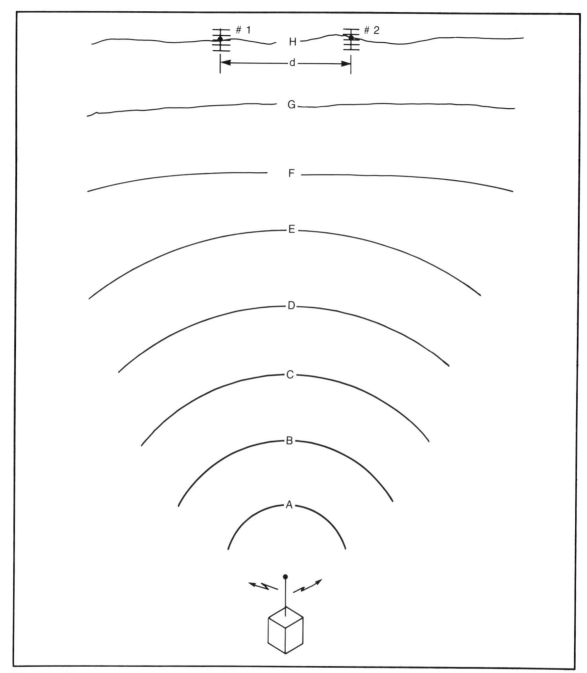

Fig. 10-4. As a radio wave travels away from the transmitter, it becomes planar and somewhat corrugated.

gamma match pointing up, and the other with the gamma match pointing down. The Wilkinson combiner of Fig. 10-5, made from 75 ohm coax (RG-59/U), matches impedances. It has very low loss and provides good isola-

tion between the antennas to minimize re-radiation effects.

To get a fix with this system, one antenna is first used alone to get an approximate bearing. The antenna is not

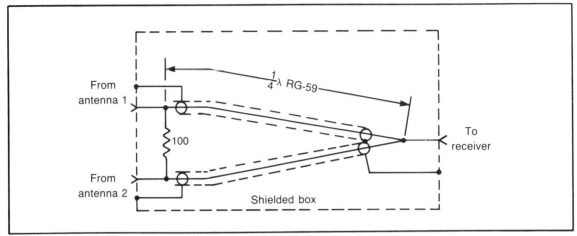

Fig. 10-5. A Wilkinson power combiner used to add signals from two antennas.

connected through the power combiner at this point. Both horizontal and vertical polarization are checked, and the one giving strongest signal reading is chosen for the interferometer measurement. This antenna is now kept fixed.

The second antenna is set up with the same polarization (but opposite phase) along a line perpendicular to the apparent bearing to the source. It is easiest if the first interferometer check is made with the second antenna less than 10 feet away. Both antennas are connected through the power combiner to the receiver, and the antenna is moved forward and backward (toward and away from the source) until the signal meter indicates a good null.

The null occurs when both antennas are on the same phase front, so at that point the ELT or other source is perpendicular to the line between the two antennas, in the direction indicated by the antennas. For more accuracy, move the second antenna further away from the first, while continuing to watch the null. If you don't do this a few feet at a time, it is possible to slip into another phase front, as they are only slightly more than eight feet apart at 120 megahertz, in the direction to and from the source. When the antennas are separated 30 feet or so, the best bearing can be taken by using a sighting compass.

Measuring Distance

As pointed out earlier, a reflection-free signal wavefront is piecewise planar when the observer is a mile or so away, but it still maintains the curve corresponding to the arc of this very large circle. We can use this characteristic to get an indication of distance to the transmitter from the interferometer setup. This is practical only under the following conditions:

☐ There must be few reflections and other local wavefront distortions to upset the circularity of the wavefront.

☐ The distance to the source must be less than two miles. At greater distances there is too little circularity to the wavefront.

☐ A wide baseline (100 to 150 feet) must be used.

Distance is measured after the bearing is taken. A marker is used to indicate the end position of the second antenna. A rope or line is stretched between the first antenna mast and the marker. The second antenna is moved to the exact midpoint of the line and set up facing the source. This second antenna is then moved away from the source until a good null is achieved. The approximate distance is then calculated with:

$$R = \frac{D^2}{8Z}$$

R = Range (distance) to the source
D = Distance between the first antenna and the marker
Z = Distance behind the line that the second antenna had to be pulled to achieve a null

Note that R, D, and Z must all be in the same units (feet, for example). While this method cannot be expected to yield super accuracy, it can give the team an idea of how well it is closing in. In the tests in New Mexico, a distance measurement of a transmitter 3000 feet away was off by only 300 feet.

Chapter 11

Weak Signal Hunting

One of the oldest ham radio adages is, "You can't work 'em if you can't hear 'em." This usually applies to T-hunting, too. Although K6BMG has shown that his dual antenna RDF unit will take bearings on signals which do not break the FM receiver squelch (see Chapter 8), there are many pitfalls with DF'ing signals that are below the threshold of detection. When is the signal on and when is it off? Is it the desired signal? Or another one? Or noise? Or sidebands from an adjacent frequency? In this chapter we discuss ways of increasing the sensitivity of our RDF system, and getting the most information from the signal we get.

WHY BOTHER WITH WEAK SIGNALS?

If the only signal you've ever wanted to hunt is a jammer running a high power amplifier, you may be wondering why it's ever necessary to dig down into the noise for a signal to hunt. The best reason is that someone's life may depend on the ability of hunters to hear his weak transmitter. The vhf Emergency Locator Transmitter (ELT), built to begin transmitting after an airplane crash, runs only a fraction of a watt. Airplanes don't always crash in populated areas, or on high hills which allow the ELT signal to be easily heard. They tend to end up in

remote, deep canyons, many miles from listening receivers. It is entirely possible that lives could be lost unnecessarily due to RDF equipment that lacks sensitivity.

Jammers can have weak signals, too. It doesn't take much signal to hold up and time out a mountaintop repeater. If the interfering signal is close enough to the receiver, it's easy for it to block out more distant signals. A few minutes with a calculator and the path loss formula will show that, given equal antennas and unobstructed paths, a half watt transmitter located one mile from the receiver will have a 10 dB signal advantage over a 10 watt transmitter located 15 miles away. The capture effect of FM allows the closer but weaker transmitter to completely obliterate the other.

Perhaps the best example of the potential of a weak jamming signal is the intermittent MCW signal that showed up on a major Los Angeles 2 meter repeater in January of 1977. At the time, the repeater had the call WR6AMD and was located atop Mt. Wilson, covering the entire metropolitan area. The jammer obliterated all but a few base stations and high-powered (300 watts ERP or so) mobiles for about four seconds at totally random times, usually about seven times an hour.

It was a real test of the perseverance and skill of dedicated T-hunters, but this automatic jammer was found 21 days later, still operational, in the San Gabriel Wilderness Area four miles from the repeater and about two miles from the nearest decent road. The 1-watt transmitter was buried, and the antenna was a ground-plane, made from a coat hanger and secreted in a very thorny bush. Estimated effort for the hunters was about 275 man-hours, much of it spent hunting a very weak signal.

Sport hunters also have good reason to want excellent weak signal performance from their gear. Smart foxes often run the least amount of power they can, just enough to be heard at the starting point. For one reason, this allows a smaller battery supply. It also means some hunters hear the hidden T at the top of the starting hill, lose it on the way down, and do not hear it again until they get quite close. If the hunter gets a poor starting bearing, perhaps due to reflections, he may never hear the fox again!

Rich Krier, N6MJ, once hid a transmitter, attenuated to a few microwatts output, and a very long yagi beam about 750 yards down the hill from the starting point in a field. The signal was very weak, and disappeared immediately as the teams left. Almost everyone except one team settled in for a long hunt and traveled dozens of miles. One team did not become suspicious until they could not hear the T from high atop Signal Hill, 17 miles away from the starting point. But Bruce and Karen Gallant (WB6DCB and WB6DCC), the team that won, snatched a tiny bit of weak signal from the side of the hill, with a different bearing from the hilltop bearing. This allowed them to close in with very little time and mileage.

On vhf, you'll find that the team with the most sensitive receiver setup has a definite advantage on many hunts. Time put into improving your system sensitivity is time well spent.

GRABBING THE SIGNAL

Below 30 MHz, the main limitation on a receiver's sensitivity is atmospheric noise. As long as the system hears plenty of atmospheric noise, the receiver is probably sensitive enough. With a small hf loop antenna, gain of the receiver may be insufficient to hear atmospheric noise well. In such a case, an rf preamplifier is just what is needed. Preamps always excel at making noise louder, but if the noise is atmospheric in origin, a preamp alone won't help the signal-to-noise ratio.

A bigger loop brings in more signal, but also more noise. To increase the signal and not the noise, a highly directive antenna is needed. Atmospheric noise comes from all directions, but the signal is (we hope) coming from only one. So a narrow pattern antenna, such as a beam, quad, or rhombic, blocks out noise in all directions except forward, allowing the signal to stand out.

At 10 or 11 meters, a rhombic is out of the question for mobile work, but a loaded yagi or quad is possible (see Chapter 12). Unfortunately, as these antennas are loaded more to make them physically smaller, their capture area diminishes rapidly, and the advantage over a loop is soon lost.

Atmospheric noise is not generally a limitation at vhf, although locally high noise levels are often generated during dry windstorms. At vhf, the signal must be strong enough to overcome the thermal noise generated in the first rf stage of the receiver. A detailed discussion of receiver noise performance is beyond the scope of this book, but suffice it to say that if your 2 meter FM receiver is rated at greater than 0.6 microvolts input for 20 dB SINAD, you will probably benefit from the addition of a low-noise rf preamp ahead of the receiver for weak signal hunting. Another indication that your system needs help is if you're among the minority of hunters that can't hear the weak signal at the start of the hunt.

Just because your owner's manual says your receiver is "hot," don't accept it blindly. Take advantage of any opportunity to test it on a good calibrated signal generator. We saw a case where the factory test technician apparently forgot to adjust one tuned circuit in the early stages of a 2 meter receiver, causing a 6 to 10 dB loss of sensitivity. While testing, make sure that sensitivity is maintained over the entire range of frequencies you wish to use, or peak it up for your segment of interest.

Any losses in the antenna/feed line chain are not recovered by preamp gain on very weak signals due to thermal noise in the receiver front end. Thus it's important to eliminate as much of this line loss as possible. Keep feed lines as short as you can. Don't use RG-174 or other miniature cable to go to receiving antennas. Use RG-58 size of cable only for short runs of less than 10 feet. Use RG-8 or larger cable for longer runs, particularly in fixed installations.

Don't trade away too much gain in your beam or quad just to get a smaller antenna for easy handling. We recommend at least four elements, and more for long distance hunting at 2 meters. Short-spacing the elements makes a compact antenna, but causes a loss of gain. Though many hams use 0.1 wavelength spacing for beams, they do not get best system sensitivity. Try to get 0.19 wavelength between directors and 0.15 wavelength from reflector to driven element.

Recent editions of *The Radio Amateur's Handbook* include a good discussion of vhf/uhf receiver sensitivity, noise figure, and types of preamplifiers. The best performing preamp designs available for the 100 to 500 MHz range use Gallium Arsenide Field Effect Transistors (GaAs-FETs). Their gain per stage is highest and cross modulation performance is by far the best, when compared to bipolar transistors, junction FET (JFET) and Metal Oxide Semiconductor (MOSFET) designs. Commercial GaAs-FET preamps also cost two to three times as much as other types.

Fortunately for your pocketbook, we don't recommend GaAs-FET preamps for most amateur T-hunting below 500 MHz for these reasons:

☐ Their superior noise figures are of little help in the typical urban electrical noise environment.

☐ They are prone to damage from disturbances on the power source.

☐ Their gain is so high that they may cause typical receivers to generate cross-modulation products in their mixer stages.

GaAs-FET preamplifiers are most effective in the high uhf and microwave region, where bipolar and conventional FET designs provide poor performance.

An inexpensive one-stage preamp is an excellent way to perk up a receiver of marginal sensitivity. JFET or MOSFET designs work well in the vhf region (below 300 MHz), while bipolar transistor designs are recommended for uhf. Excellent preamps are commercially available from companies such as Lunar Electronics and Advanced Receiver Research Company. If you like kit building, Hamtronics has a line of popular preamp kits which can be constructed to cover frequency segments from 27 to 650 MHz with about 10 percent bandwidth. Circuit Board Specialists also offers bipolar preamp kits usable on the 146, 223 or 440 MHz bands.

Deciding where to install the preamp can be a dilemma. For highest sensitivity and best system noise figure, the preamp should be at the antenna, driving the feed line. This is mechanically awkward and creates the risk of accidentally transmitting into the preamp. Even very low power levels destroy MOSFET and GaAs-FET preamps; others fare only slightly better. Preamps with automatic power sensing relays built in are available, but the relay loss may negate the advantage of antenna mounting.

Most hunters find that the best compromise for 2 meter hunting is to mount the preamp in the receiver in line between the antenna transfer relay and the first rf stage (Fig. 11-1). With short feed lines, very little degradation is noticed.

If one preamp is good, cascading two would be better, right? Wrong. The first preamp establishes the system noise figure and the second gives no benefit. Sometimes even one external preamp is too many. If the specifications for your vhf or uhf receiver are state-of-the-art, and if the receiver meets them, a preamp probably won't help and may hurt. Too much gain ahead of the receiver's mixer stages can result in overload, birdies, and intermodulation products if a strong in-band signal such as a repeater is nearby. Since these undesirable effects are more likely with the added gain of a preamp, it might be wise to put connectors on the preamp or even add a bypass switch.

A BUILD-IT-YOURSELF PREAMP

A simple one-stage preamp is not a difficult project to build from scratch. It can be built into a shielded enclosure or mounted uncased inside the receiver. This 2 meter design is an adaptation by Gary Frey, W6XJ, of a popular JFET circuit. It is stable and requires only one bias resistor.

Figure 11-2 includes construction information. The metal can U310 is preferred, because the gate lead is connected to the can. The part is soldered into a hole drilled in the circuit board ground plane, providing excellent input-to-output circuit isolation. The plastic cased P310 can also be used if care is taken to keep device leads, particularly the gate, very short. Minor supply voltage variations cause no problem because the FET is inherently

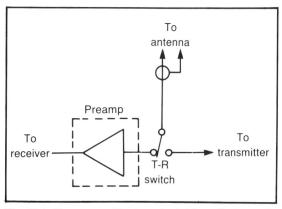

Fig. 11-1. The preamplifier is protected from the transmitter if it is connected between the antenna relay and the first rf stage.

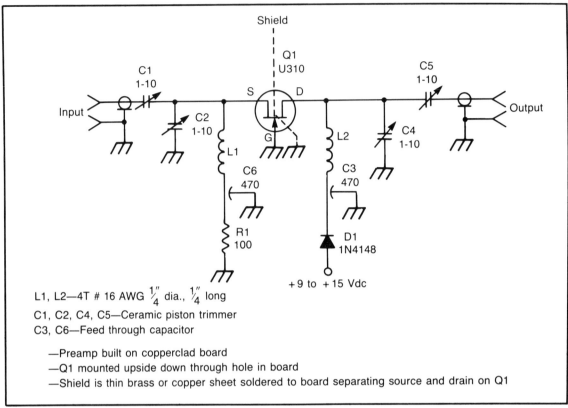

Fig. 11-2. Schematic of the JFET preamplifier.

L1, L2—4T # 16 AWG $\frac{1}{4}''$ dia., $\frac{1}{4}''$ long

C1, C2, C4, C5—Ceramic piston trimmer

C3, C6—Feed through capacitor

—Preamp built on copperclad board
—Q1 mounted upside down through hole in board
—Shield is thin brass or copper sheet soldered to board separating source and drain on Q1

a constant-current device. Diode D1 provides protection from accidental reverse supply polarity.

When power is first applied, measure the supply current. Change the value of R1 as required to get between 8 and 13 mA. The preamp may be tuned with a weak signal if a signal generator is not available. C1 should be tuned for best signal-to-noise ratio, not just maximum gain. If good test gear is available, C1 should be adjusted for a noise figure of 1.8 dB or better.

C4 and C5 are tuned for maximum gain, about 13 dB. If C1 or C4 peaks at the end of the range, adjust the inductance of L1 or L2 accordingly. In some receivers, there may be some interaction with the first receiver rf stage, producing instability. If this occurs, try putting a 100 ohm resistor across the output of the preamp. Use a quarter-watt carbon composition resistor and keep the leads short to avoid inductive effects.

Best performance of this preamp is over about a 2 MHz range, although performance does not deteriorate quickly outside this range. It is suitable for mounting inside a receiver and being peaked up in the center of the band (Fig. 11-3). We recommend putting it between the antenna relay and the first stage, as shown in Fig. 11-1.

Be very careful if you mount it in the antenna line of a transceiver. Although it might take a watt for a few seconds, there's no guarantee. Set the rig in the low power position, as you probably do already to protect your resistive attenuator. Better yet, disconnect the microphone when hunting.

USING FM QUIETING

Peaking a weak tone or noise modulated AM signal is fairly easy—just listen for the loudest apparent detected signal. Peaking a weak FM signal, on the other hand, is not as simple, particularly in a moving vehicle. When the signal is very weak on an FM receiver, the meter does not show any indication. How can an accurate bearing be taken with a directional antenna under these circumstances?

The high i-f gain of an FM receiver produces a loud rushing sound when no signal is present and the squelch control is open. As signal strength increases, the noise goes down, until the signal fully quiets the receiver. Weak FM modulation may not be understandable until a considerable amount of quieting has occurred. The squelch circuit uses this noise quieting characteristic of FM, turning on the audio stages when a reduction of the noise is sensed.

When the FM signal is too weak to read on the S-meter, use the squelch to get a more accurate bearing than by guessing from the noise. The squelch control should be set at its threshold as for normal operation. Point the antenna for apparent best signal. Turn the antenna in one direction until the signal just squelches out and write down the bearing. Let's say it's 42 degrees. It doesn't matter if the reading is with respect to north or to our vehicle heading, as long as we're consistent.

Now swing the antenna back through the peak until it squelches out on the other side; say, 84 degrees. The correct bearing is halfway between, or the arithmetic mean (average) of these readings. Add the numbers and divide by two. In this case the bearing to the bunny is 63 degrees. A word of warning: This method will be prone to error if the beam or quad has a non-symmetrical pattern.

Some minor juggling will be necessary when going through the 360 degree point. If the left and right squelch points are at 347 and 31 degrees, for example, the correct bearing is 9 degrees. Just add 360 degrees to the lower number, take the average, and subtract 360 from the result if it is greater than 360.

This method may be impractical in a moving vehicle, or if there is severe flutter or fading on the signal, as aircraft or atmospheric inversions can provide. It can also be inaccurate due to local noise conditions. If the beam is swung to point to a strong noise source, such as a noisy power line, the noise may squelch the receiver even if there would normally be enough signal to keep

Fig. 11-3. The JFET preamplifier is mounted in a small Pomona Electronics metal box. You will probably want to use BNC connectors instead of the SMA connectors shown here.

it open. Sometimes leaving the squelch open and aiming the antenna for best signal quality is the best way to get a bearing in a noisy location. The best overall strategy in these situations is to take frequent bearings from different locations.

A NOISE METER FOR FM RECEIVERS

Since the reduction of noise in an FM receiver is a direct function of signal strength, the strength of a weak signal can be measured by how much the receiver noise is being quieted. In fact, having such a noise meter gives you a definite advantage over your fellow hunters who must guess the direction of best quieting.

A meter that is directly calibrated in percentage of quieting isn't required. All you need is one that reads relative noise from about zero to 90 percent quieting. By the time the signal strength rises to about 90 percent quieting, the regular S-meter begins to move upscale.

Figure 11-4 is the schematic for such a noise meter. Q1 and associated components form a high input impedance gain stage fed by a two pole high pass filter. Loading on the receiver audio circuits is negligible. The high pass filter allows the meter to respond to only the high frequency noise components in the audio, and to ignore any voice or tone modulation. It is convenient that noise in a typical receiver discriminator has the greatest energy content at high, even supersonic, frequencies.

The second stage provides 8 dB voltage gain to raise the level of the noise. If you cannot get full scale noise

indication on M1 with R8 set to minimum, increase the gain of this stage by raising the value of R7 to 1 k or 2 k ohms as required. If R7 is changed, check the dc voltage on Q2 collector and change R5 as necessary to set that voltage to about 6 volts. At the other extreme, your receiver may have enough noise output to permit deleting the second stage. In this case Q2, C4, and R5-7 are eliminated and the C5 input point is connected to the collector of Q1.

The high frequency noise is rectified and fed into a dc amplifier (Q3) to drive the meter. Control R8 is used to adjust the meter deflection to full scale with no signal present. As the input signal increases, causing quieting, the meter indication rapidly drops, due to the non-linear characteristics of the noise rectifier diodes. It is a good idea to mount R8 on the front panel near the meter. Many FM receivers have variations in i-f gain and discriminator output as their dc supply voltage varies, so you may need to readjust R8 for mobile-in-motion versus engine-off operation.

Input for the circuit must come from unprocessed receiver audio ahead of the squelch gate. The discriminator output is best if it is of sufficient amplitude. If the top of the squelch pot is not part of the dc gate circuit, it may be a good tap-off point. The output of a noise amplifier in the squelch circuit can be used if the squelch pot does not control the gain of the noise amplifier stage. Use shielded wire unless there is a capacitance loading problem.

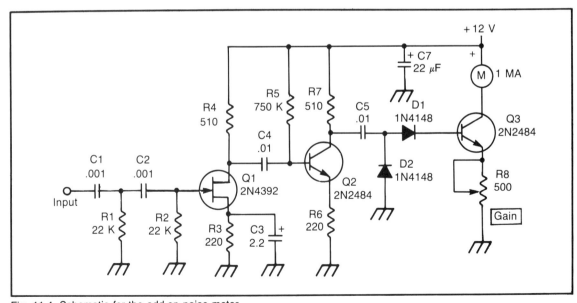

Fig. 11-4. Schematic for the add-on noise meter.

Fig. 11-5. An external S-meter, noise meter, and DSB product detector are built into this sloped front enclosure.

Figure 11-5 shows the circuit built into a typical T-hunting meter box. The noise meter is on the right. R8 is next to the miniature toggle switch. The other meter is a remote S-meter. Construction follows standard audio frequency practice and can be on perf-board. Inexpensive replacement devices can be substituted for the part numbers shown here.

Use of the noise meter is straightforward. Set R8 for full scale (but not pinned) with no signal present. The meter reading will decrease as the signal strength increases. By the time the reading gets to zero, the S-meter should be indicating.

One hunt of a weak signal with the noise meter is worth a thousand words of explanation. With care, very accurate bearings can be obtained on weak signals with the noise meter. However, atmospheric noise and modulation products can cause errors for the unwary. Use caution in high electrical noise areas when the weather is dry, as power line noise can appear to the noise meter to be unquieted receiver noise. The difference can often be discerned by listening to the receiver audio. Also, if the S-meter swings up but there is little quieting, severe external noise is present.

Experienced hunters usually keep the receiver squelch open when hunting a weak signal. The noise may be annoying, but it permits hearing electrical noise or adjacent channel signals which can upset both noise meter and S-meter readings.

Voice modulation does not normally show up on the noise meter, but tones from square wave oscillators, which hiders love to use, may have enough high frequency audio harmonics to make the noise meter needle bounce around. They can affect S-meters too, particularly if the receiver bandpass is narrow, as we'll discuss later. Use a meter with as little damping as possible, so these effects can be noticed and mentally compensated for.

SIDEBAND DETECTORS FOR AM AND FM HUNTS

Newcomers to amateur radio may not have thought about why their voice communications in the hf range are primarily single sideband (SSB) with some occasional amplitude modulation (AM), while in the vhf/uhf bands there is mostly frequency modulation. FM is convenient because it's easy to tune in and does not require oscillators with as much stability and drift resistance as SSB does.

The capture effect and quieting characteristics of FM make broadcast quality audio possible. Though such quality is seldom heard on the ham bands, broadcasters use FM almost exclusively for remote and mobile pickups on vhf and uhf, and for uhf studio-to-transmitter links. Carrier-operated repeaters, sub-audible tones, and telephone dialing tones all work well on FM but are not practical for amateur SSB.

For DX voice work, however, SSB is the mode of choice. A weak SSB signal can be copied when an FM signal of equivalent strength shows no noticeable quieting. Weak signal work on the vhf/uhf bands, such as tropospheric ducting, meteor scatter, and moonbounce is all done on CW or SSB, but T-hunting on these bands is all FM or AM.

Why, then, would a ham want a SSB receiver to hunt an FM signal? Because it's unexcelled for hearing weak carriers. The SSB receiver won't demodulate the FM signal, but the carrier is plainly audible as a tone when the receiver is tuned a few hundred hertz away from zero beat. The modulation imparts a raspiness to the tone, making it disappear into a jumble of sound at high modulation levels. The more signal, the louder the tone.

With today's typical amateur equipment, an SSB detector can hear an FM signal carrier when it's about 10 dB below the threshold of copy with an FM detector. Hunters with trained ears will have no trouble hunting such a weak signal by sound. If you're a competitive hunter, imagine the advantage of being able to hear and hunt signals that are 10 dB lower than others can detect! Think about what you'd have to do to your antenna system to get 10 dB more gain. Next time you leave the starting point on a high hill and the bunny's signal disappears in your receiver, visualize yourself switching over to SSB reception and continuing to get bearings!

Now that you're convinced (we hope!) that having SSB provisions will give you a competitive edge, let's examine the alternatives for detecting the signal in this way. There are many multimode tranceivers available for the most popular vhf/uhf bands, 2 meters, 1-1/4 meters, and 70 cm. The USB or LSB position on these rigs gives the extra sensitivity. When the signal is strong, the FM detector is used. Typically, these all mode sets cost 50 to 100 percent more than FM-only units.

When you're not T-hunting with your multimode rig, you'll enjoy vhf/uhf SSB communications. Range is considerably more than FM simplex. Signals take up less spectrum space. Such a rig can also involve you in the fun of OSCAR satellite communications. The activity you'll find on the bands depends on where you live, since SSB is not nearly as universal as FM on these bands.

Unfortunately, commercial AM/FM/SSB rigs were not designed with T-hunting in mind. Though they have a definite advantage in sensitivity, they suffer from two problems, both having to do with the S-meter. Although you won't be using the S-meter when listening to a weak signal in the SSB position, you'll want to use it in the FM mode when the signal gets strong.

Today's multimode rigs usually have separate i-f strips and detectors for FM and SSB. The SSB i-f is much narrower than the FM strip—the SSB i-f is usually about 2.7 kHz wide, while the FM i-f may be 8 to 15 kHz wide. The narrow i-f on SSB is an advantage for hearing weak signals, as less noise passes through the narrow filter with the bunny's carrier.

Manufacturers usually put the S-meter circuitry in the SSB section of all-mode receivers. There is no problem with SSB signals, but highly deviated FM signals have energy outside the SSB filter bandwidth. The result is that FM signal modulation, even on strong signals, may make the S-meter bounce around. A T-hunt tone box and rig with sufficient deviation may make the S-meter bounce so much as to be useless as a direction indicator in a quad/attenuator setup. The authors gave up trying to use a Kenwood TS-700A for sport hunting because of this characteristic.

A second problem is that many multimode rigs use automatic gain control (AGC) feedback in the SSB portion, which tends to cause the S-meter to have a logarithmic characteristic. As explained in Chapter 5, this can have a disastrous effect on apparent antenna directivity. One rig we tested (the Icom IC-260) had this characteristic plus such heavy damping that it was impossible to get bearings on the internal meter while in motion, and very difficult to get them when stopped. With an added external meter circuit, the readings were much better, though the narrowband problem remained.

Before purchasing a multimode rig for T-hunting, check it and the specifications very carefully to be sure it's suitable. Borrow one and try it if you can. Be sure it is stable in mobile operation and has a good receiver incremental tuning (RIT) system.

If the rig has the S-meter associated with the SSB i-f, as described above, the best way to use it for T-hunting is to add an external S-meter to the FM i-f section. See Chapter 5 for appropriate circuit ideas.

AN ADD-ON SIDEBAND DETECTOR

It's not necessary to buy a new vhf transceiver to take advantage of sideband detection for increased receiving sensitivity. This external adapter does that job very inexpensively. While it does not allow transmitting SSB, and does not provide the noise reduction of the narrow SSB i-f strip, it gives you that 10 dB sensitivity edge in hunting FM signals, and lets you eavesdrop on sideband activity in your area. You can continue to use your present S-meter and the FM noise meter described earlier in this chapter.

Figure 11-6 shows a simplified block diagram of a typical SSB-only receiver. The local oscillator (LO) may be tunable, crystal controlled, or synthesized. Sometimes it is called a heterodyne oscillator. The beat frequency oscillator (BFO) is at the same frequency as the suppressed carrier of the SSB signal in the i-f; hence, it is occasionally called a carrier oscillator. The SSB detector mixes, or beats, the SSB i-f signal with the BFO signal to produce the sum and difference. The difference product is the detected audio signal, which is amplified and fed to a speaker or earphones.

Because this mixing is actually a multiplication technique, the detector is called a product detector. In the SSB-only receiver, the BFO is set about 300 hertz outside the i-f passband, so that only the lower (or upper)

300 Hz to 3000 Hz sideband energy in the signal passes through the 2.7 kHz wide filter and is detected.

An external product detector is shown connected to an FM receiver in block diagram form in Fig. 11-7. It uses the same product detection principle, but the i-f filter is, of course, wider (12 to 15 kHz). The SSB signal goes straight down the middle of the i-f filter, unless the BFO is offset with the RIT control. The BFO is set in the middle of the filter passband for sideband reception, so both sidebands are demodulated, making this unit a double sideband (DSB) detector.

RIT (Receiver Incremental Tuning) is accomplished by tuning the BFO. It can be set up to plus or minus 5 kHz of the center of the i-f to accommodate off-frequency carriers and to give a pleasing tone when hunting FM signals.

The add-on DSB detector described here is designed to be used with receivers having a 455 kHz i-f strip. Most amateur vhf FM receivers have a first i-f at 10.7 MHz or other convenient frequency in that range, followed by conversion to 455 kHz, because good, inexpensive ceramic filters are available for FM bandwidths at that frequency. If your receiver has a different i-f frequency, the BFO can be modified for it.

Building the Detector

The schematic diagram and parts list for the detector are given in Fig. 11-8. All parts should be readily available. The ECG 222 can be used at Q2. Q1 is RS 276-2062 and AR1 is RS 276-706. Perf-board construction works fine. Using a board with a ground plane on one side is recommended. Supply voltage is not critical. The BFO is stable to plus or minus 20 Hz from 10 to 15 V input, so no voltage regulation is used on the oscillator.

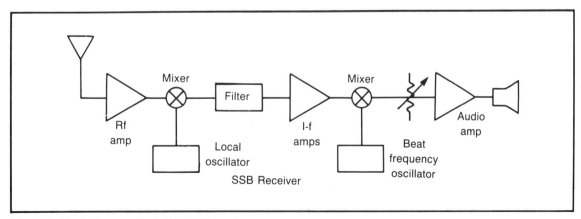

Fig. 11-6. Block diagram of a typical SSB receiver.

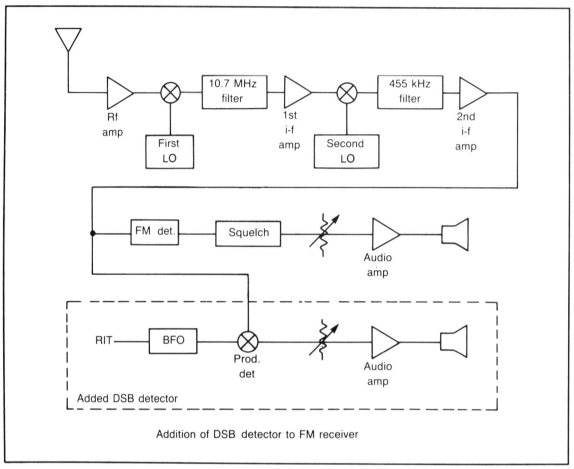

Fig. 11-7. A typical vhf-FM receiver with an added product detector.

An unusual oscillator configuration called a Seiler oscillator is used in the BFO to achieve its excellent stability. C4 and C5 are very large compared to capacitances of the FET and strays. Changes of device capacitances due to temperature are small compared to C4 and C5, so they are masked. Stability is also ensured by the use of a good high-Q powdered iron coil form (not ferrite), and mica or polystyrene capacitors at C2 through C6.

To change the BFO frequency for other receiver i-fs, tank capacitors C2 and C3 should be scaled. C1 is chosen for ± 5 kHz range. Values of C4 and C5 are not critical but should be equal and much larger than C2. If they are too large, the circuit won't oscillate.

The i-f signal and the BFO output are mixed at Q2, a dual-gate MOSFET. Audio transformer T1 couples only the difference frequencies to the audio amplifier AR1. T1 is an interstage audio transformer with about a 1:3 step-up turns ratio. C10 is chosen to tune the secondary winding of the transformer to about 1 kHz. This peaks up the carrier tone at that frequency for ease of hunting.

AR1 is an audio power amplifier to drive the speaker, with 34 dB gain. Use of a socket is recommended. To prevent oscillation, the eight grounded socket terminals are soldered directly to the ground plane. Leads of C12 are kept very short, with the capacitor connected directly to pins 14 and 7. Likewise, R8 goes directly to pins 2 and 6. If oscillation occurs despite these precautions, the manufacturer recommends adding a 2.7 ohm resistor and 0.1 microfarad capacitor in series from pin 8 to the ground plane (R9 and C15).

Though the bandwidth of the LM380 is specified as 100 kHz, it still has plenty of gain at the 455 kHz i-f frequency. Any BFO signal radiated or conducted into the stage will be amplified and may cause instability. The only problem we had with the prototype was pickup at R6, since R6 and C1 were mounted close together on the front panel. The solution was to run a shielded lead (RG-174 coax) from the oscillator to C1, and shielded leads to the top and tap of R6. A grounded shield plate was put between R6 and C1. Double-sided PC board works perfectly for such a shield. Figure 11-9 shows this shield in place. These precautions may not be necessary if the two parts can be mounted on the board at separate locations.

The circuit fits on a board small enough to go into the enclosure that houses your remote S-meter and other devices. Since super fidelity isn't needed, a 2-inch or smaller speaker will work well and produce plenty of sound. If there isn't room on the front of the box, mount the speaker on the back. This assumes that your box is mounted on the top of the dash, since most hunters don't like to look or fumble under the dash while driving.

Connecting To Your Receiver

The signal should be taken from the receiver at the earliest point in the i-f chain that will give sufficient audio output on detected signals. This minimizes the effects of limiting on strong SSB stations. The collector or drain of the first i-f amplifier after the i-f filter is a good place to start. This is the same point where the S-meter detector of Fig. 5-1 tapped into the i-f.

If the DSB detector is built into the receiver, gate #1 of Q2 can be connected directly to the i-f tap point through a 10 to 100 picofarad capacitor, if the lead is kept quite short. In most cases this isn't practical, and a coax lead from rig to detector is desired. Figure 11-10 shows the schematic of a simple untuned buffer stage to drive the coax line. It has high impedance input and moderate gain at 455 kHz. It was used successfully in a trunk-mounted Drake UV-3, driving 15 feet of RG-174 coax to the product detector mounted in the front.

The buffer can be made small enough to squeeze into the circuit area where the tap is made, as shown in Fig. 11-11. The output can then be taken in coax through the rig's accessory socket and over any convenient length to the DSB detector. When not hunting, the detector can be unplugged from the accessory socket.

A frequency counter and an oscilloscope are useful for initial setup of the detector. Connect the counter to the collector of Q1 through a 100 pF capacitor. Apply power and adjust L1 for 455.0 kHz on the counter with the plates of C1 half meshed. Adjusting C1 over its range should move the BFO from about 450 to 460 kHz.

Observe the waveform at Q1 collector with the scope,

using a 10X probe. If clipping at either top or bottom is noted, or if the midpoint of the BFO sine wave is not between 4 and 8 volts, change R1 as necessary. If there is clipping on both top and bottom, decrease the oscillator output by making C3 smaller in value. Retune L1 as required. Check for 455 kHz signal in the output (pin 8) of AR1. It should be less than 0.5 V peak-to-peak over the range of R6 with the receiver i-f input disconnected. If not, check the shielding and ground paths as discussed earlier.

Now apply the receiver i-f signal to the mixer stage. On-channel FM signals should produce a zero beat with C1 at midpoint. When listening to a carrier and tuning the BFO, there should be a broad response peak in audio output around 500 to 1000 Hz. If the peak is not where you want it, change C10 to suit.

Using The Detector

You'll find that FM signals too weak to hold the squelch open produce a tone in your DSB detector. Set the BFO tuning control (C1) for a tone that stands out to your ear the best, as you will be pointing your antenna by ear. Anything that affects the frequency or stability of the local oscillator in your receiver, such as synthesizer noise, subaudible tones connected to the VCO, or temperature/voltage sensitivity, shows up as a change in the quality of the tone. So do instabilities in the received signal. Modulation affects the tone, but usually does not stop you from getting bearings.

When the signal gets loud enough to hunt with an S-meter or noise meter, the DSB detector can be turned off using the switch mounted on R6. Since the speaker is not needed once the signal gets strong, it can be shared with an audio S-meter (see Chapter 5) using a DPDT switch if space is limited.

SIGNATURE ANALYSIS

No, we're not talking about hunting down forged checks. Every transmitter has certain identifying characteristics about its signal which can help separate its signal from others on the same frequency. These characteristics make up its signature. While very sophisticated methods are available and used by authorities to identify the source of signals, some characteristics show up readily on an FM signal when monitored on an SSB or DSB detector. In a jamming situation, and on some hunts, there may be many signals on the frequency from time to time, and the DSB detector may be useful in identifying and separating them out.

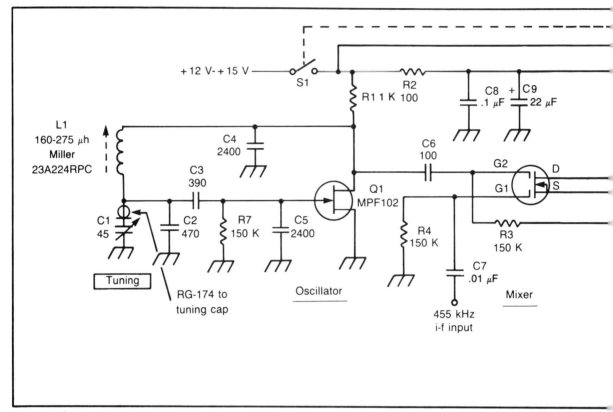

Fig. 11-8. Schematic of the add-on DSB detector.

Fig. 11-9. Use shielded cables and a shield made from copperclad board to prevent radiation of the BFO signal into the audio circuits.

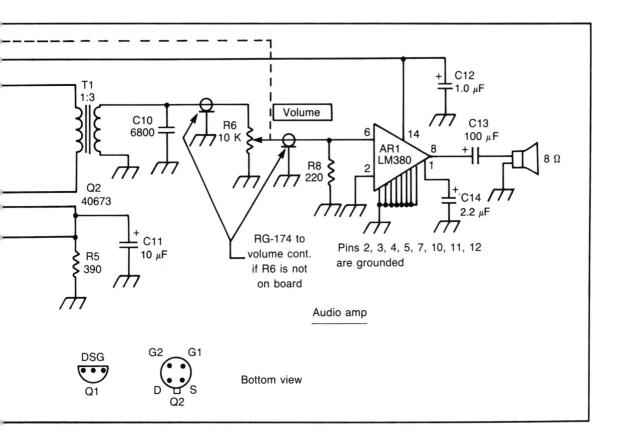

Audio amp

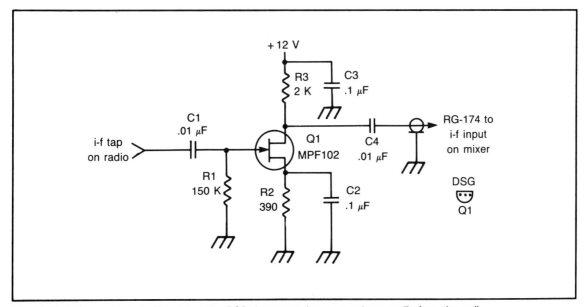

Fig. 11-10. Using a simple buffer allows the DSB detector to be mounted externally from the radio.

Fig. 11-11. The buffer board can be made quite small and squeezed into the radio.

Assuming the local oscillator in your receiver is stable, all changes in pitch of tones at the detector output are due to changes in frequency of the incoming signal. There will be slight differences in pitch of the tones from each signal that comes on. Here are some characteristics to look for when trying to identify a signal:

□ Absolute frequency. Many signals can be identified because they are several hundred hertz away from others.

□ Frequency change at key-up. Many rigs have distinctive frequency changes within the first couple of seconds after key-up as the local oscillator settles down.

□ Instability or modulation. A base station may have a distinctive 60 Hz hum that warbles the DSB detector output. Such warble could also be a subaudible tone encoder output. A mobile station may vary frequency slightly as the engine speed changes, often accompanied by alternator whine in the FM detector output. Some rigs have noise or random frequency variation. Battery operated rigs may have slow frequency changes over long transmissions as the batteries drop in voltage.

While you may not be able to compete with the FCC at identifying signals, you'll find SSB/DSB detection a big help when you want to know if the carrier you're hearing is the one you should be hunting.

Chapter 12

Sniffing Out the Bunny

You may have the finest mobile T-hunt setup in town, but it won't get you to the bunny if it's 500 feet off the road, and although your attenuator may have enough resistors for 120 dB, it will do no good if the rabbit's signal gets so strong it goes right through the receiver's case. In this chapter, we'll deal with the problem of closing in on a well-concealed transmitter and antenna, mostly on vhf.

Since the earliest days of amateur T-hunting, special devices for close-in hunting have been called sniffers. You may also hear them called fox-boxes or other creative names, depending on where you live. After we discuss the various methods of DFing close in, we'll show you how to build one.

HOMING DF UNITS

Switched antenna RDF units can be made very lightweight and portable for sniffing. Pattern switching units such as the L-Per and Happy Flyers work well at close range if the antenna cables do not pick up large amounts of rf directly. The BMG Engineering unit is even better in this respect because the stronger signals from the antenna effectively suppress any direct rf input on the cables.

The biggest drawback with the BMG unit is that it gives only directional information. Unless the receiver has an S-meter, which is not the case on many synthesized vhf handhelds, there's no way to tell whether you're two feet or two miles from the hare. The S-meter mode in the L-Per gives an indication of closeness, if you don't mind a lot of switching back and forth. Even better is a meter on the automatic attenuator of the Happy Flyers deluxe unit.

Other problems with switched antenna units are discussed in detail in Chapter 8. The problem of reflections is even more severe on foot because bearings taken while stopped may be in error due to reflections, and taking bearings while walking or running may be impractical or even dangerous.

If the signal is not vertically polarized, reflections are emphasized, and it may be impossible to get a correct bearing by holding the unit in the normal manner. Despite these drawbacks, many hams use homing RDF units for sniffing and swear by them (others swear at them). But now let's look at some alternatives.

THE BODY FADE

Your beam and attenuator have been working fine

up until now, but that blankety-blank jammer must be running a kilowatt! You know you're close, but the S-meter is pinned in every direction, and all the attenuation is in. He's getting into the receiver directly. That rig's no good any more. Let's get out of the car and look around. No antenna in sight. Now what?

If you have a handheld rig, there's a way to continue. It's not elegant, but it can work. Take the antenna off the handheld, and hold the unit down close to your chest. Now pirouette around and find the direction where the signal sounds the weakest. Although this is easiest with an S-meter, you can also do it by finding the maximum noise direction on an FM receiver. If the signal is full quieting, detune the radio 5 or 10 kHz until some noise is heard with the signal.

When you have found the direction of weakest-sounding signal, the signal is coming from behind you. This is the direction from which your body provides the most attenuation. If you don't mind the chuckles and belly laughs of observers, you can sometimes walk backwards all the way to the transmitter in this fashion. After the first time, however, you'll want something better.

SEALING UP THE RECEIVER

Few receivers are built with totally rf-tight cases. Control shaft openings, meter holes, and painted seams all let rf in. Plastic-cased handhelds have no shielding at all. If your battery-operated receiver has an S-meter and you'd like to try on-foot hunting with it, you can build an rf-tight enclosure for it that allows you to get much closer to the hidden T without overload.

Figure 12-1 shows an enclosure made from copper clad PC board material. The antenna, power, and audio are brought into the box through fittings in the rear. The power and audio leads are filtered with bypass feed-through capacitors, both mounted in the little box soldered into the corner. Metal extension shafts and bushings are used for the controls.

The enclosure and cover plate should have no holes larger than 1/4 inch. A small (one inch square) area could be covered with copper screen to allow viewing of the S-meter, provided that the screen is soldered in place all around. Do not use a two-piece minibox for this application, as it does not have good overlapping seams. The cover should be secured with screws no further than 2 inches apart, and the mating surfaces should not be painted.

The shielded case shown had the cover sealed with finger stock soldered to the lid (see Fig. 12-2). Even with this seal, the box needed to have all of the external con-

Fig. 12-1. The receiver can be put in a shielded enclosure to increase its resistance to rf overload.

nections removed and be wrapped in aluminum foil to be really rf-tight. With the foil wrapping, the rubber duckie antenna of a keyed handheld transceiver could be waved all around the box without opening up the squelch on the Drake TR-22C FM receiver inside.

Another effective way to seal the cover is to solder a half-inch strip of copperclad board all the way around the outside lip. Take care when soldering the strips so

Fig. 12-2. Finger stock can be used to seal the lid on the box.

that they are flush with the top of the lip. Buff the copper surface with steel wool until it's shiny at the mating surfaces. The lid can be attached with screws and nuts. If the flange look isn't pleasing to you, the strips can be put on the inside; however, it's harder to solder inside strips flush to the top. This method requires tapping screws or nuts soldered against the copperclad board on the underside.

Anything connected to the antenna input picks up signal, so the shielding must be carefully continued up to the attenuator. If the attenuator is external, double shielded (RG-223/U) or semi-rigid (RG-402/U) coax should be used for the attenuator-to-enclosure run, and it should be very short, perhaps only two or three inches, with the attenuator box mounted to the outside of the enclosure (Fig. 12-3).

Building the attenuator into the enclosure (Fig. 12-4) works even better. The cable from the outside connector to the attenuator must be double-shielded or semi-rigid to prevent radiation of strong signals from the coax into the inside of the box. The attenuator cells must be sealed, just as in the attenuator of Chapter 6, to prevent any unwanted signal radiation.

PRIMITIVE SNIFFERS

Modern handheld sets have a very sensitive receiver, capable of hearing signals that produce a fraction of a microvolt of signal at the antenna jack. This is exactly what we do *not* want at the end of the hunt. We want a low sensitivity receiver. So how about a crystal set? It has just an antenna, a germanium diode, and a meter. Headphones can substitute for the meter if the signal is

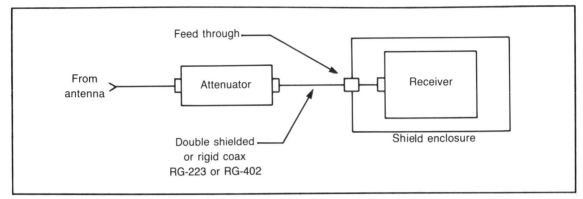

Fig. 12-3. The shield box with an external attenuator.

AM. Figure 12-5 shows such a unit, called a field strength meter (FSM).

Small FSMs can be bought commercially. Since there are no tuned circuits, no tuning is required, and they may work up to 400 MHz. Some SWR meters also have a whip antenna input for this purpose.

Without amplification, these units are very insensitive and quite ineffective for sniffing. Two we tested, the Monarch model FSI-1 field strength meter and the Henry Radio model HRB-3 SWR meter, required over 250,000 microvolts for full-scale indication. On 2 meters the whip antenna on the FSI-1 had to be 3 feet away from a 1 watt handheld to get a quarter scale reading, the minimum necessary for effective DFing.

Unamplified FSMs such as these are of little use except with very high power transmitters. In one local 2 meter hunt, a 50 milliwatt transmitter was concealed inside a portable outhouse at a construction site, with a quarter-wave antenna on the roof. One hunter attempted to sniff it out with his SWR meter and finally gave up after many minutes of frustration for himself and uncontrolled laughter for the onlookers. Even with a four-element quad, the meter didn't indicate anything until it was one foot away from the transmit antenna! In this particular case, his nose would have been a better sniffer.

A numb FSM can be souped up by adding a transistor dc amplifier. Figure 12-6 shows a typical circuit. Experiment with the transistors in your junk box to find one with best gain in this application. PNP types can be used if the 1N34A diode and the battery supply are reversed in polarity. This is a step up from the unamplified FSM, but the transistor base-emitter threshold voltage limits the sensitivity of this circuit.

THE SNIFF-AMP

An operational amplifier integrated circuit provides far more gain and can be adjustable over wide gain range. The unit described next, called the Sniff-Amp, has as much dc gain as you'll ever need for this purpose. With

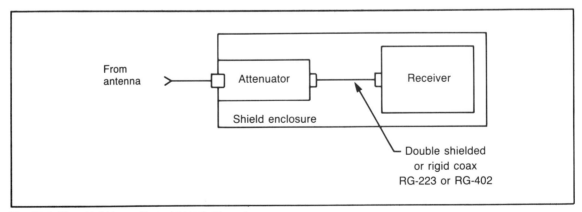

Fig. 12-4. The shield box with an internal attenuator.

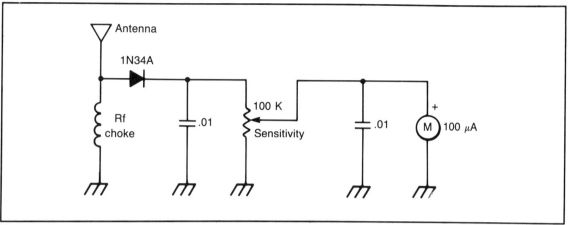

Fig. 12-5. A simple untuned unamplified field strength meter.

its three-position gain switch and gain control pot you can lower the gain smoothly to prevent meter pinning as you close in on the target. After a bit of practice it will be easy to guess the approximate distance to the hidden T by the gain range and control setting for full-scale indication.

The circuit is designed for maximum sensitivity by using a high gain amplifier with a high input impedance to load the detector diode very lightly. An inexpensive high sensitivity germanium diode is used for the detector (D1). Others, including microwave detectors (1N21) and Shottky types, were tried, but the lowly 1N63 outperformed them all in this application. A 1N60 was almost as good. See what the diodes in your junk box will do. The half-wave circuit gives just as much sensitivity as a two diode full-wave detector and loads the tuned circuit less.

With the circuit shown in Fig. 12-7, a 350 microvolt signal is huntable at maximum gain. With a full sized four-element quad and this sensitivity, we could get bearings on a 2 meter repeater 1.5 miles away. Full scale sensitivity on the three ranges measured 650, 1900, and 6000 microvolts with the gain control set at maximum.

With the gain control at minimum on the least sensitive range, full scale indication is about 500,000 microvolts. If the hidden transmitter still pins the needle at that setting—and you haven't found it—you can still hunt, as we'll show later. Although tuned circuit infor-

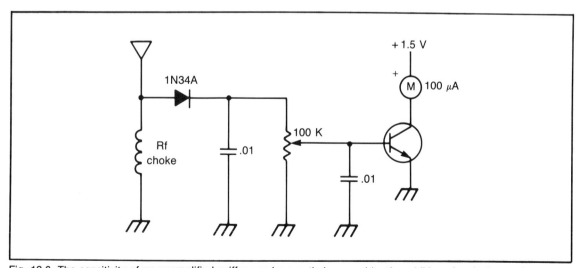

Fig. 12-6. The sensitivity of an unamplified sniffer can be greatly improved by the addition of a single transistor.

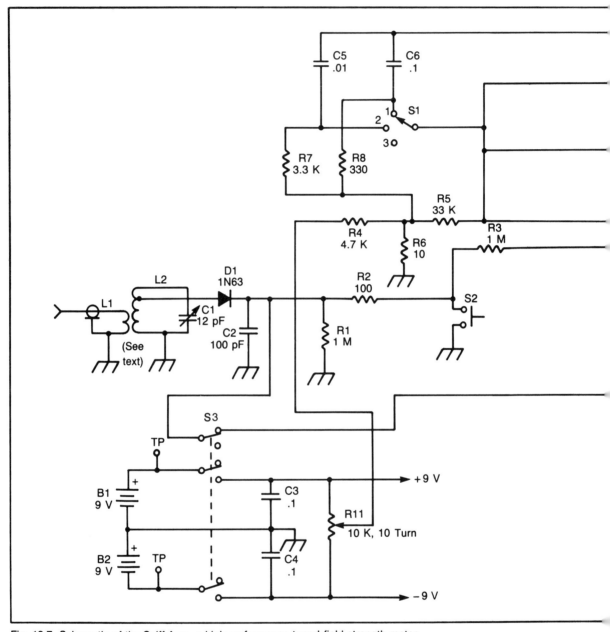

Fig. 12-7. Schematic of the Sniff-Amp, a high performance tuned field strength meter.

mation is for the amateur 2 meter band, these components can be changed to put the Sniff-Amp on any band up to 500 MHz. It should work fine hooked to the output of a loop antenna for hf with an appropriate tuned circuit.

At maximum sensitivity, the operational amplifier (AR1) provides 90 dB of dc gain, about the most that can

be had without severe drifting of the zero setting. For ease of adjustment, a ten-turn pot is recommended for the zeroing control. The circuit as shown zeroes almost any operational amplifier IC available.

You will find that zeroing is critical only on the highest sensitivity range, and is no problem at all on the

170

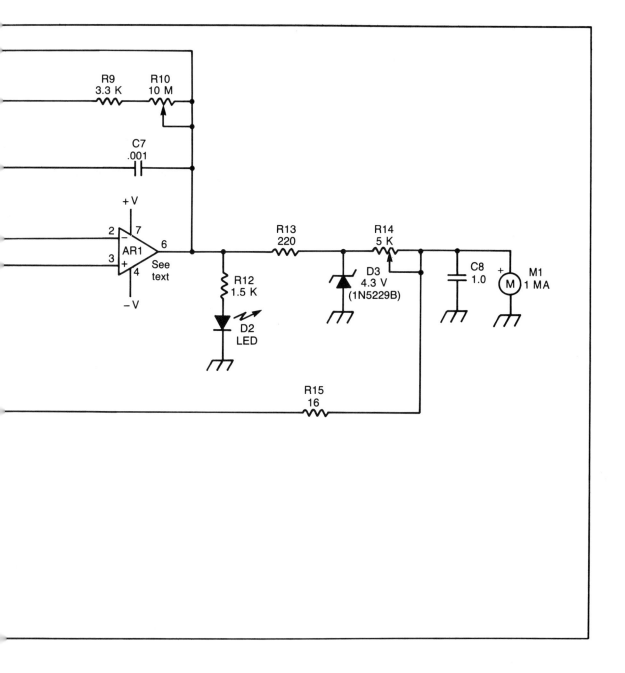

other ranges. Fortunately, you very rarely need high sensitivity. In most hunt situations, the other ranges provide plenty of sensitivity. Still, you may wish to experiment with zeroing circuit modifications to get less touchy zeroing with the particular type you use. Try connecting the zero control (R11) between +9 V and ground, or –9 V

and ground, instead of +9 V to –9 V. You can also add fixed resistance in series with R11, or raise the value of R4, to spread the range. To prevent a battery voltage slump from causing drift through the zeroing circuit, regulate the voltages applied to R11.

Capacitors C5, C6, and C7 roll off the amplifier response to prevent oscillation. With them, it doesn't matter whether or not the op amp you use has internal compensation. A number of different MOSFET and J-FET input op amps were tried in this circuit and all worked, but some were better than others. The LF355N is a JFET type available from Digi-Key at low cost. The RCA CA3140, with a MOSFET input, worked better but is harder to find. Use a socket and don't be afraid to try several different types, selecting the one with least drift and highest sensitivity.

Remaining components are non-critical. The 10 megohm (R10) pot may be hard to find. It may have to be pulled from a piece of surplus equipment. If you can't find one at 10 megs, use the highest value available. R13 and D3 limit the meter deflection, both positive and negative, to give maximum sensitivity without hard pinning. Values are changed appropriately if you use a meter more sensitive than the inexpensive 1 milliampere one called out. R14 is adjusted for full-scale deflection with enough input signal applied to provide clamping at D3 cathode. C8 is a meter damping capacitor and may be changed as required to suit your needs and the meter you obtain.

Other circuit features include S2, a momentary switch to short the input signal while zeroing the amplifier. You'll need it to zero the amplifier in the presence of a strong continuous signal. R15 and third pole of S3 provide a very low gain sniffing option, connecting the diode directly to the meter. When the signal is too strong even with the amplifier at lowest gain, switch off the power and hunt with the diode/meter combination. You'll be able to sniff right up to the transmit antenna. Full-scale sensitivity in the power off mode is 1,550,000 microvolts. (That's right, over a volt and a half!)

Figure 12-8 shows the Sniff-Amp as built into a 5 × 4 × 3 inch (HWD) aluminum box. A two-pin jack at the top of the box provides a means of quickly checking the battery voltages with a VOM before the hunt to prevent a possible unpleasant surprise during the hunt. Although a rotary switch can be used for S1, a center-off toggle switch was used to conserve space. The center position becomes the least sensitive range (position 3).

Parts placement is generally not critical and perfboard construction works fine. All components except the controls, meter, and L1/L2/C1 are mounted on the perfboard. As with any high gain circuit, keep input leads separated from output leads. For two meters, L2 is five turns of AWG 16 tinned bare wire, 1.25 inches long, 1/2 inch diameter, connected directly across the ceramic trimmer, C1. The high impedance amplifier allows a high tap (one turn from the ungrounded end in the prototype) on the coil for D1. Experiment to find the most sensitive tap

Fig. 12-8. The Sniff-Amp as built.

point. L1 is two turns of insulated wire wound between the first two turns at the grounded end of L2.

It is very important to use a metal box, and to have its assembly screws well tightened. If any part of the case is allowed to float, erratic field strength readings result. The circuit board ground should be connected to the chassis at only one point, preferably the ground end of L1/C1, with a short lead. R14 is mounted on the perf-board.

To check out the completed Sniff-Amp, set R10 and R14 to maximum, and S1 to the high sensitivity range. Turn on the power, and set the zero control for maximum upscale indication. Adjust R14 such that the meter barely pins. Now verify that R11 zeroes the amplifier. Response is slowest on the high sensitivity range. Set S1 to the lowest sensitivity range and tune C1 for maximum deflection at the frequency of interest, using a signal generator fed into the input jack. If you don't have a signal generator, put a whip antenna on the sniffer and use an on-the-air signal from a nearby transmitter.

ANTENNAS FOR SNIFFING

Some hams sniff on vhf using only a whip antenna. It's easy to wonder how they ever find the transmitter. You can wear out a lot of shoe leather guessing which way to walk when your antenna has no directivity. Either a null or peak indicating antenna is suitable, but gain is desirable to increase sniffing range.

At 2 meters and above it's hard to beat a yagi or quad for gain and directivity. You merely disconnect your beam from your mobile receiver and attenuator and hook it to the sniff amp and start walking. But a four element 2 meter quad will seem quite heavy after a few hundred feet. Trees love to snag wire quads. And then there's the problem of how to hold the heavy quad mast with both hands and still have a hand free to hold the sniffer box. A smaller, lighter antenna is needed for closing in.

It's possible to build a quad with elements about one half their normal circumference and still get excellent directional performance. Only two elements, a driven element and a reflector, are needed for the Incredible Shrunken Quad. While it doesn't have the gain of a four element full-size quad, and we wouldn't recommend transmitting through it, it's just right for sniffing in a strong signal area with the high gain Sniff-Amp.

The volume (airspace) of this antenna at 2 meters is only one eighth that of a full sized four element quad, which makes it also a fraction of the weight and less likely to be eaten by tree branches. It can be kept in even a small car trunk without disassembly. Figure 12-9 shows the combination of the Shrunken Quad and Sniff-Amp ready to hunt, capable of being held and maneuvered with one hand.

Figure 12-10 shows element dimensions and construction information. Capacitors are 1-10 picofarad trimmers. Use sealed piston trimmers if you can find them, as they will be less affected by rain and moisture. The balun is needed to keep the feed line from destroying the directivity.

The theory we used to design the 2 meter Shrunken Quad has been detailed by Roger Sparks, W7WKB, in QST for April 1977. (No, it was not the annual April Fool article!) He did his work at 40 meters, but the principle applies to either hf or vhf. This antenna can be scaled as required for other frequency ranges, such as the aircraft band or even the Citizens Band.

The small size of this antenna is made possible by the capacitors and wires connecting the maximum volt-age nodes of each element. This is a loaded antenna, and just like any other loaded antenna, such as a 75 meter mobile whip, the Q is high and bandwidth is narrowed. This is at the same time both advantageous and disadvantageous. The disadvantage of this high Q system is apparent only if you like to hunt over a wide range of frequencies. While a full-sized quad on 2 meters works well over three megahertz or so, the gain and directivity of the small antenna is much more sensitive to frequency.

Figure 12-11 illustrates on- and off-frequency performance. At resonance (a), a full scale signal shows near zero meter indication off the sides and back, easily showing the exact direction to the transmitter. A half megahertz away from resonance (b), the forward response is down 3 dB and there is no complete null in back—still huntable. One megahertz away (c) there is a definite lobe in the back that is only about 3 dB down from the front lobe, and things can get confusing.

Fortunately, most southern California competitive hunts are on the coordinated frequency of 146.565 MHz. But there's no telling where a jammer or stuck radio may show up. It's best to retune the antenna when changing the hunt frequency by more than 200 kHz. With a little

Fig. 12-9. The Sniff-Amp and Shrunken Quad can easily be held and operated with one hand.

173

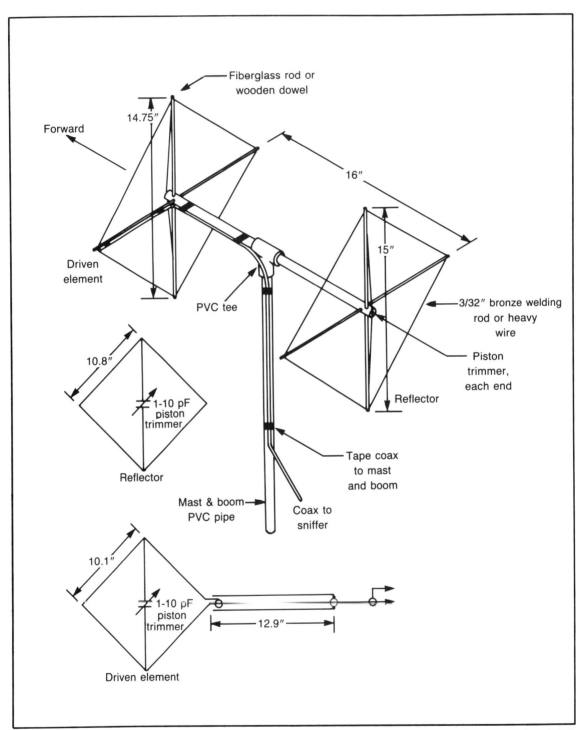

Fiberglass rod or
wooden dowel

14.75″

Forward

16″

Driven
element

15″

PVC tee

3/32″ bronze welding
rod or heavy
wire

Piston
trimmer,
each end

Reflector

10.8″

1-10 pF
piston
trimmer

Reflector

Tape coax
to mast
and boom

Mast & boom
PVC pipe

Coax to
sniffer

10.1″

1-10 pF
piston
trimmer

12.9″

Driven element

Fig. 12-10. Construction details for the Shrunken Quad. It is shown vertically polarized, but both elements can be rotated 90 degrees for horizontal polarization.

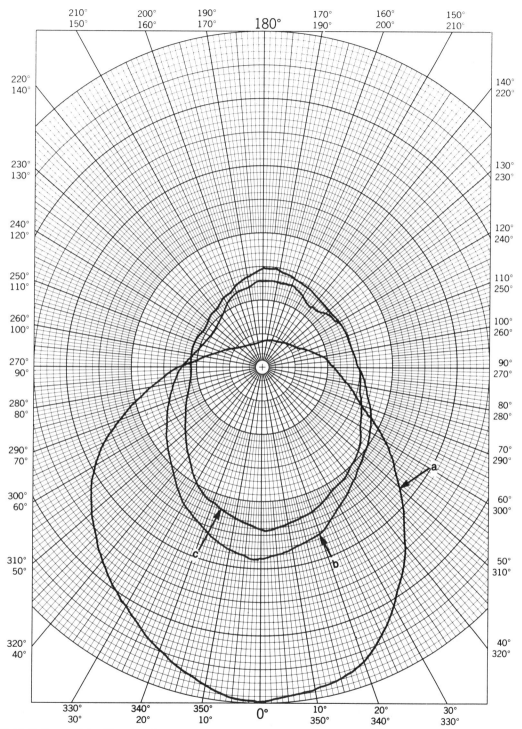

Fig. 12-11. Patterns of the Shrunken Quad with a Sniff-Amp. At (a) a good front to back ratio is achieved at resonance. It degrades at 0.5 MHz off resonance (b) and is somewhat bidirectional at one MHz off resonance (c).

practice, this can be done very quickly, following the procedure described later.

The narrow antenna bandwidth is actually an advantage in most situations, since it adds selectivity to the Sniff-Amp, which has only one tuned circuit. On one Sunday afternoon fun hunt in a regional park, ordinary sniffers with standard four element quads were confused by a nearby business band transmitter while hunters were trying to get bearings from the parking lot on the Bunny Box's 35 milliwatt signal (see Chapter 15) five hundred feet away on a hillside. The Shrunken Quad and Sniff-Amp, however, pointed right to it, thanks to the narrow antenna bandwidth.

Before tuning the antenna, make sure the tuned circuit (L2/C1) in your Sniff-Amp is set for the middle of the range of interest. One setting of this circuit suffices for the top 2 MHz of 2 meters. It doesn't need to be reset every time you tune the antenna. You need a plastic tuning tool with a small metal bit to tune the antenna trimmers, and also a source of signal on the frequency of interest. The Bunny Box hidden T is ideal, but a handie-talkie in the hand of a willing helper also works. The Shrunken Quad should be in a relatively open area away from cars and other pattern-distorting objects, and at least 50 feet away from the transmitter.

Tuning can be done in just a couple of minutes. Remove the reflector and hold the antenna so that the boom points toward the source. Carefully tune the driven element capacitor for a peak meter reading. You'll find hand capacitance will detune the loop, so move your hand away after each adjustment to check the reading. (You may even have to over-tune past resonance in one direction or another so that the signal is peaked when you remove your hand.) Now put the reflector on, turn the antenna to point in the opposite direction of the source, and adjust the reflector capacitor for the deepest null possible, again being careful to avoid hand capacitance effects.

Check the tuning by rotating the antenna and observing the front-to-back ratio. With the gain control set for about 3/4 scale meter reading pointing at the antenna, you should get a deep null (near zero) off the back. If not, repeak the driven element pointing toward the source, with the reflector attached this time. Then re-null the reflector with the antenna pointing 180 degrees away as before. Continue this procedure until you are satisfied with the pattern. Large objects near you could cause reflections, preventing you from seeing the null during tuning. Check with your full sized quad first, and move your location if necessary.

SYSTEMATIC SNIFFING

Good sniffing is a practiced art, no matter what equipment you use. Get someone to hide a small transmitter (your handie-talkie?) in various places around your house and yard and practice finding it, learning to judge distance by the strength of the reading.

Don't get so wrapped up in watching the meter that you ignore some obvious clues. For example, it was Clarke Harris's (WB6ADC) observation skills, not his sniffing technique, that helped him win the Fake Pole Pig Hunt. Crafty hider Ken Diekman (WA6JQN) built a model power distribution transformer out of a big metal can, along with a bogus crossarm and insulators to hold the horizontal two meter antenna. He climbed up a telephone pole (without any power lines on it of course) and mounted the whole assembly to the top. Six hundred feet of small coax led from the antenna down the pole and underground to a borrowed nondescript van.

The low level horizontally polarized signal, coming from a pole that happened to be next to a very large metal water tank, gave the hunters plenty of trouble getting into the general area. But then they had to try to sniff out a weak signal that they didn't know was 30 feet up in the air. At night there was absolutely nothing unusual-looking about it. We don't know how well the Sniff-Amp and quad would have done—the hunt was before this combination was built—but WB6ADC says it was his eyes, not his ears, that did the trick. While everyone was stumbling around the water tank, Clarke checked out the van with his flashlight and noticed a car battery inside through the rear window. That prompted him to confront the amorous occupants of the van, who turned out to be accomplices.

Most of the time the antenna is more accessible than that, and this sniffing system can lead you right to it. Don't just try to guess where the signal's coming from. Follow the meter to it and be systematic. Perform mental triangulations as the bearings swing. Remember that sometimes there may be more than one source of signal. For example, if the transmitter and antenna are widely separated, you may get radiation directly from the transmitter as well as from the antenna.

Hunt the peak, not the null. This way you will be less likely to be fooled by reflections, which can eliminate the null or cause multiple nulls. As you approach the hidden T, the signal increases dramatically (as does your pulse rate, no doubt). Keep reducing the gain, as you will lose directivity if you allow the meter to reach full scale, causing protection zener D3 to conduct. When you know you're close, sweep the unit up and down like a wand, as well as around. The fox may be up in a tree (or pole) or down on the ground.

Before you begin walking, verify that the sniffer bearing agrees with the one obtained with the vehicle antenna. If you suspect that the hidden T may have a horizontally polarized antenna, check by rotating the antennas for horizontal polarization to see if the meter reading increases. While it's possible to sniff a cross-polarized signal with this unit, you'll find it harder to separate the reflections and you may get less apparent null on the back side.

Take any opportunity to triangulate to help determine the distance to the transmitter. Figure 12-12 shows how multiple bearings along the road pinpoint the distance to the transmitter. This triangulation was necessary on one actual hunt because the transmitter was using only two milliwatts. This was too low to see from a distance with amplitude sniffers, and there was a lot of multipath from the fence to confuse homing DFs.

After a few hunts using this sniffing system you will agree that having a physically small directional antenna and strength indicator has important advantages over sniffing with a switched antenna system in some circumstances. On one Rio Hondo Amateur Radio Club (CA) outing, a half dozen hunters converged on an amusement park in Buena Park where a strong signal appeared to be coming from the middle of a duck pond. Circling the pond, the hunters found that the signal appeared to move, but always stayed in the same general area. The rules of the hunt called for a 15-second transmission every two minutes, and it seemed like every transmission resulted in a new bearing.

While some of the hunters were eyeing the paddling ducks very carefully, K0OV noticed a woman pushing a baby carriage back and forth along the walk next to the pond. The carriage seemed heavily loaded with blankets and junk, but only a doll was sitting in the seat. He nonchalantly walked up beside her and let the driven element of the Shrunken Quad dangle down next to the carriage, with the amplifier set for minimum sensitivity.

Sure enough, as soon as the time for a transmission came, the needle went to full scale. The "first in" slip was quietly obtained from the carriage pusher and a retreat was made before any of the other hunters even noticed. Some of the other participants never did successfully sniff out WA6JQN's ten watt mobile rig and car battery under the blankets before time was up for the hunt.

DELUXING YOUR SNIFFER

Successful transmitter hunters are always looking for ways to improve the performance and convenience of their systems. Here are some easy ways to improve your sniffer.

Preparing for Night Hunts

If you can find a lighted meter, you'll find it a big help on night hunts. If you can't, there are other ways to find your way to the hidden T in the dark. That's the

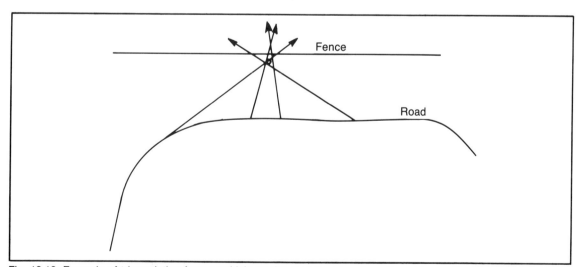

Fig. 12-12. Example of triangulation from a vehicle on the road to determine off-road transmitter location.

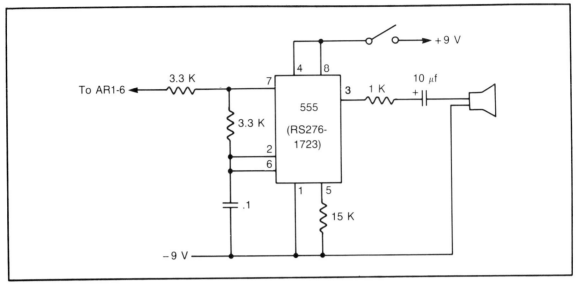

Fig. 12-13. Audio S-meter for the Sniff-Amp.

purpose of the light-emitting diode (LED) across the amplifier output (D2 in the schematic). It begins to glow at about one quarter scale and gets brighter as the meter reading increases.

You can sniff without looking at the meter at all if you add a tone output as shown in Fig. 12-13. It is similar in operation to the audio S-meter described in Chapter 5. Note that there is no connection to circuit ground in this version, as the return is connected to −9 V. The resistor from pin 5 of the 555 IC to −9 V modifies the internal threshold of the IC so that the tone begins at a very low frequency near zero meter reading and goes to several kilohertz at full scale. The supply bypass capacitors (C3 and C4) in Fig. 12-7 prevent interaction between the oscillator and the sniffer dc amplifier.

Battery Minder

The flasher circuit shown in Fig. 12-14 provides a reminder to the operator that the unit is on. The Zener diode also makes it a battery monitor. The LED stops flashing when the positive battery gets down to about 5.5 volts. The batteries should be replaced at that point to prevent inability to achieve full scale readings. The LM3909 (RS 276-1705) is readily available at local parts stores.

Better Selectivity

Although the high-Q Shrunken Quad and tuned cir-

cuit in the sniffer provide good rejection of out-of-band signals, it may not be enough in some cases. Hiders have been known to put low power transmitters near 100 kilowatt FM stations, or airports, or hilltop vhf two-way radio facilities.

If you want to be prepared for these possibilities, you'll need a small, very high Q filter for the input of your Sniff-Amp. Companies such as Piezo Technology will

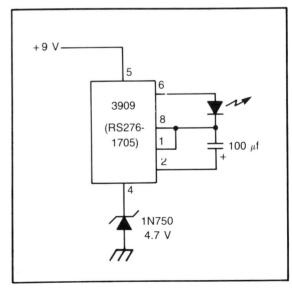

Fig. 12-14. Battery monitor and power-on indicator for the Sniff-Amp.

178

gladly sell you a 13.5 kHz wide crystal filter, with skirts going down to better than – 20 dB at 60 kHz either side of center, in the 138 to 170 MHz range (2133 VBP series). Unfortunately each filter will cost you about a hundred dollars and be good for only one hunting frequency.

Cavity resonators also have high Q, but it's so high they must be retuned for each frequency. They are also too cumbersome for our purposes.

A helical resonator, on the other hand, is perfect for this requirement. It is quite small, yet can achieve a Q

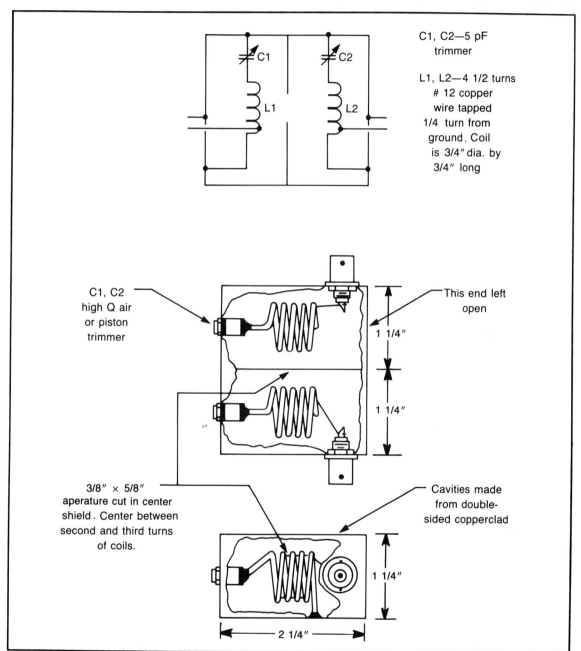

Fig. 12-15. Construction details for a simple helical resonator for the two meter band.

of 1000 if carefully built. Two resonators, coupled and stagger tuned, provide an effective filter that does not require tuning for any reasonable segment of the 2 meter band.

Figure 12-15 shows construction details for the simple sniffer filter. The walls are built from double-sided copper clad circuit board. The non-perforated boards are soldered along the seams on the inside. This may require an extension on your soldering iron tip. Use AWG 10 wire for the coil, soldered on both inside and outside of the box at one end. Capacitors are 1-10 pF piston trimmers, as used in the miniature quad, soldered through the top of the filter enclosure. High quality air variable capacitors may be substituted.

One end of the filter box can be left open. BNC connectors allow easy installation and removal of the filter between the antenna and the Sniff-Amp. The almost-complete filter is shown in Fig. 12-16. The front wall has yet to be soldered in place. The board should be well cleaned before assembly for best filtering action.

Fig. 12-16. The helical resonator before installation of the top wall. The end can be left open without affecting performance.

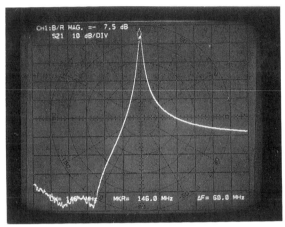

Fig. 12-17. The response of the helical resonator over an 86 to 206 MHz range. Rejection at the FM broadcast band is around − 90 dB and in the business band − 40 to − 50 dB.

This method of construction is quite a compromise. For high performance in a sensitive receiver, you wouldn't want to use these non-elegant construction techniques. For best Q the enclosure should have been seamless and silver plated. Silver-plated wire would improve it even more. In this case, with the sniffer hunting relatively strong signals, the method described works just fine.

Tuning can be done using the Sniff-Amp as an indicator, and a sweeper or rf generator as a source. Both tuned circuits can be peaked at the same frequency for maximum sharpness, or stagger tuning can be done to allow a wider bandpass for hunting several frequencies. Figure 12-17 and Fig. 12-18 show a typical filter's response when tuned for hunting on the 2 meter band.

It does no good to have the helical filter get rid of that FM station's signal on the coax, only to have it get into the sniffer or receiver through the case. Be sure the sniffer amplifier or other detecting device is well shielded and sealed to prevent direct pickup of out of band signals.

If you want to build a filter for another frequency range, or design a high performance version for some other use, there are many good articles on helical resonator design. Recent editions of *The Radio Amateur's Handbook* have detailed design information for these filters. Helical resonator filter kits are available for the 144, 220, and 432 MHz amateur bands and nearby frequencies from Hamtronics in Hilton, New York.

LISTENING TO THE SIGNAL

Although T-hunting on 2 meters began in the AM days, most vhf/uhf amateur hunting is now done with

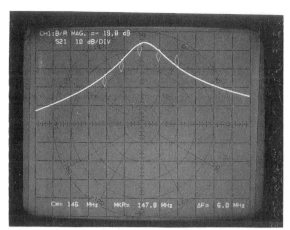

Fig. 12-18. When peaked at 146 MHz, the resonator has −15 to −20 dB response at the band edges.

NBFM signals. (An interesting exception is the Southern California Amateur Television Club, which has conducted hunts using a TV signal on 434 MHz.) The simple sniffer circuits described here do not demodulate FM signals, so most hunters carry a handheld rig to verify that the fox is on the air, and to hear any instructions from him while sniffing.

Aircraft band stations, ELT's, and CB operators use AM, as do some hf amateur radio T-hunting groups. The simple detector of Fig. 12-5 provides a demodulated AM (not SSB) output, but the Sniff-Amp has too much rolloff to allow listening to an AM signal. If you'd like to try AM demodulation on the Sniff-Amp, change the rolloff capacitors (C5, C6, and C7) to much lower values and con-

nect headphones through a blocking capacitor to pin 6 of AR1. It is quite possible that AR1 will oscillate without the rolloff, and additional compensation measures will then be needed, depending on the IC you use.

OTHER SNIFFER USES

You may find some other interesting uses for your sniffer around the shack. Couple it to a shielded ferrite rod antenna and use it to find sources of pickup and re-radiation of your 75-meter base station. Pipes or ac power wiring that re-radiates rf can be located even if they're inside a wall.

Connected to a suitable sense antenna, it makes an excellent field strength meter for antenna adjustments and pattern tests. It even sniffs noise at vhf. At a recent Amateur Radio demonstration at a southern California amusement park (the same park where the duck pond was located), severe intermittent electrical interference was noted at a display station on 10 and 15 meters, preventing any operating on those bands. While K0OV was walking through the shopping area just outside the park, he suddenly began to hear the same interference on signals on his 2 meter handheld rig.

Out came the Sniff-Amp and Shrunken Quad. In a few minutes, the source had been pinpointed to a single fluorescent lamp starter in a bank of three dozen lamps inside a gift shop. Turning off the lamp eliminated the QRN. This interference hunt drew quite a crowd of park-goers who stood in a circle watching the sniffing effort intently. Apparently they thought it was part of the park entertainment!

Chapter 13

Planning for Hunts
in Your Community

A regular fox hunt can become an important technical and social event for your local radio club. It can help keep the club going in the summer months after Field Day when interest and attendance normally fall off. This chapter offers some hints to get your group going, increase participation, and keep things interesting.

WHEN AND WHERE

Saturday evening is the most popular time for serious hunters, with some hunts lasting into the wee hours. Sunday afternoon hunts, planned to end before sundown, add a measure of safety and encourage family participation. They're even more appealing if there is a picnic or refreshments at the end. Saturday morning after the club breakfast at a local restaurant may also prove popular.

The starting point for vhf hunts should be higher than the surrounding terrain in all directions, to allow for best accuracy in bearings. It should be in an electrically quiet rural or residential area, away from noisy power lines. The local radio or TV tower hill is not a good place if strong signals cause desensing or intermodulation in the hunters' receivers. For hf hunts, height is not as important a factor, but electrical noise, if present, will be more of a problem. A well-lit starting area provides added safety

for night hunt (Fig. 13-1).

Some groups do not have a formal starting point, but allow the hunters to start anywhere within the hunt boundaries. This precludes exact mileage taking, so time is usually the only factor in determining the winner. Approximate mileage can be computed and used as a factor by having the hunters declare their starting point on the air at the beginning, and then computing air mileage to the finish with a good map. To be sure, luck plays a large part in this hunt, as the winner is often the team that happens to start from the closest point.

WRITING THE RULES

In our area there are about ten regularly scheduled hunts. No two hunts have the same set of rules. Each group has chosen rules that suit the locale, the abilities of the hunters, and their preferences. For successful hunts, a good set of written rules is a necessity. If the rules are ambiguous, or unfairly applied, conflicts occur which can result in hurt feelings and reduced participation.

When the rules are decided on, they should be written up and a copy given to each potential hunter. Make sure some provision is included to facilitate future

Fig. 13-1. Saturday night, 8 PM. Well lit hilltop starting location for the Fullerton (CA) Radio Club monthly T-hunt.

changes. Spell out who shall have a voice in the change process—all club members or just active hunters. If the latter, define who qualifies as an "active hunter."

A sample set of T-hunt rules is given below. They are presented as a guide to issues that must be faced when rules are written, and should be modified as required to meet the needs of your group. Several considerations should be kept in mind when you write your rules.

FOXFINDERS AMATEUR RADIO CLUB TRANSMITTER HUNT RULES

Date:	Third Saturday of month
Time:	8 PM
Frequency:	146.565 MHz FM
Starting point:	Top of Crest Drive, at Lover's Lane parking lot
Boundary:	15 air mile radius from starting point
Winner:	Hunter or team with lowest driven mileage
Hider (Huntmaster):	Winner of previous hunt
Refreshments:	At nearby restaurant after the hunt

1. The signal must be copyable at the starting point.
2. The output of the transmitter must be continuous and must maintain constant power throughout the hunt. The type of audio modulation must remain the same except for identification. Peak deviation shall be not

less than 2.0 nor more than 4.5 kHz. Any antenna configuration or polarization may be used, but must not be changed during the hunt.

3. The antenna and transmitter shall be within 250 feet of access by standard passenger car. Reasonable care must be taken to ensure that hunters on foot safely reach the hidden T. It must be in a location which is publicly accessible and with no charge for admission.

4. If all cars have not arrived by 11 PM, the hunt will be ended, unless the hider and remaining hunters agree to continue.

5. The huntmaster shall determine the point(s) at which ending mileages are taken. It is recommended that hunters not be penalized for long distance "sniffing," and that any reasonable stopping point be permitted.

6. The huntmaster shall monitor another frequency, repeater or simplex, during the hunt. Any team dropping out of the hunt shall immediately notify the hider.

7. Each hunter shall provide evidence of odometer calibration. This may be from the Auto Club (AAA), a commercial company, or a witnessed rolloff on the measured 10-mile course on highway 43 west of Podunk. Calibration shall be valid for six months, provided that no changes, such as new tires, have been made. Hunters without calibration evidence may participate, but are ineligible to win unless a witnessed rolloff is done immediately after the hunt.

8. Starting mileages shall be radioed to the huntmaster prior to the start of the hunt.

9. In case of bad weather, the hunt shall be held, but the hider shall take extra care to ensure the safety of hunters, by arranging for shelter at the end point and remaining in a paved area.

10. In the event of a first place tie, the next hider shall be chosen by lot.

11. RULE MODIFICATIONS as follows are permitted provided that they are previously announced by the huntmaster in the club newsletter and on the ARES net the week prior to the hunt. No more than one modification per hunt is allowed, except for the bearing contest.

- *Boundary extension*, up to 10 miles
- *Intermittent signal*, at least 10 seconds per minute
- *Time penalty*, up to 0.2 miles per elapsed minute
- *Off-road distance extension*, up to 500 feet
- *Fixed station bearing contest*, concurrent with mobile hunt

12. PROHIBITED PRACTICES as follows may cause disqualification at the discretion of the huntmaster:

- Backing up, except where impossible to turn around
- Any traffic citation
- Use of outside assistance, such as bearings from fixed stations, except when a bearing contest is part of the hunt

13. ADDITIONAL RESPONSIBILITIES of the huntmaster:

- Obtain permission from property owners as necessary
- Cover "No Trespassing" signs if posted area is used by permission
- Bring rotating trophy and award to winner
- Select restaurant

14. These rules may be amended by a majority vote of the active hunters at any regular club meeting. Active hunters are those participating at least once in the past three months.

Winning Criteria

Time versus mileage scoring probably generates the most controversy. Proponents of time-scored hunts like the fact that they most closely simulate a hunt for a jammer or ELT. Every second counts, and the ability to make split-second decisions is a valuable asset. But some experienced hunters have told us they will not participate in any hunt where time is a factor, saying that such hunts encourage reckless driving and hunch hunting. The techniques of accurate bearings and triangulation are difficult to learn in the frantic pace of time-only hunts.

Mileage-only hunts, on the other hand, can be much more leisurely. It's possible to stop for gas or even a snack. Occasionally the last team in is the winner, which keeps the suspense level high at the end. But mileage-only hunts have their problems, too. The hider may find himself having to stay on the air for what seems like an eternity waiting for the last hunter to arrive. Then there are the odometer calibration controversies. Inaccuracies of 20 percent are not unheard of when oversize tires are used. Variations of plus or minus 5 percent often occur in stock vehicles.

It is possible for two teams to take exactly the same route and end up with a tie. Also, the mileage-only hunt gives no incentive to good sniffers. Teams who find the bunny quickly after leaving their cars are not rewarded over those who don't sniff well or can't sniff at all.

The Fullerton (CA) Radio Club has found a successful compromise between the two types of hunts. The mileage of each team is increased by 0.1 for each elapsed minute between the start of the hunt and the time it locates the hare. This keeps the hunt moving and serves as a tie-breaker when routes are nearly identical. It also rewards teams that sniff efficiently. This combination rule has been very well received. Usually the first team in also has low mileage. A review of the FRC monthly hunt results over a one year period showed that the winner was first in 10 times and had low mileage 9 times.

Avoiding Arguments

In a time-only hunt, it's easy to define the winner as the first person to actually point out or touch the transmitter or antenna. In a mileage hunt, the rules should state that the transmitter must be located before mileage counts. But heated controversy can occur over where the end point is, for the purpose of mileage taking.

As an example, assume the hidden T is in the middle of a public park. The hider expected the vehicles to all enter the park on the main road and park in a lot 100 yards from the bunny. All except WA6JQN do just that. Ken, however, comes in on the back side of the park, leaves his car on a public street, and walks 250 yards into the park to find the bunny. His mileage is 0.3 lower than the next lowest, and his is the first vehicle to arrive.

By the rules of some groups' hunts, WA6JQN would be required to drive his car around the park to the parking lot and have his mileage recomputed. In this case, he would end up losing the hunt. Is this rule fair? Or should the rules allow a hunter to leave his car as soon as his sniffer or portable gear can detect the bunny?

In a time/mileage hunt where time is a penalty, hunters should certainly be allowed to park anywhere that's legal. If a team parks too far away, it is automatically penalized by the long walking time. Parking anywhere should also be allowed on mileage-only hunts, as it encourages hunters to develop sensitive sniffing equipment. Hunters can shave off miles by stopping short if they can sniff earlier. Your group may disagree, or may want to have the hiders decide and state the rule at the start of the hunt.

Odometers

The suggested sample rules show an effective way to avoid odometer calibration arguments. Still, compli-cations can arise when a car's cumulative odometer does not read tenths of a mile and the trip odometer must be used. Orange County (CA) T-hunters, one of the oldest hunting groups in the country, faced this situation and developed some useful guidelines.

Trip odometers with tenth mile indication can be used, but should not be reset to zero before the hunt. If the odometer has been partially or completely reset, the vehicle should be driven for at least 0.1 mile indication before starting mileage is recorded. This ensures that the gears have re-meshed. Both the cumulative and trip odometer readings should be recorded at the beginning and end of the hunt, and compared to prevent resetting during the hunt.

For cars without tenths indication on either the cumulative or trip odometers, the cumulative odometer is used. If the outcome is close, a special calibration is done after the hunt. The subject car is rolled off from the end point next to a car with a calibrated odometer reading tenth miles. The number of tenths of a mile to the next even mile on the subject car's odometer is determined. This is subtracted from 1.0 mile and added to the end point elapsed mileage of the subject car by the huntmaster.

When this last method is used, it is to the hunter's advantage, before starting mileage is recorded, to drive around the starting point until the odometer just advances to the next mile. If this is not done, the team can be unwittingly penalized up to 0.95 mile in scoring.

Miscellaneous Rules

Boundaries should encompass enough area to provide a challenging hunt. They should include varied terrain, if possible. Conversely, making the area too large scares off fledgling hunters and those with a limited gas budget. The starting point should be near the center of the area. If the boundaries are complicated, include a map with the rules.

Decide on whether a continuous carrier hunt or intermittent transmission hunt is desired, or whether to make it the hider's option. Shorter transmissions tend to favor Dopplers and other quick-to-read devices. They are easiest when held on a repeater input so hunters can use the output to tell when the T is on the air.

Some measure of signal quality at the starting point should be specified. For repeater hunts, it might be wise to require a minimum level of quieting (say 50%) at the repeater input. Of course, the huntmaster has to have well-calibrated ears!

Rules should define any areas that are off limits for

hiders and hunters. This usually includes places where admission is not free, posted "No Trespassing" areas, and military or other secure areas. To simply state that private property is off limits is not a good idea, as this precludes many publicly accessible areas such as amusement parks. While advance permission from the owner may not be necessary for hiding in a commercial parking lot, it is almost mandatory for rural private property.

Traditionally, the winner of the hunt gets to be the hider next time. New hunters usually like this rule, as they seek the honor of getting to hide, and it's an impetus for them to win. Conversely, many old timers are far more interested in hunting than hiding, and some will admit to planning a "controlled second place" to avoid having to hide. Other options are possible:

☐ The second-place finisher can be required to hide

next. This greatly increases the incentive to win for many veterans at hunting.

☐ The hider can decide which place hunter will hide, declaring this at the beginning of the hunt.

☐ The next hider can be selected by lot at the end of the hunt, regardless of who wins. The hunters draw straws, with the winner getting the first draw. The current hider can either prepare the straws or be required to draw with the others.

Other Hunt Ideas

There are several rule variations your club can try for a change of pace:

☐ Multiple bunnies transmitting sequentially every thirty seconds or so. The rules should specify whether or not hunters must find the bunnies in sequence.

☐ A moving bunny, in a vehicle driving a preset

Fig. 13-2. Gary Holoubek, WB6GCT, as a moving "bunny." Note the rubber duck transmitting antenna in the rear basket.

course. This could range from a car driving a five-mile street pattern to a cycle being ridden around a public park or parking lot (Fig. 13-2).

☐ A bearing competition for fixed stations, allowing even shut-in hams to get in on the fun of DFing.

Plenty of advance notice should be given to hunters for any of these variations. Any special rules should be carefully thought out and announced ahead of time.

NOVELTY HUNTS

Dedicated fox hunt fans are always looking for new challenges for their equipment and their skills. To keep the sport fresh and exciting, and to provide additional opportunities for fun, they have developed some unusual types of hunts. Here are some examples. While they may not closely simulate the search for a jammer or a distress call, they still provide keen competition and practice at refining one's gear and hunting techniques.

The All-Day Hunt

This traditional southern California quarterly event should actually be called the all-weekend hunt, because it often lasts that long. The transmitter may be in any terrestrial location in the continental United States that can be heard at the starting point.

Other than its length, it is similar to most other mileage hunts. Since bearing accuracy is of extreme importance due to the distances involved, there is no time factor. Odometer calibration is required if any team comes within ten percent of the apparent winning reading. Winning mileages in the hundreds of miles aren't uncommon.

The hunt is very unhurried. Hunters usually plan for a complete weekend outing, including meals and camping gear. End points are often near campgrounds, and a social "happening" frequently results.

If such a hunt is only for stalwarts, then there are a lot of stalwarts in southern California. You can expect to see a dozen or more cars and campers lined up at 10 am when the hunt begins. This type of hunt is ideal for the region due to the wide variety of terrain, including flatlands, mountains, canyons, and deserts.

The Requested Transmission Hunt

First done in Maine on 10 meters in the 1950's, this is a lowest-mileage hunt with an unusual twist. After an initial transmission, the hidden T operator remains silent until he is requested to transmit by one of the hunters. Each time a team asks for a transmission, it is penalized one mile.

To ensure fairness, all transmissions should be timed to be exactly the same length. The northern Maine group used 30 seconds. The huntmaster keeps track of the penalties, computes the mileages, and determines the winner.

Hunters must be very careful to get the maximum amount of information from each bearing opportunity, and be ready at all times to take advantage of a transmission asked for by another hunter. They must constantly face the dilemma of whether to ask for a transmission, causing a penalty and giving a "free" bearing to other hunters, or to remain silent and risk driving the wrong way or overshooting the target.

Time to find was not a factor in the AM hunts in Maine, but a time penalty could also be included if the hunts get very long due to hunters' reluctance to ask for transmissions. We suggest instead the following solution to that problem. Let the hunter requesting a transmission specify how long he wants it to be. The penalty should vary with the length requested, say perhaps only 0.1 or 0.2 miles per five seconds.

With a little practice, it's not hard to get a rough bearing in five seconds with a hand rotated antenna. Ten seconds should be the maximum needed if the hunter is stopped and ready. The penalty is small for a short transmission, and it may not aid the other hunters if they are in bad locations or are not prepared. Such a modified rule is an incentive to hunters to learn to take bearings as quickly as possible.

It would appear that having a Doppler or other electronically rotating antenna type DF system would give an advantage under these conditions because it continuously reads out and rapidly updates. It should be possible to get a bearing on even the shortest transmissions. But remember that Dopplers work best when bearings are taken continuously while moving, and are eyeball-averaged. This can't be done when transmissions are short and intermittent. So the requested transmission hunt is an equalizer in that respect.

The No-Holds-Barred Hunt

For the truly dedicated hunter, this hunt approaches the maximum challenge. For the creative hider, this is the chance to try everything in his bag of tricks. He is free to do things that would be disallowed in most other hunts, such as changing power and antenna parameters, or deploying multiple simultaneous transmitters, perhaps some mobile. He can be as accessible or inaccessible as he dares.

There are as few or as many rules as the hider wants. He sets the time and the boundaries, if any. He also chooses how the winner is determined. This has resulted in some bizarre requirements. One huntmaster taped cups of water to the hunters' car hoods and gave the prize to the team with the most water left in his cup at the end point. Fortunately, most hiders are more conventional.

One NHB hider chose to begin his hunt Saturday night at midnight. He rode his motorcycle up and down the Los Angeles flood control channels, which were dry at that time of year. By listening to the hunters and transmitting only infrequently with his handheld rig, he was able to successfully keep away from them for several hours. As the hunters closed in, he rode from the channel up into the storm sewer system and kept transmitting using a "rubber duckie" antenna on the end of a mast, pushed up through a small hole in a manhole cover in the middle of an intersection. Because of his infrequent transmissions, it was after dawn Sunday morning before WB6ADC tied a note to the duckie, shoved it back down the manhole cover, and won the hunt.

The Mount Wilson Repeater Association uses its infrequent NHB hunts to generate revenue for the treasury. Teams bid for the privilege of hiding. Members and supporters of the teams pledge money to the club treasury, and the team with the most pledged at the cutoff time gets the go-ahead. After this auction, the successful team has one week to collect all the pledges or forfeit. When all pledges are turned in, the hunt is scheduled.

FINDING YOUR BATTING AVERAGE

Who is the best T-hunter in your radio club? Every hunter is interested in how he stands, and whether he is improving, but a single count of wins does not give a complete picture.

Here is a way to grade each hunter or team based on the placing in each hunt, taking into account the number of competitors hunted against. A score, similar to a box score in baseball, is assigned to each hunter as a function of his placing according to the following formula:

$$ \text{SCORE} = \frac{n - p + 1}{n} $$

n = number of vehicles
p = position (1 = first, etc)

For example, in a four-car hunt, the winner gets 1.000, second place gets .750, third gets .500, and fourth gets .250. Note that no one gets .000 unless he gives up or is talked in. Table 13-1 is a listing of scores given by this formula for up to 15 teams.

These scores can be averaged over time to give each hunter a long-term batting average. This method is used to keep track of hunters in the Fullerton (CA) Radio Club (FRC). At the end of the year, the scores are averaged and a ranking of the hunters is published in the club bulletin. The summary includes the average, the ranking, the

Table 13-1. Scoring Chart for up to 15 Teams, One Hunt.

TEAMS	1ST	2ND	3RD	4TH	5TH	6TH	7TH	8TH	9TH	10TH	11TH	12TH	13TH	14TH	15TH
1	1.0														
2	1.0	.500													
3	1.0	.667	.333												
4	1.0	.750	.500	.250											
5	1.0	.800	.600	.400	.200										
6	1.0	.833	.667	.500	.333	.167									
7	1.0	.857	.714	.571	.429	.286	.143								
8	1.0	.875	.750	.625	.500	.375	.250	.125							
9	1.0	.889	.778	.667	.556	.444	.333	.223	.111						
10	1.0	.900	.800	.700	.600	.500	.400	.300	.200	.100					
11	1.0	.909	.818	.727	.636	.545	.455	.364	.273	.182	.091				
12	1.0	.916	.833	.750	.667	.583	.500	.416	.333	.250	.167	.083			
13	1.0	.923	.846	.769	.692	.615	.538	.461	.384	.307	.230	.153	.076		
14	1.0	.929	.857	.785	.714	.643	.571	.500	.429	.357	.286	.214	.143	.071	
15	1.0	.933	.867	.800	.733	.667	.600	.533	.467	.400	.333	.267	.200	.133	.067

Use the row corresponding to the number of teams and read across for the score of each team. For example, in a seven-car hunt, the third place team score is .714.

number of hunts, the number of wins in the past year, and rankings from previous years. Statistics from a very small database have little meaning, so the summary charts should cover at least eight hunts, and only hunters who participated three or more times should be included.

You may be surprised to find that the hunter with the best average may not have the most wins. In one year's FRC summary chart, Gary Holoubek (WB6GCT) was the highest scorer, even though he did not actually win any hunts. He placed second nearly every time he came out. Could it be that he successfully avoided winning to prevent having to hide the next time?

One thing that is almost always shown by the tables is that those who hunt most often tend to have the best scores. This should encourage the new hunters to keep at it and watch their scores improve.

GETTING GREATER ATTENDANCE

To increase participation, the results of each hunt should be detailed in the club newspaper, with the help of the club's most creative writer. Readers should be made aware that they're missing a lot of fun if they're not out hunting regularly. Novices and unlicensed members should be reminded that they can hunt without transmitting under the rules of many hunts.

Find out who the potential hunters in your club are and invite them out with you to ride along on a hunt. Let them see for themselves how much fun it can be. Don't worry about giving away any of your trade secrets to a future competitor—better that than to have no one to compete against. Remember that the goal is to develop more and better hunters, to increase the level of competition, and to ensure that the group is ready if jamming or an emergency occurs.

Another way to generate more hunters is to hold "clinics" at club meetings showing how to build home brew gear such as quads, loops, and attenuators. Many southern California hams have gotten into the fray after attending club meetings and seeing Ray Frost, WA6TEY, and others show how easily a PVC pipe and wire quad can be strung together in a few minutes.

Even better is to have a club RDF clinic where members can get together and actually build and test the gear they'll be using, under the guidance of the group's most experienced members. Here are some hints for setting up your clinic:

☐ Plan for a day long session during good weather. Advertise it well in advance.

☐ Have materials and parts available for purchase on the spot, or give attendees in advance a list of materials to bring with them.

☐ Some work can be done in advance in batches, such as cutting sets of copper clad board for attenuator boxes.

☐ Limit the initial clinic to antennas, mounts, and attenuators, plus other basic information and possibly S-meter help. Later clinics can be scheduled for advanced topics such as sniffers or Dopplers.

☐ Plan a hunt soon after the clinic, to get everyone going.

Amateur photographers should be encouraged to take and share pictures of unique setups and hiding spots. A home movie or videotape can help show the fun of hunting and even be good PR. Videotapes by Larry and Patti Curtis, WA6LPI and KA6OTX, have been shown at both ham club gatherings and on local cable television. They did it on their own, but you might be able to get your local cable access club to help you with equipment or editing. Who knows, maybe we can even get T-hunting into the next ARRL promotional film or tape about amateur radio.

Chapter 14

You're the Fox

Sooner or later (probably sooner than you expect) you'll get your chance to hide the transmitter. Now's your chance to get even, right? You carefully scrutinize the rules, looking for a good loophole. You wonder: "Is it a no-no to hang it from a helium weather balloon? How about putting it in a duck decoy in the middle of a lake?"

Experience, both as hunter and as hider, is your best teacher at finding hiding spots that are challenging. There are plenty of examples in this book to give you ideas. This chapter starts out with some suggestions on how to go about determining when you've found a good spot. Then we move on to the important technical considerations of equipment, antenna, power, and audio.

FINDING THE PERFECT SPOT

Don't wait until it's your turn to hide before starting to think about good locations. Pay attention to your surroundings. Be on the lookout for changes in your area as you do your daily driving. A bridge out or road closure for repair may suggest a hard-to-reach hiding place. Knowledge of a new road, bridge, or freeway access point is useful both for hiding and hunting. Check into the free guides to new housing in your town, which give an idea where new roads, perhaps not on the maps, may exist.

(These guides are available at savings and loan institutions and real estate offices.) New roads can also be found by getting the latest maps and comparing them with older ones.

Finding the perfect hiding spot is an intuitive process, and instructions for finding one would be impossible to write, given the wide variety of terrain around the country and the different characteristics of each ham band. As you hunt and hide more, you see which tricks to confuse hunters work and which don't. Try new ideas each time, and soon you'll be enjoying your ability to cause consternation among the top-notch hunters in your area.

Check the route from the starting point to any potential hiding spot. If there's a direct arterial street going right between, chances are most hunters will take it and ties will result. Try to find a location that spreads them out.

Some hiders scour their maps looking for a trap for the hunters, where the direct path leads the hunters to a river, a rural dead-end road, a washed-out bridge, or other uncrossable obstacle. The hunters are then forced to choose the minimum time or mileage method to backtrack out of the trap. Of course the best (luckiest?) teams avoid the trap in the first place.

Keep it safe and legal. Don't pick a spot which could

easily result in an injury or cause hunters to run afoul of the law. People think hams are weird enough without us giving them cause. Be prepared to explain the purpose of amateur DF experimentation simply and politely to passers-by, pointing out the potential public service benefits. Take advantage of any opportunity for good public relations for our hobby.

When the bunny's location is finally chosen, there is still more to do. Check the signal path ahead of time. It's very embarrassing to find at hunt time that the power and antenna you're using isn't enough to get a usable signal to the starting point. Conversely, you don't want to put in too much signal and reduce the challenge.

Be prepared for degraded conditions. It's not uncommon for a non-direct path to have more or less loss at hunt time than it does when you check it out. Perhaps an inversion layer or clouds during the daytime checkout provides some ducting or signal reflection. If it is not there at night when the hunt starts, more power or a better antenna may be needed. The converse can also happen.

New housing areas with incomplete homes usually have special private security forces. Be sure to get permission before hiding if required. Don't wait until the last minute and then ask the first security guard you see. He will probably be unwilling to give you access on the spot without an OK from his supervisor or even the property owner. It is better to seek out the owner or developer well ahead of time and have him give instructions to the security subcontractor. (Having the private patrols on your side can be very useful. One even let us use an unfinished house and garage as a place to hide ourselves and the car while we watched the hunters trying to find the transmitter, which was in a construction outhouse on the street outside.)

Notifying the local police department isn't usually necessary, but might be desirable for some spots. A T-hunting group in St. Paul, Minnesota, discovered this the hard way when they were suddenly confronted one night by anxious authorities with guns at the ready. It turned out that they were across the street from a new bank, ready to open but with the money not fully secured. A night security officer had thought people standing around vehicles with strange equipment on them might mean trouble, and had called in the alarm.

PREPARING THE T

Once you've found the perfect hiding place, it's time to think about what equipment to use. For many hunts, simplest is best. Just drive to the spot with your mobile station and start transmitting when the time comes. If the hunt calls for an intermittent signal, use your watch for timing, and make taunting remarks to the hunters in your brief transmissions.

After a few hunts, you'll begin to think of ways to make your hunt more challenging by concealing the transmitter setup and watching from a distance as the hunters try to close in. It's hard to describe the jubilation you'll feel watching hunters drive up or walk up within feet (or even inches) of your transmitter and walk away still confused. With planning, the ideas on the next few pages, and a little luck, you can have that kind of success.

THE TAPE RECORDER

Talking continuously is probably not what either you or the hunters want on a non-intermittent carrier hunt. When the first hunters arrive, you'll want to spend time chatting with them, not manning the transmitter. A tape recorder provides a simple continuous audio source for you. A battery powered cassette unit is ideal.

What you put on the tape is limited only by your imagination and the FCC rules prohibiting broadcasting (with exceptions) and music transmissions (with no exceptions). Sound effects recordings are popular, as are Westlink amateur radio bulletins. Just use something distinctive so there is no doubt it's the hidden transmitter.

John Moore, NJ7E, tells of how a remotely controlled transceiver was hidden in the middle of the Arizona desert with a carefully contrived tape recorder and timer mechanism. It came on with one minute of live-sounding chit-chat every four minutes, but the cunning hiders had taped their patter at a busy downtown intersection. The street sounds were baffling to searchers who saw the bearings crossing far from any urban area, and led more than one to distrust his DF gear.

You'll want a direct connection between the recorder and the transmitter input for two reasons. Audio quality from a speaker-to-mike transfer is far from good. Also, the open microphone may pick up local sounds that give away your location.

A quick and dirty direct connection method is to patch the external speaker jack on the recorder to the transmitter microphone input. Don't do this unless you have a separate receiver, such as a handheld, to listen to your transmitted signal. The plug will probably disable the recorder's internal speaker, and the setting of the recorder's volume control for proper modulation is critical. This adjustment has to be repeated each time you hook the recorder to the transmitter.

A much better way to hook a recorder to the transmitter is to use the separate audio output jack found on

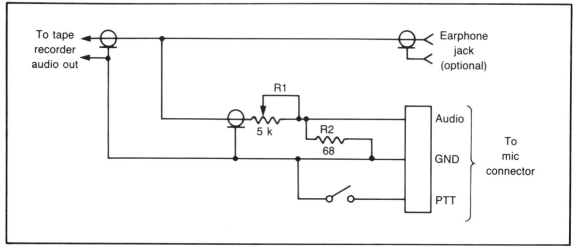

Fig. 14-1. Circuit connecting a portable tape recorder to a transmitter.

many recorders, which is unaffected by the volume control. The level is about one volt. If your recorder does not have such a jack, you can add one by tapping the ungrounded end of the volume control. The circuit of Fig. 14-1 is intended for use with such an audio output jack, but can be hooked to the external speaker output in a pinch. For some rigs, R2 may have to be a higher value.

Some pointers:

☐ Use a tape long enough to keep you from having to change it often.
☐ Put callsign identification on your tape at least every ten minutes.
☐ Make sure the batteries in the recorder will last the length of the hunt.

A SIMPLE TRANSMITTER CYCLER

For an unattended transmitter on an intermittent carrier hunt, you'll need a way to turn the transmitter on and off. Figure 14-2 shows a very simple method. The 555 IC timer is connected as an astable multivibrator with a one-to-three on/off ratio. The output of the IC drives a small sensitive relay, K1. The contacts of K1 can either be connected to the push-to-talk terminals on a transceiver, or placed in series with the hot power supply lead on low power transmitters.

With the parts shown, the relay is closed for about fifteen seconds out of every minute. The ratio of R2 to R3 determines the on/off ratio with this formula:

$$\frac{T_{on}}{T_{off}} = \frac{R3}{R2 + R3}$$

The time for one complete on/off cycle is determined by the value of C1. Any leakage or capacitance tolerance variation in C1 affects cycle time, so you may need to experiment to get the desired timing. Put a number of small electrolytic capacitors in parallel at C1 to get the desired value if one doesn't hit it on the nose.

All parts are readily available at local parts stores and construction is non-critical. The relay has a 12 volt, 1000 ohm coil (RS 275-003). The unit may be built into a small box or incorporated into a noisemaker unit. The circuit draws about 25 milliamperes total maximum with the LED and specified relay. The LED (D3) indicates that the transmitter is keyed, and is optional. Diode D2 protects the IC from the inductive kick of the relay coil, and is not optional.

Switch S1 interrupts the timing cycle and turns on the transmitter for giving clues or station ID. Closing S1 resets the 555 IC and keys the relay. When S1 is opened, the timing starts again with an off period.

R4, R5, and D1 are optional components which prevent the first off period from being up to 35 seconds longer than those following due to the complete discharge of C1 when power is off or S1 is closed. With the added components, C1 is kept charged to just below the trigger threshold, and the first off period is only about three seconds longer than normal. If you don't mind an initial period that's extra long, delete these parts.

TONE BOXES

A distinctive electronic sound source helps the hunters by instantly identifying the signal as the hidden T. It also helps the hider by permitting a completely unattended transmitter (provided FCC station identification requirements are met). A single continuous audio tone works but most hunters have more ingenuity. An electronic telephone ringer, a solid-state chirping bird ornament, and the guts from a kid's space gun noisemaker have all been used.

At the opposite extreme are elaborate units which randomize the rate and sequence of tones to produce a pattern that seems to never repeat exactly during a hunt. Construction of such units is simplified nowadays by the availability of sound generating ICs. Some clubs build a tone box which is passed around to each hider, so that the sound is always the same and the hunters don't have to build their own boxes. The club callsign can be used in the ID circuit.

The Un-Music Box

This tone box answers the need for a simple and inexpensive noisemaker. The non-melodious tone frequency steps occur every quarter of a second, with over four minutes between repeats of the complete pattern. It can be made to work with most FM rigs, even a handheld if it has an external mike input. The components are readily available, and it's so easy to build that it's a good first digital construction project. Yet it has enough features to make it the only tone box you may ever need, including:

☐ CW identification. It IDs automatically at ten minute intervals, or the ID can be manually started at any time.

☐ Programmable ID'er. The callsign is easily programmed and changed. The box can be passed around from hider to hider in the local club, each using his own call when he hides.

☐ Intermittent keying. A 15 second on, 45 second

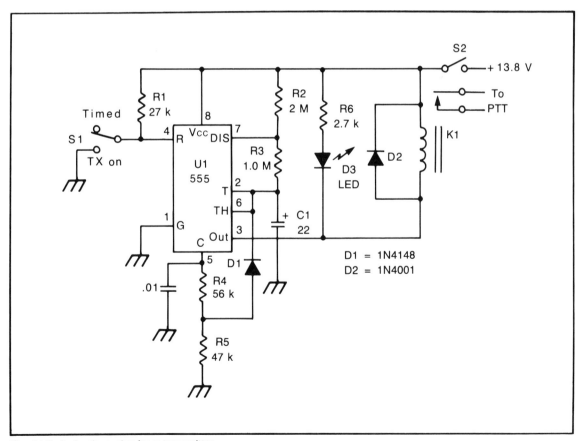

Fig. 14-2. A simple cycler for a transmitter.

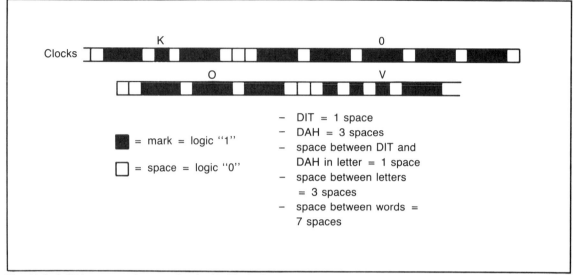

Clocks

K   0

O   V

■ = mark = logic "1"

□ = space = logic "0"

- DIT = 1 space
- DAH = 3 spaces
- space between DIT and DAH in letter = 1 space
- space between letters = 3 spaces
- space between words = 7 spaces

Fig. 14-3. A CW programming map for the Un-Music Box.

off carrier control mode is available, or the carrier can be on continuously.

□ Battery operation. Power can come from an internal battery or the radio's 13.8 V source. Current drain in operation is only about five milliamperes.

Most popular commercial ID boards use either a PROM chip, which must be programmed on a special machine; a diode matrix, which has to be carefully worked out using mapping techniques for each message; or dip switches, which can get expensive. This ID generator uses a shift register to hold the callsign, making it simple to program and re-program. A shift register is analogous to an old time firemen's bucket brigade. In this case there are 128 cells (buckets) in a row. Each is either full (logic 1), representing a mark (tone), or empty (logic 0), representing a space (no tone).

Each time a clock command is given, all buckets are shifted one position to the right. One bucket comes to the end, is checked to see if it is empty or full (read), and then passes back into the input. After 128 clocks, the buckets are back in their original position, and the entire message has been read. The message remains in the register as long as power is applied to the circuit.

The bucket brigade is programmed by filling or not filling each bucket. A sample pattern of a CW message ready for programming is shown in Fig. 14-3. The actual programming is actually very simple. Once you get the hang of it, you can program a callsign in less than one minute without writing anything out.

Circuit Description

Figure 14-4 is the complete schematic of the tone box digital section. The gated tone oscillator is one quarter of a four-section Schmitt trigger NAND gate IC (U8-D), with C5 and R12-R19 in the feedback path.

The Schmitt trigger (Fig. 14-5) is ideal for simple logic oscillators. In this circuit, developed by O. H. Schmitt in 1938, the threshold for negative-going inputs is lower than for positive-going inputs. This property is called hysteresis. The charge and discharge of the capacitor between these thresholds, taking place through the feedback resistance, gives an astable output with fast rise and fall times.

Analog switches U4 and U5 select the combination of feedback resistors for a particular tone frequency as commanded by U7, the twelve-stage ripple counter. U7 is clocked at a 4 hertz rate by U8-A, another gated Schmitt trigger oscillator. Besides driving the analog switch, U7 controls the 15/45 second timer with its Q7 and Q8 outputs, and starts the IDer when both Q10 and Q12 go high at 640 seconds.

Two 64-bit static shift registers, U2 and U3, form a first-in first-out memory for the ID. Assuming that the message has been loaded, it is read out by setting the reset pin on U6 high, either manually with S5 or automatically from U7. This enables the astable ID clock (U1) by pulling its reset pin low.

U1 clocks the shift registers and counter U6 simultaneously, so when Q8 of U6 returns to high 128 clocks later, the message is complete. During the ID cycle, U8-B and U8-C key the tone oscillator from the shift register output. The 12.5 Hz rate of the U1 clock sets the message rate at 15 words per minute.

When ID is initiated by the U7 timer, the CW tone pitch is determined by the value of R19, as Q10 is the only high output of U7 driving U4/U5 at 640 seconds. If ID is started with S5, the CW pitch will be set by whichever outputs of U7 are high at that moment. When ID is complete, U7 is restarted from zero through C7 and R8.

To program the IDer, the ID clock is put into a manual mode with S1, and S3 switches the shift register from recirculating to input from S4. Pushbutton S2 advances U1 as programming progresses. R1 and C1 circuits debounce S2 so that extra clocks are not generated when S2 is released.

Supply current drawn by CMOS ICs is primarily a function of the number of gates changing states at a time, and the rate at which state changes are taking place. All oscillators in this unit operate at very low frequencies, which keeps the supply current drain to a very small amount. The greatest drain is in the push-to-talk keying circuit, which need not be powered from the internal battery.

D1 and D2 remove the load from the 9 volt battery when external power is applied. Zener regulator D6 keeps the logic supply voltage (VDD) at about 8.5 volts maximum, protecting the CMOS from automotive system transients and keeping the tone pitch constant. Even when using external power, it's a good idea to keep a battery in the unit. This prevents the ID from losing its program if the external source is accidentally disconnected during the hunt.

To conserve the 9 volt memory hold-up battery when the unit is not in use, set S1 to STEP, then press ID START (S5) once. This stops all oscillators to minimize battery drain, provided that the first ID bit is a space. The circuit draws less than one microampere with oscillators off, so the battery should last its shelf life. Returning S1 to SEND starts the unit cycling again.

The push-to-talk (PTT) control circuit of Q1 and Q2 is designed for rigs which ground the PTT lead on the microphone connector to transmit and have positive voltage on the PTT lead. For other rigs, minor changes may be needed. Measure the unkeyed PTT voltage and keyed PTT current and redesign accordingly.

A relay can be added (Fig. 14-6) to key almost any radio by grounding the PTT lead or putting the contacts in series with the B+ lead of board level transmitters. The V+ end of the relay goes to the 13.8 volt transmitter source, not the 9 volt Un-Music Box battery. Coil re-

sistance should be greater than 100 ohms. The RS 275-003 is okay here.

If input voltage supply polarity to the unit is accidentally reversed, the logic will not be damaged, but nearly the full supply voltage will be across R11. Make R11 a two-watt resistor and it will survive polarity reversals.

Figure 14-7 illustrates a typical connection for the box into the mike and PTT lines of Drake 2 meter equipment. It also works for Kenwood rigs having four-pin mike plugs. Use connectors mating with the ones on your particular rig and wire accordingly. The mike lines are low level and shielded wire is recommended.

Switch S7 selects a TALK mode, where the mike and its PTT button operate the radio normally, or TONE mode, which applies tone, cuts off the mike, and transfers PTT control to S6. The mike cutoff is quite important. You wouldn't want any local sounds to give away your hiding place, would you?

Construction

The prototype unit (Fig. 14-8) was built using wire wrap construction, but that isn't mandatory. Sockets for ICs are recommended, to prevent static damage to the CMOS ICs during wiring. The oscillator pin on the 4047 (Pin 3 of U1) does not have as effective static protection as all the others, so special care should be used in handling that IC. Keep ICs in conductive foam when not in the circuit.

The CMOS ICs are packaged with a variety of numbering variations. RCA's part numbers begin with CD (CD4066AE), while Motorola has an equivalent 14000 series (MC14066B). Catalogs usually list only the 4000 number without the prefix. Either A or B suffix parts can be used at this low voltage and frequency. All are very inexpensive.

The values shown for R12-19 are just one of the many possible configurations, each of which gives a different tone pattern. They were chosen with a BASIC computer program that computed all possible tone frequencies and their differences, given the resistor values. As you go up the musical scale, each note is about 6 percent higher in frequency than the one below, so each tone frequency step should be at least that great.

Total tone range from lowest to highest is limited by the audio response of your transmitter. Our values give a tone range from 160 to 4000 Hz. Experiment to see what you can do with the resistors in your junk box.

It's a good idea to mount R12-19 on two four-pin dual in-line (DIP) headers, so the values can be easily changed. You can even make up a few personality headers with

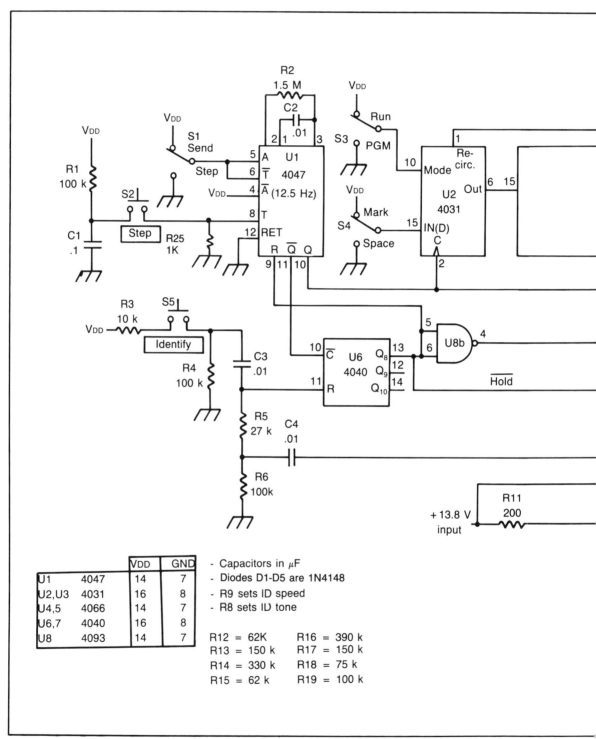

Fig. 14-4. Schematic and parts list for the Un-Music Box.

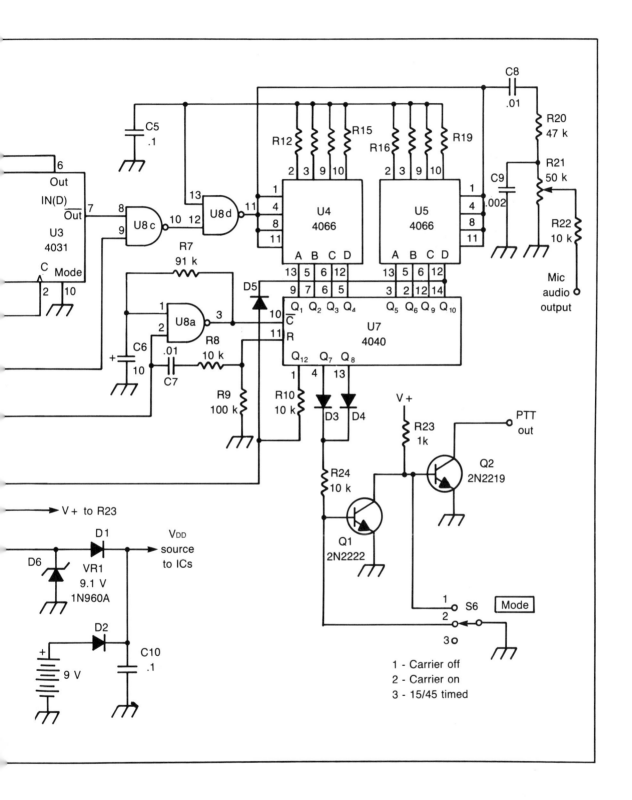

197

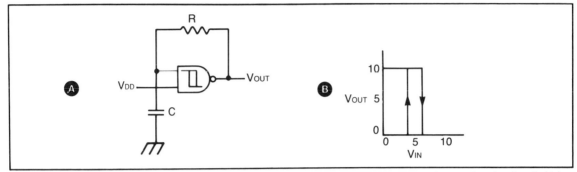

Fig. 14-5. The basic Schmitt trigger oscillator circuit (A). The Schmitt trigger has hysteresis, causing the threshold for negative-going signals to be lower than for positive-going signals.

Fig. 14-6. Relay keying addition allows for universal keying.

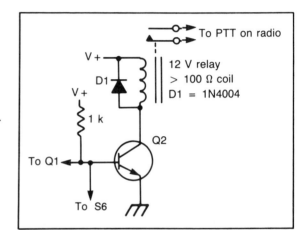

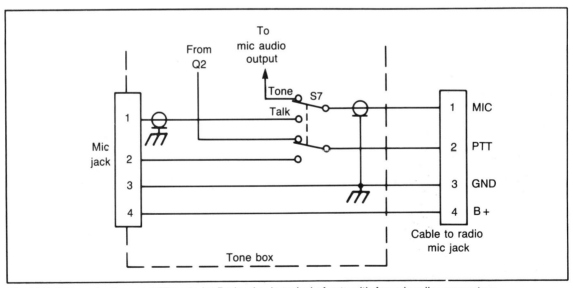

Fig. 14-7. Wiring of mike and PTT leads for Drake rigs is typical of sets with four pin mike connectors.

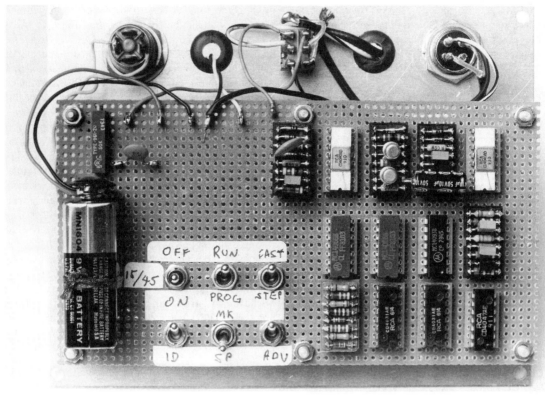

Fig. 14-8. Tone box board mounted inside box lid.

varied resistor combinations so your tone box doesn't always sound the same on each hunt. Each pair of headers gives four possible tone patterns by reversing and exchanging them between the two sockets.

While the unit can be built into a very small box, we made no attempt to miniaturize the prototype. It was mounted in a 6-3/4 × 5-1/4 × 2-1/4 inch plastic box (Keystone #701). By mounting the board to the lid, the box can be removed completely with no dangling wires for ease of ID programming (Fig. 14-9). Only S5 and S7 need to be mounted through the lid. The other switches can be mounted on the board to simplify wiring. They are used only during ID programming and when setting up the unit for the hunt.

Check-Out and Programming

After wiring up the IC sockets and other components, check your work carefully. Then power up the board without the ICs in place. Current from the supply should be zero when 9 volts is connected to the battery input. The Zener dropping circuit will draw about 15 milliamperes

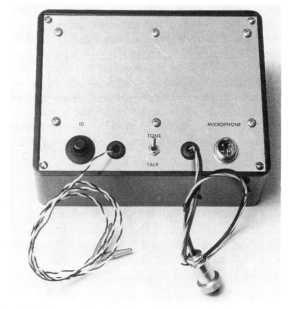

Fig. 14-9. Complete tone box with lid installed.

199

when +12 volts is connected to the external power input. Check to make sure that the proper voltages and polarities are present on each IC supply/ground pin. Now remove power and install the ICs. Set the switches as follows before applying power for test:

S1 to SEND
S3 to RUN
S4 to SPACE
S6 to CARRIER ON

Connect an earphone to the tone output to monitor the sounds. If the output to the rig's mike input doesn't give enough volume in an earphone, connect it to the output side of C3. With power on, you should hear the rhythmic but non-musical sound of stepping tones. If an ID is initiated, ten seconds of CW-ish garbage will replace the tones.

To program the IDer to send something useful, plot the mark and space values for your message, following the example of Fig. 14-3. You have 128 bits to play with, which is long enough for the longest possible US ham callsign (something like KJ0JJJ), plus DE and some space at the beginning and end. For proper-sounding Morse, a DIT is one mark bit, a DAH is three mark bits. Spaces between DITs and DAHs in a letter is one space bit, between letters is three space bits, and between words is 7 space bits.

When your message is all plotted out, set S1 to STEP and S3 to PROGRAM. S4 should be set to SPACE. Press ID START (S5) once. Press S2 twice to put in some initial spaces. Now enter your CW message by setting S4 for mark or space as appropriate and pressing S2 to enter each bit, following your plot.

At the end of the message, set S4 to SPACE and set S1 to SEND. This automatically fills the remainder of the register with zeros. Wait a few seconds until the sequential tones restart, and set S3 to RUN. Programming is now complete, and you will hear the CW message when S5 is pressed.

Use care when setting R21 for your transmitter. Listen on a separate receiver and bring up the level slowly. It is easy to overdeviate the transmitter and cause adjacent channel QRM from the high frequency tones if the level is set too high. If the transmitter has a low-impedance microphone input, R22 may have to be made smaller to get sufficient modulation.

Improvements

The 4031 shift registers are inexpensive (about $2.50

by mail order) and it's easy to add more for a longer message. They should be chained as shown in Fig. 14-10, which illustrates how four ICs are hooked up to get 256 bits. As more are added, the hold signal output on U6 must be moved to correspond with the number of total bits. For example, take the output from Q9 instead of Q8 if four register packages are used. Use Q10 and eight shift register ICs for 512 bits.

You can also use your creativity in the selection of counter outputs driving the FET switch ICs. We didn't use Q7 and Q8, for example, because those outputs would be changing the pattern when the transmitter is off in an intermittent-keyed hunt. They could be used instead of Q9 and Q10. Try taking the U4-13 input directly from the 4 Hz oscillator at U8-3 instead of from U7-9. This produces a rhythmic output which is quite interesting. It sounds almost polyphonic, with an echo every 16 notes.

Lest you be concerned that use of this tone box constitutes transmission of music, have no fear. The dictionary says that music has varying rhythm (the box doesn't). Music also is "structurally complete and emotionally expressive." The only emotion you'll feel listening to this box is pride in its creation. The only emotion the hunters will feel is eagerness to find the hidden T.

ANTENNAS FOR HIDING

Antenna selection is the area where hiders can truly unleash their creativity. It is also the area that causes hiders the most frustration. After all, how can one make an invisible antenna with 20 dB gain and a perfectly unidirectional pattern?

If you need a lot of gain to cover a long path, it's probably best to forget unobtrusiveness and use a conventional gain antenna. If gain isn't important, you can make the antenna hard to find by doing one or more of the following:

☐ Conceal it. We once obscured a four-element 2 meter beam behind a chain link fence and some dead branches, and watched as three hunters drove up to it, shined headlights right at it, and drove away. An antenna in a tree is also very hard to spot.

☐ Make it small. A rubber duck fastened to the side or bottom of a car with a magnetic mount in a row of cars in a parking lot is very hard to find. How much can you shrink your antenna and still be heard at the starting point?

☐ Make it look like something else. The best example of this we've ever seen is the sprinkler ground plane, designed by Walt Brackmann, WA6SJA, and shown in Fig. 14-11. There are four pipe radials and a

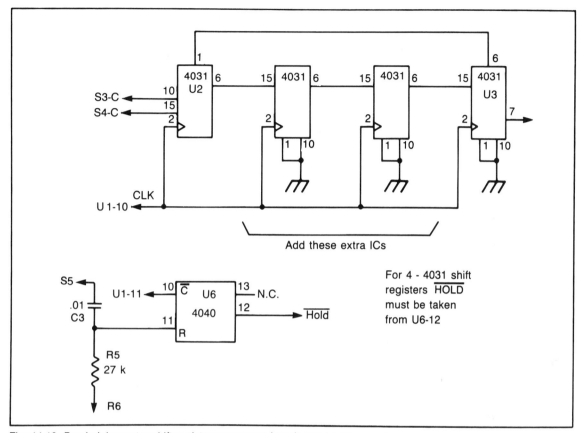

Fig. 14-10. By chaining extra shift registers, message length can be increased.

19-inch vertical driven element pipe with a PVC insulator, making it load up perfectly on 2 meters. He even arranged for it to be dripping water during the hunt.

□ Make something into the antenna. Load up a fence, a flagpole, a bridge, or something else that's big. This is easier to do at hf than at vhf, because transmatches are common and gamma matching to grounded objects is simpler. But vhf transmatches are possible to build. Other matching methods can also be used, as will be shown later.

□ Divert attention from the antenna. If there's a large metal object nearby, such as a water tower, concentrate your vhf signal towards it and make it into a reflector which appears to be the antenna. K0OV did this once at a well known large church in Garden Grove, California, which is all glass with a framework of metal tubing. The signal came from a 2 meter quad located low, outside the property in a cul-de-sac, and behind a fence. It was pointed up and into the structure and effectively lit it up with ten watts of rf. Only if they read this book

will the attendees at the wedding there that Saturday night know why the strange people with weird electronic devices were wandering through the church grounds and staring up at the building with puzzled expressions.

You don't need fancy commercial antennas for hiding. Anything that radiates well will do if you're not worried about directivity. Figure 14-12 shows two examples. The spike and cookie sheet ground plane can be put on top of something, or the cookie sheet can be buried just under the surface of the ground. Similar antennas can be made for other bands. The floppy antenna works well for hanging from high places. See how creative you can be.

Don't overlook long wire antennas. A span of wire many wavelengths long has very interesting directive lobes. Clarke Harris, WB6ADC, well known for using unusual antennas for hiding, has successfully used a rhombic, with dimensions similar to those in Fig. 14-13. Such an antenna, supported on small wooden sticks or dowels about a foot above ground, has a very directive pattern,

Fig. 14-11. An antenna disguised as a sprinkler head. It even dripped water!

with high gain and horizontal polarization. It is as effective as a 14-element beam at the same height, but is very hard to spot at night. It is easiest to set up in a large level area. Take precautions to avoid having hunters tripping over it.

The rhombic in the figure is six wavelengths per leg, about the biggest practical size for this application. The quarter wavelength parallel line section and balun match the 800 ohm impedance of the rhombic to 50 ohm coax. The load dissipates about one third of the transmitted power, and should be made of carbon or other non-inductive type resistors. To make a rhombic for other vhf bands, scale all dimensions, except the angles and the parallel line matching section spacing, by the ratio of the frequencies.

Most vhf transceivers have internal protective circuits which reduce or shut off the transmitter output when sig-nificant amounts of apparent reflected power are sensed. It is likely that your unusual antenna will present an unsatisfactory SWR to the rig and cause a shutdown. Traditional matching techniques, such as L-C tuners, coax stubs, and quarter wave sections can be used to match the antenna to the line and lower the apparent SWR. If you don't mind a reduction of power to the antenna, you can keep the transmitter happy by putting a resistive attenuator, often called a pad, between the transmitter and the antenna.

A 7 dB pad effectively protects most transceivers from anything, even an open or short circuit. This is because forward power is attenuated by 7 dB going toward the antenna, and any reflected power is attenuated by another 7 dB getting back to the transmitter. The minimum return loss of the antenna system is thus 14 dB, which results in a worst-case SWR of 1.5:1 at the transmitter.

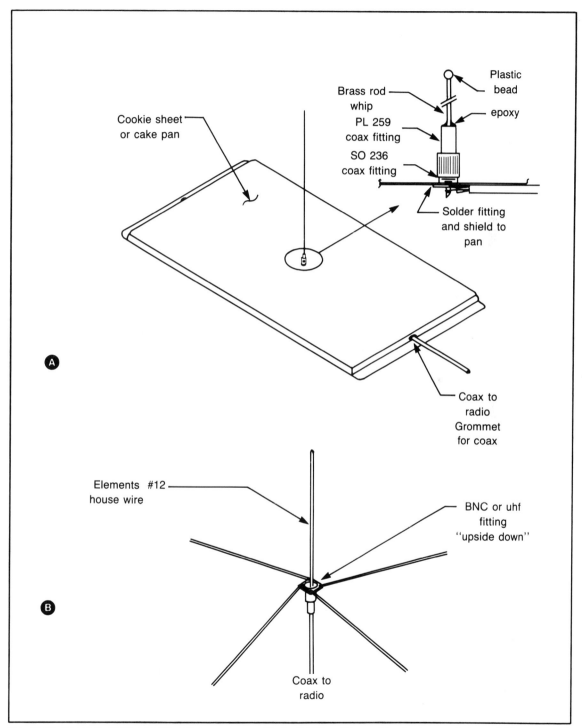

Fig. 14-12. These customized antennas are very versatile for T-hiding. The cookie sheet ground plane of (A) can be put on top of objects or buried. The floppy ground plane in (B) can be suspended from anything handy.

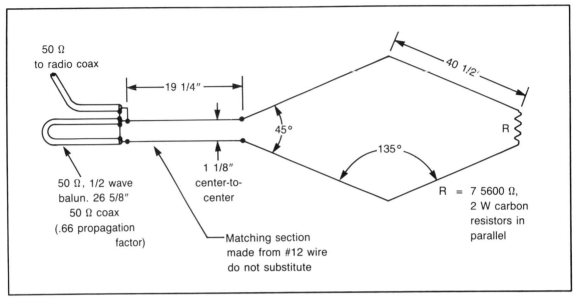

50 Ω
to radio coax

← 19 1/4" →

40 1/2'

45°

R

135°

R = 7 5600 Ω,
2 W carbon
resistors in
parallel

50 Ω, 1/2 wave
balun. 26 5/8"
50 Ω coax
(.66 propagation
factor)

1 1/8"
center-to-
center

Matching section
made from #12 wire
do not substitute

Fig. 14-13. A 2 meter rhombic gives a very narrow pattern.

That can be accommodated by most rigs.

The advantage of the pad over other matching methods is that absolutely no tuning is required, although tuning might help the antenna to radiate better. No SWR meter is needed for setting up, either. The disadvantage of this method is that only 20 percent of the transmitter output power is present at the output of the pad. For many hunts where the hidden T is in the clear or the distance is short, this poses no problem.

A well-remembered Fullerton Radio Club 2 meter hunt is a good example of the effective use of a pad. A 10 watt rig and a car battery were hidden underground in a telephone junction vault in a new housing area. Since the area was atop a high hill and line of sight to the starting point, a 20 dB Narda high power attenuator was placed in line with the coax to the antenna, which was a two foot piece of #28 wire wrapped around the stem of a handy dead mustard plant. One hapless hunter stopped his car right on top of the vault and spent quite a while trying to get a bearing.

Calibrated medium power fixed attenuators are available from Narda and other companies. They work from dc to the microwave region, and are used for test setups in the industry. Such pads are too expensive to be bought new for amateur use, but are occasionally found in the industrial surplus market. They can readily be home built to work on the hf and vhf bands. Figure 14-14 shows a 7 dB T pad for 50 ohm coax which handles the output of a 10 watt rig. It should be built in a Pomona Electronics

#2391 or similar small metal box with appropriate connectors. Do not use wire-wound resistors, as their inductance prevents a good match to the transmitter. Keep resistor leads as short as possible. You may want to put fins of some sort on the box if it gets too hot.

There is an alternative to the attenuator box which is particularly suitable for higher powered transmitters at vhf/uhf. A suitable length of small coax will provide a calibrated loss. The use of the coax as an attenuator for SWR reduction has the additional advantage of allowing the transmitter to be remote from the antenna.

Tables 14-1 and 14-2 give data on common coaxial cables which will help you determine their suitability for matching or attenuating the hidden T signal. Listings are in descending order of loss at 2 meters. Attenuation is given for four vhf/uhf bands in dB per 100 feet of cable. The loss in dB at any frequency is directly proportional to the length. For example, 100 feet of RG-8/X has 4.5 dB typical loss at 2 meters. A 200 foot length has 9.0 dB loss in that band. For 100 watts into the 100 foot length, 35.5 watts will reach a matched load at the end of 100 feet, and 12.6 watts is left at the end of 200 feet.

To figure out how much RG-8/X it would take for 7 dB attenuation at 2 meters, divide the table value by 100 to get loss per foot, then divide that number into 7 dB. The answer is 155 feet. A 7 dB pad of tiny RG-174A/U is more practical, as only 52 feet is needed at 146 MHz if the cable has maximum loss.

A few words of caution about these tables is in or-

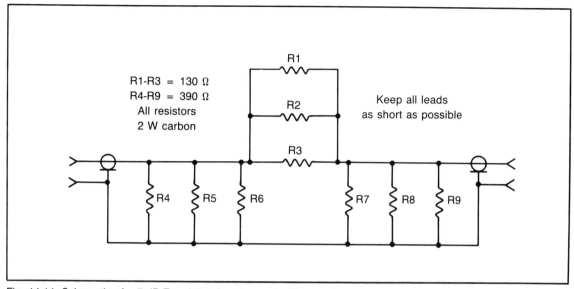

R1-R3 = 130 Ω
R4-R9 = 390 Ω
All resistors
2 W carbon

Keep all leads
as short as possible

Fig. 14-14. Schematic of a 7 dB T pad that handles up to 10 watts through the amateur 2 meter band.

Table 14-1. Mil-spec Coax Characteristics.

COAX TYPE	Max. Loss per 100 ft, in dB				Dc resistance per 100 ft.	Nominal Impedance (ohms)	Propa-gation Factor	Max. Power @ 146 MHz
	50 MHz	146 MHz	223 MHz	440 MHz				
RG-174A/U	6.6	13.5	17.0	26.5	9.670	50	0.66	42
RG-58C/U	4.0	8.6	11.7	18.0	1.480	50	0.66	118
RG-223C/U	4.8	7.7	9.3	12.4	0.897	50	0.66	160
RG-62A/U	2.8	4.8	6.0	8.5	4.400	93	0.83	198
RG-59B/U	2.2	4.6	5.9	9.6	4.800	75	0.66	240
RG-214/U	1.7	3.1	4.0	5.8	0.173	50	0.66	740
RG-11A/U	1.4	2.7	3.6	5.6	1.480	75	0.66	580
RG-213/U	1.5	2.7	3.4	5.1	0.173	50	0.66	640

Table 14-2. Commercial Coax Characteristics.

BELDEN PART NO.	COAX TYPE	Nominal Loss per 100 ft, in dB				Nominal Impedance (ohms)	Propa-gation Factor
		50 MHz	146 MHz	223 MHz	440 MHz		
9258	RG-8X	2.5	4.5	5.7	8.5	50	0.78
9251	RG-8A/U	1.6	2.8	3.4	4.9	52	0.66
8214	-	1.2	2.3	3.0	4.4	50	0.78

der. Table 14-1 is for cables built to military specifications, and will help you use that big roll of government surplus coax you found at the last flea market. The data comes from MIL-C-17F, which is a government specification. The attenuation values are specification maximums, meaning that your cable will have less attenuation than these values, provided it has not deteriorated with age or wear.

Old coax, particularly if it does not have non-contaminating jacket material, may have even more loss than the table value. If the exact value of attenuation is important, the coax should be checked for loss with a power meter by measuring power at the input and the output of the line with the output terminated with a matched dummy load. When the loss is 7 dB, the power at the end of the line is one fifth the power going into it. The measurement must be made in the same frequency band as will be used for the hunt.

Notice that there is no current specification for RG-8/U cable, which has been replaced with RG-213/U for military applications. Any RG-8/U cable made today is not controlled by military specs, and you never know what you're getting when you buy cable marked RG-8/U or RG-8A/U unless it is identified in some other way, such as Belden #9251. Table 14-2 gives data on some popular 50 ohm non-military coaxes from the Belden catalog, including RG-8/X. The attenuation numbers in Table 14-2 are nominal, not maximum.

Other information in the tables may be useful in T-hiding. The propagation factor is important for calculating the length of fractional wavelength matching or phasing lines. The dc resistance is of interest if you want to use your surplus coax as a long dc power cord. RG-214/U and RG-223/U have two shield braids, and will have less signal leakage along the length of the cable than types with single shields.

OTHER ANTENNA POSSIBILITIES

A discussion of antennas for hiding would be incomplete without mentioning some of the avant-garde aerials that advanced hiders have considered. We don't mean to imply that this is all of them, as someone is always trying something new. Avid T-hunters always look at new antenna products and ideas with an eye toward their applications for creative hiding. Be sure to check the rules of the hunt to be sure these or any other oddball radiators you have in mind are allowed.

The Distributed Antenna

Hunters expect the antenna to be in one place. Even a multi-element antenna such as a ZL Special gives a single bearing to a sniffer. But what if the elements are spaced many yards apart? Consider an antenna system made up of several individual concealed dipoles or quarter-wave spikes, spaced at 25 to 50 foot intervals in

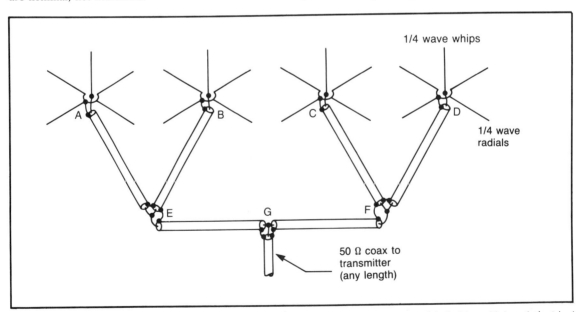

Fig. 14-15. The distributed antenna consists of several radiating elements spaced out and fed with multiples of electrical quarter wavelengths of 75 ohm coax.

a line or some other pattern, as shown in Fig. 14-15. No difference will be evident at the starting point, but the multiple sources will cause lots of grief when the hunters get close enough to be inside the antenna system.

Dual antenna RDF systems which depend on phase or time-of-arrival information may become completely confused when nearly equal strength signals are received from widely different directions. Null-seeking RDFs will find there is no null to seek. Beam or quad sniffers will probably fare best, but will have to contend with multiple bearings.

Four ground planes in parallel are not a good match to the transmitter, so the feeders for each ground plane (AE, BE, CF, DF) are made of 75 ohm coax such as RG-59, and are odd multiples of an electrical quarter wavelength. At feeder junctions E and F, each whip impedance has been transformed to be in the neighborhood of 100 ohms. When paralleled, each pair (AE/BE and CF/DF) looks like close to 50 ohms again. This is also true of pairs of pairs, so the whole four element system looks like 50 ohms at junction G to the single coax from the transmitter. Another four element block could be connected to form an eight element pyramid by using two more 75 ohm matching sections, and so on, and so on . . .

At 2 meters (146 MHz) the quarter wavelength driven elements and radials are 19-1/4 inches long. A quarter wavelength of RG-59/U is two thirds this length, due to the velocity factor of the coax (per Table 14-1), so the feeders can be any odd multiple of 12-3/4 inches, such as 12-3/4, 38-1/4, 63-3/4 inches, up to 28 feet 8-1/4 inches or even more. Even multiples won't work.

The Lossy Coax Antenna

An even more cunning form of the distributed antenna technique, the lossy coax antenna has no discrete radiators. A long piece of coax is terminated at the far end, and then made to leak signal along its entire length. Depending on how much radiation is required, this can be done by spreading or slitting the braid at intervals or continuously. Some inexpensive coax has such poor shield coverage that no modification may be required!

Higher transmitter power may be required with this system, because it is a very ineffective radiator. In wooded areas, stringing the coax through trees overhead allows the system to get out better than if the coax lies on the ground.

This antenna setup may be even harder to sniff out than the distributed antenna, provided that it is arranged so that as the hunters approach it from the most likely direction, they are equidistant from almost any point on

the coax. If the coax is laid out in a straight line and a hunter approaches from the direction of the far end, he will probably find it easily. The transmitter should be buried or well shielded, or else there may be more radiation from its case than from any point on the coax, which would lead the hunters directly to the transmitter.

Antenna Switching

Instead of radiating simultaneously from several widely spaced antennas, how about switching among them? If the antennas can be separated by some natural boundary—a river or creek for example—it is much harder for the hunters to close in. Signal amplitude variations can be masked by synchronizing the switchover with changes in the tonebox audio.

Several surplus merchants sell four-way coax relays with BNC fittings, suitable for vhf use. The relay has four coils, one for each output. Relays with 26 volt coils often are usable at 13.8 volts, and draw about 45 milliamperes at that voltage.

A schematic diagram for connecting such a relay to the Un-Music Box is given in Fig. 14-16. The transistors can be any silicon type capable of driving the relay current. The four coils are energized in sequence, with changeover occurring every few seconds, coincident with a tone sequence change.

Two cautions:

☐ Do not attempt to "hot switch" high levels of rf power with these relays.

☐ The transmitter should be capable of handling momentary high SWR during the switching.

In this section we have dealt specifically with the application of various antennas and matching systems to transmitter hiding. A number of publications cover the more general techniques of building and feeding antennas, and can guide you if you're new to antenna making. Look into them. (They may also give you some more ideas!)

POWERING THE TRANSMITTER

To decide what battery power system to use for your hidden T, first make an educated guess of the length of the hunt. (Typical fun hunts run about three hours maximum, but much longer hunts are possible, depending on the rules, the distance, and the skill of the hider and hunters.) You also have to know the rig's power drain and other power needs.

Let's try an example of a typical Sunday afternoon

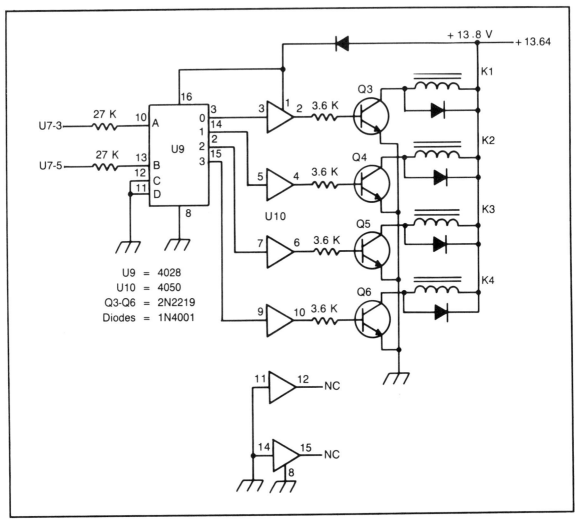

Fig. 14-16. Schematic of relay driver for antenna switching to be used with the Un-Music Box tone generator.

repeater input fun hunt with an intermittent (fifteen seconds per minute) tone transmission from a one watt portable set. Most rigs of this type, whether handheld (such as the Icom IC-2) or shoulder carried (such as the Kenwood TR-2200), require about one half ampere while transmitting and 130 milliamperes (mA) when receiving unsquelched. Since the hidden T will be transmitting one quarter of the time and listening to the repeater the remainder of the time, we can compute the average current drain.

$$500 \times .25 = 125.0$$
$$130 \times .75 = \underline{97.5}$$
$$222.5 \text{ mA}$$

Add to that 25 mA for the tone generator and timer for a total average drain of about .25 ampere. Multiply that by the length of the hunt to get close to three quarters of an ampere-hour (Ah). Lct's put in a 25 percent safety factor to allow for chatter before and after the hunt, and to be sure that the output power doesn't drop radically at the end of the hunt. The result is one Ah. That's about double the amount available from the typical AA size nickel-cadmium (Ni-Cad) battery packs for these rigs.

In this case, some form of external power source is required. There are many possibilities. We'll discuss the most promising, including applicability to various types of hunts.

Nickel-Cadmium Batteries

Ni-Cad batteries are today's most popular form of rechargeable power. Almost everyone has some gadget that uses Ni-Cads. They pack a lot of capacity in a small volume, can support high discharge currents, and last for hundreds of recharges if properly used.

Although multiple cell packs are available, it's easy to assemble your own using standard cell sizes, which range from 1/2-AAA to D and larger. In our example, a pack of industrial C cells, typically rated at 1.8 Ah, give plenty of reserve power for our T-hunt. Be aware, though, that the Ah rating is based on a ten-hour discharge cycle. A 2.0 Ah cell lasts for about ten hours at 200 mA, but only about 9/10 hour at 2.0 amperes.

The rated terminal voltage of a Ni-Cad cell is 1.2 volts. This is lower than the 1.5 volts of a flashlight cell. Ten cells in series are needed for a 12-volt transmitter, as compared to eight flashlight cells. After a full charge and a half-hour open circuit rest, the standing voltage of a cell is 1.35 volts, but through most of the discharge period the voltage is approximately 1.2 volts. When the cell voltage quickly drops to 1.0 it is discharged, and the load should be removed immediately.

Avoid the blister pack Ni-Cads that are sold to replace throw-away batteries in radios and toys. Both the C and D cells of these battery lines have only 1.0 to 1.2 Ah capacity, compared with up to 1.8 Ah for C and up to 4.0 Ah for D industrial Ni-Cads. Table 14-3 lists the part numbers of popular consumer Ni-Cad batteries.

Table 14-4 details the typical capacities of the more desirable industrial types. There are many sizes of cylindrical cells, and the table will help you find the capacity of unmarked surplus cells which frequently can be found at attractive prices. Some manufacturer's cells have more capacity than listed, and some have less. Be prepared to pay a premium for new industrial Ni-Cads, which sell at net for well over twice the price of consumer cells.

Use care in soldering the cells into packs. Get batteries with soldering tabs if possible. Keep soldering heat to a minimum and do not solder over the small hole in the positive terminal. It is a vent; if it's capped, cell rupture could result.

Ni-Cad batteries have higher self discharge rates than lead-acid types, particularly at high temperatures. A typical pack loses about 50 percent of its charge in three months. Charge your pack right before the hunt to avoid an unpleasant surprise.

The biggest killer of Ni-Cad batteries is cell reversal. If one cell in your pack goes to zero volts before the others during the hunt, the load current into the transmitter tries to charge it in reverse polarity. The result is usually a dead shorted cell. Though these shorts can sometimes be "zapped" out, the full capacity probably won't be restored. That cell is likely to short again during a future discharge cycle.

The charging of Ni-Cad batteries can be a complex topic, and there is a lot of information, and misinformation, in print about it. The easiest way to get a fast charge and protect them is with a voltage regulated current limited power supply, such as a bench supply. The open circuit voltage should be set at 1.43 times the number of cells in series, and the current limit at up to about 3C. (The expression 3C means three times the Ah rating of the individual cells.) This current limit determines the initial charge current and depends on the quality of the batteries and how fast a charge is needed. The current should be monitored and charging stopped when it falls to less than 0.1C.

For occasional T-hiding use, slow charging works well. Most wall chargers sold for Ni-Cad packs give a 15 hour slow charge at 0.1C. Do not confuse this with trickle charging, which is only meant to sustain a charge and is done at even lower rates. The voltage should be monitored during slow charging, and the charging stopped when the voltage reaches 1.43 times the number of cells. This is particularly important when charging a pack that is only partially discharged.

The basic circuit for slow charging is given in Fig. 14-17. V_{in} comes from a dc source that is several volts higher than the pack voltage (V_{bat}). The desired charge

Table 14-3. Part Numbers of Consumer Ni-Cad Batteries.

SIZE	CAPACITY (Ah)	RAY-O-VAC	GE	DURACELL	RADIO SHACK	EVEREADY
AA	0.5	615	GC1	NC15AA	23-125	CH 500
C	1.2	614	GC2	NC14C	23-124	CH 1.2/C
D	1.2	613	GC3	NC13D	23-123	CH 1.2/D

Table 14-4. Capacities of Industrial Ni-Cad Batteries.

SIZE	CAPACITY (Ah)	LENGTH (in)	DIAMETER (in)
1/3-AA	.1	.705	.588
1/2-AA	.25	1.28	.588
AA	.5	1.953	.588
1/3-A	.15	.657	.657
A	.60	1.965	.657
1/2-SUB-C	.55	1.047	.906
SUB-C	1.2	1.646	.906
2/3-C	1.0	1.303	1.036
C	1.8	1.858	1.036
1/2-D	2.2	1.465	1.303
D	4.0	2.346	1.303
F	7.0	3.488	1.303

current in amperes (I_{chg}) determines the value of R by the formula:

$$R = \frac{V_{in} - V_{bat} - 0.7}{I_{chg}}$$

For example, for an eight cell pack of AA cells, charged at 0.1C from a 13.8 volt base station supply:

$$V_{bat} = 8 \times 1.2 = 9.6 \text{ V}$$
$$R = (13.8 - 9.6 - 0.7)/0.05 = 70 \text{ ohms}$$

Power in the dropping resistor R is given by $P = I_{(chg)}^2 \times R$. In this case R dissipates only a fraction of a watt normally, but if the leads to the battery are ever shorted, the dissipation is almost 2.5 watts. A resistor capable of handling this power should be used.

You may have heard that repeated shallow discharging results in memory in Ni-Cads, but don't worry about it. The odds are overwhelming that you'll never encounter memory from that cause. But if you overcharge the battery for a long time, a premature voltage depression occurs upon discharging that appears to be just like the memory effect. For this reason be sure to monitor the slow charge and stop it when complete. Finally, *do not* deliberately discharge your pack to zero volts every time before charging, as some articles recommend. It doesn't do any good, and you'll risk damage from cell reversal.

Build a Fast Charger

One simple way to set up a miniature vhf hidden transmitter is to use a handheld. The problem is how to power it externally for a long hunt. This regulated fast charger powers the HT continuously in either receive and transmit mode from an external vehicular supply. When not transmitting, it applies a tapered charge to the batteries that brings a dead pack to full charge in less than an hour, and then keeps the battery charged and ready to go.

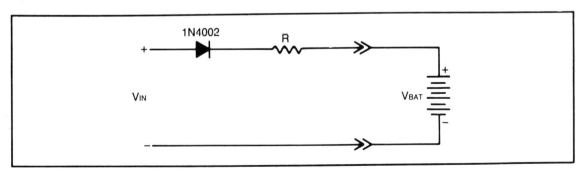

Fig. 14-17. Circuit for slow charging batteries.

The charger is designed for sets like the Tempo series which use eight cell packs of AA or half-AA size cells with an external charging connection like that shown in the schematic (Fig. 14-18). Other configurations and voltages can be accommodated, but may in some cases require rig modifications. The current limit protects the charger from the short which inevitably occurs when it is plugged into the rig, while the diode prevents the battery from being shorted by the plug or discharged into the charger when power is off.

An unusual pnp transistor (Q1) with very low V_{CE} at saturation allows proper regulation with input voltages only a fraction of a volt higher than the output voltage. Such a situation occurs when trying to charge or maintain the pack from a car battery with the engine off. Q1 is available from Motorola MRO (formerly HEP) distributors, and should not be substituted if you need the low offset voltage feature of this regulator. It should be mounted on a heat sink with at least 8 square inches of surface area, and more if a higher voltage input is used.

You must use the 14-pin DIP packaged 723 regulator IC at AR1 instead of the 10-pin can version, which is missing an important internal connection. Luckily, the DIP version is most commonly available (RS 276-1740).

The other parts, except for Q1 and M1/shunt, should be easily obtained.

The meter movement is 100 mA full scale, with a shunt made of nichrome wire or low-value resistors in parallel to give a switchable 1 ampere range. For best regulation and minimum voltage offset, combined meter and shunt internal resistance should be less than 2 ohms on the high current range. A good surplus military type meter is recommended.

The short circuit protection diode in the S1 battery pack drops 0.71 V at 25 mA. R1 should be set for (1.43 × N) + .71 volts unloaded, measured at Q1 collector. In the Tempo S1 case the number of cells (N) is eight, so the output should be 12.15 volts. If you don't have a digital voltmeter, you can set R1 by another method: Charge the pack with the slow charger supplied with the rig for 14 hours, then unplug it. Set the output of the regulator for about 10 V, then connect it to the rig in the taper charge mode and slowly adjust R1 upward until the current meter reads about 15 mA.

Current limit can be checked by hooking a 10-ohm power resistor, 10 watts or more, to the regulator output and reading the current meter. It should be 750 mA plus or minus about 100 mA. R2 can be changed appropriately

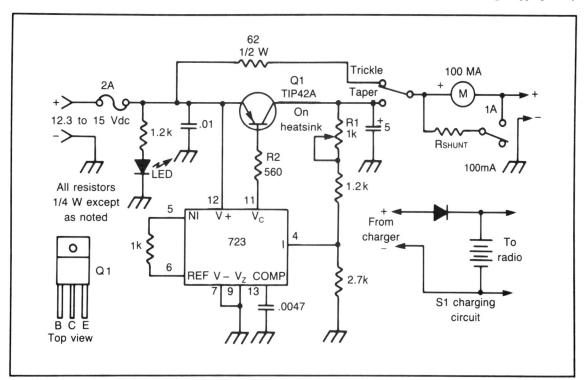

Fig. 14-18. Schematic of fast charger for small Ni-Cad packs, with Tempo handheld charging circuit.

211

to get the current limit into this range. Exact current limit value isn't critical.

Depending on how you package the unit, you may need to take measures to keep rf out of the regulator. If regulation problems occur when transmitting, try a .01 microfarad bypass capacitor to ground right at the collector of Q1. A capacitor of the same value directly between pins 12 and 7 of AR1 may also help.

For use with handhelds other than Tempos, be sure there is a diode between the regulator and the battery pack to prevent damage to the regulator. If not, add one at the charger output. Without the diode, the battery discharges back into Q1 and AR1 when input voltage is not present. Use a one ampere or greater device such as the 1N4001.

For Icom 6-, 7-, or 9-cell packs, R1 and R3 may have to be changed to set the voltage per the formula above. The higher voltage pack requires an input voltage to the regulator of at least 13.7 V. At the lower output voltages, check to be sure Q1 doesn't overheat during fast charging, as it will dissipate more. The charger should be connected through the output diode to the two screws on the bottom of Icom battery packs (watch polarity) instead of any charging input, to bypass the current limit resistors. The screws go directly to the battery terminals, and may make a do-it-yourself charging stand practical.

Don't charge your Ni-Cad batteries in extremely hot or cold temperatures, as damage to the cells may result. As with any Ni-Cad charger, it's best to disconnect it after charging is complete unless the radio is in use, to prevent voltage reduction due to the sustained overcharge effect mentioned earlier.

If you power your handheld rig for a transmitter hunt with this charger, remember that for a long period of time it presents the transmitter with a voltage equal to or higher than the maximum it will ever see from a fully charged battery. Is there adequate heat sinking on the rig's final stage transistors? If in doubt, use the radio's low power mode for the hunt.

Lead-Acid Batteries

Ni-Cad battery systems are best suited for hunts which require 4 Ah or less of battery capacity. Above that they become quite expensive, and lead-acid batteries become quite attractive. If you stay in your car or don't mind lugging a 35-pound car battery around, you can have enough power for a weekend-long hunt with a one watt rig or a day-long hunt at 10 to 25 watts without recharging.

Regular automotive batteries provide up to 100 Ah

when in good condition. Chargers are available at auto parts stores or are easily made. But besides being heavy, car batteries contain acid, which can cause holes in clothing and skin burns if you don't use a lot of care. Ordinary car batteries are not intended for many full charge-discharge cycles. If you want the full battery capacity for long hunts often, consider one of the special deep discharge batteries sold for marine or recreation vehicle use (at a higher price, of course!). For a smaller system, look into motorcycle batteries.

Sealed lead-acid (SLA) batteries are an attractive alternative to automobile batteries. A 10 Ah pack can be bought for about forty dollars, and may suit most of your hiding needs. Such a pack is about a fourth of the volume of a standard car battery and weighs less than ten pounds. Because it is sealed, it can be carried about easily, and mounted in any position. If you're careful not to crack the case or severely overcharge it, you'll never have to worry about acid spills. Also, deep cycle discharge performance is much better than car batteries.

SLA batteries are available from a number of companies. The electrolyte is gelled in Globe Gel/Cell® and Gould Gelyte® batteries, but liquid in most others. Yuasa type NP batteries are available in sizes from 1.2 to 24 Ah at some local parts houses. Panasonic has almost 40 different SLA battery models, in various form factors, with capacities from 1.2 Ah to a whopping 120 Ah unit that weighs 115 pounds. Cyclon® cells by Gates are made in a cylindrical form from 2.5 Ah (D size) to 25 Ah. Information on these batteries and how to charge and use them is available by writing to the manufacturers (addresses are in the back of this book). Ask for their battery application manuals.

The constant voltage current limited bench supply mentioned in the Ni-Cad section also makes a good fast charger for SLA packs. Open circuit voltage is set to between 2.4 and 2.5 volts per cell, resulting in 14.4 to 15.0 volts for a 12 volt pack. Current limit should be 0.25C to 0.4C. The various manufacturers disagree somewhat on these values, so be sure to check the specifications for the brand of cells you buy. For instance, Gates Cyclon cells are capable of initial charge rate of 4C or higher in a constant voltage charge system.

Fast charging must be done with care in packs with cells in series to get even charging. Do not use heavy duty automobile battery chargers with small (20 Ah or less) SLA batteries. They may cause overheating and permanent damage. Slow charging can be done with the same scheme as for Ni-Cads (Fig. 14-17). The nominal voltage is 2.0 volts per cell in lead acid batteries. Remember to allow plenty of safety factor in the rating of the dropping resistor.

Although some SLA batteries can be stored in a discharged condition without damage, others cannot, so it is a good idea to always recharge your pack promptly after the hunt. Panasonic tested its SLA battery design by discharging one at 122 degrees Fahrenheit for four days, then letting it sit at that temperature for four weeks. After a constant voltage charge-up, the battery was discharge tested to determine how much of its capacity was recovered. It was good for 90 percent of rated capacity. Automotive type batteries don't fare nearly that well. One accidental long term discharge probably won't spell the end of your SLA pack, but don't make it a habit, because the life is shortened each time you abuse it this way.

Unlike automobile batteries, self discharge on the shelf is not a problem with most SLA batteries. Yuasa states that their SLA packs self discharge less than 0.1 percent per day at 20 degrees Celsius. But there is a rapid increase in self discharge at higher temperatures. Don't keep your SLA pack in the garage in the summer. It might even be wise to keep it in the back of your refrigerator if you don't expect to be hiding again for several months. (Oh come now—you're a better hunter than that!) Check and recharge as required every six months or so of storage and before use.

SLA and Ni-Cad batteries are quite safe normally, but store a lot of potential energy. As with any battery capable of high currents, take extra care to avoid shorts across the terminals. Put an in-line fuse right at the positive terminal of the pack. Without such a fuse, a wiring short could cause the cable to the rig to burn. Tape the terminals up so that batteries cannot fall over and short, or allow something metallic to short them out. Lead acid batteries should never be operated in a tightly closed container, as hydrogen gas may seep from the vents and create an explosion hazard.

Primary Batteries

The rechargeable batteries we have been discussing are often called secondary batteries. Non-rechargeable batteries, on the other hand, are primary batteries. Primary batteries may be useful for transmitter hiding, particularly if you hide transmitters only occasionally.

Ordinary carbon-zinc (Leclanche) batteries leave a lot to be desired. Their main disadvantage is their significantly reduced life at high currents. We should not use C size Leclanche cells for the T-hunt in our previous example because the terminal voltage falls too rapidly with the 500 mA transmitter drain. Even though there is some

recovery during periods when the transmitter is off, the cells will not last through the hunt. Even conventional D size cells would be marginal for this application.

Because of this tendency for decreased life at higher loads, carbon-zinc cells are not given ampere-hour ratings. Representative data on carbon-zinc batteries is given in Table 14-5. The table can't be used to determine exact life because the number of hours of use per day is a factor, but it gives you a general idea of their suitability for your planned hunt. Because the terminal voltage falls off steadily during use, a rated load resistance is stated instead of a rated current.

The heavy duty cells have a slightly different electrolyte and provide up to double the life at a given current drain compared to conventional cells. Performance of all Leclanche cells falls off rapidly at low temperatures, so don't use them in freezing weather. Buy them fresh before the hunt, because their shelf life is limited to 18 to 24 months, and is much less at higher temperatures.

Alkaline batteries, which use potassium hydroxide electrolyte and different construction, are superior to conventional cells in almost every way. The terminal voltage remains high throughout the discharge period. They operate down to about −25 degrees Celsius. They maintain 90 percent of their capacity after a year on the shelf. Best of all, they provide two to ten times the life of conventional cells when used in high or continuous current applications, such as powering transmitters.

Table 14-6 shows the typical performance of alkaline batteries. In this table, cells are considered discharged when the terminal voltage falls to 0.9 volts per cell. For our sample hunt we could use a pack of nine C size cells for 13.5 volts initially, falling to 8.1 V at end of life. However, the transmitter output will probably become unacceptably low at 10.5 V or so. This can be compensated for by monitoring the voltage and adding more cells in series as the hunt progresses.

Since this alkaline pack costs between eight and eleven dollars retail, it won't take many hunts before rechargeable cells begin to pay their way. By the way, in most stores C and D size alkaline batteries are sold at the same price, so get the D size cells to get more transmissions for your money.

REMOTE CONTROL

Suppose you want the advantage of a completely unattended transmitter, but the hunt rules call for you to be able to turn it on and off at will, or be able to talk on it. A remote control setup to do that sounds complicated, but can be done quite simply. You can conceal yourself

Table 14-5. Primary Battery Capacities.

SIZE	LIGHT LOAD			HEAVY LOAD		
	R (load) (OHMS)	DUTY HRS/DAY	LIFE HRS	R (load) (OHMS)	DUTY MIN/DAY	LIFE MIN.
N	- - -	- - -	- - -	15	10	200
AAA	- - -	- - -	- - -	15	10	320
AA	10	1	4.8	5	30	150
AA HD	10	1	5.8	5	30	195
C	75	4	145	4	60	275
C HD	10	2	22	2	30	200
D	40	4	155	4	60	720
D HD	10	4	54	2	60	540

a hundred feet or so from the bunny and watch all the fun.

R/C Gear For Control

If you have or can borrow the transmitter, receiver, batteries, and servos for radio control of models, you can make an excellent short range control system for simple commands. Modern R/C gear output is proportional shaft movements, not relay closures, so an interface unit must be constructed to get the switch closures needed for hidden transmitter control. Figure 14-19 shows how it's done at the hidden T end of the R/C link. The servos are mounted on a wooden frame so that their movement actuates microswitches.

You can use as many servos as you have channels in the R/C equipment. In this case there are three. Four switches are used to control The Un-Music Box. Three of them are shown being controlled by two servos in the closeup of Fig. 14-20. Channel one actuates two switches in parallel with S6, selecting either OFF, CONTINUOUS ON, or 15/45 SECOND INTERMITTENT transmitting. One switch is actuated by the channel two servo. It is used to manually ID the transmitter, wired in parallel with S5. The switch on the channel three servo is a backup, turning the whole setup on and off.

The R/C transmitter doesn't have to be on for the duration of the hunt, and it's not necessary to hold an R/C joystick in one position for long periods. The R/C transmitter is turned on only long enough to set the servos for the mode of operation desired, or to ID the transmitter. It can then be turned off until needed again. The servos remain in position when the R/C signal disappears.

Use of R/C gear need not be limited to controlling the hidden transmitter. For one memorable night hunt a remotely controlled decoy was built. A second noisemaker, identical to the one providing audio for the hidden transmitter, was connected to an audio amplifier and speaker and placed about fifty feet away under a bush. A fake antenna was stuck alongside. Any time a hunter began to sniff near to the real bunny, the R/C link was used to turn on the audio-only bunny to lure him away. Hunters who knew their sniffing gear well weren't fooled for long, but several others took the bait.

Table 14-6. Alkaline Battery Capacities.

SIZE	RATED DRAIN (ma)	R (load) (ohms)	CAPACITY (hrs) into R (load)
N	25	50	19
AAA	25	50	27
AA	130	10	12
C	300	4	12.5
D	320	4	31

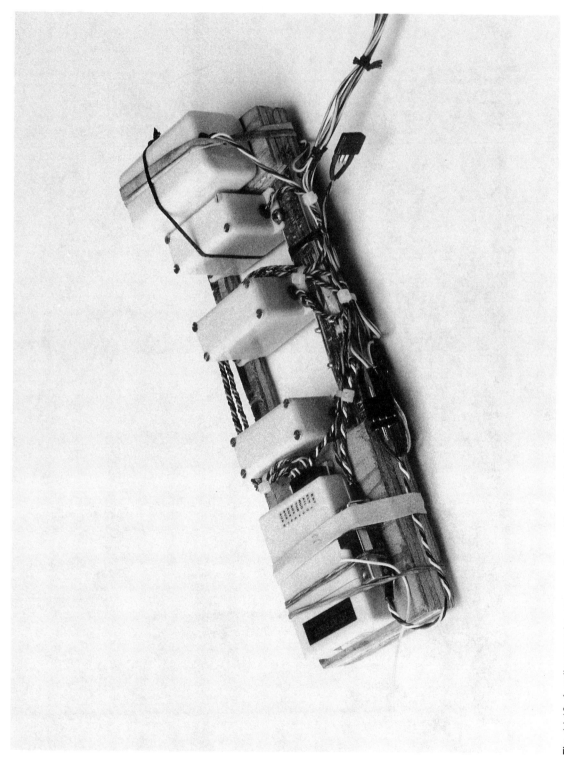

Fig. 14-19. A radio control system can be used to control the hidden T transmitter, tone box, or antenna connections.

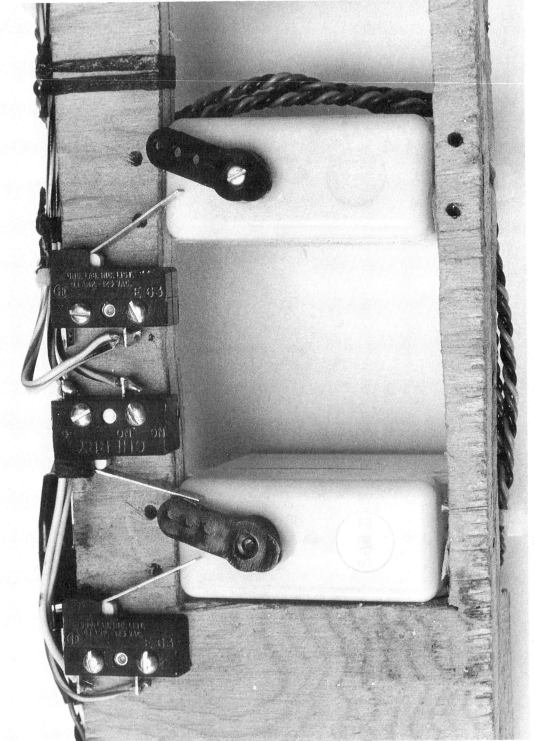

Fig. 14-20. The radio control servos are used to open and close individual micro switches.

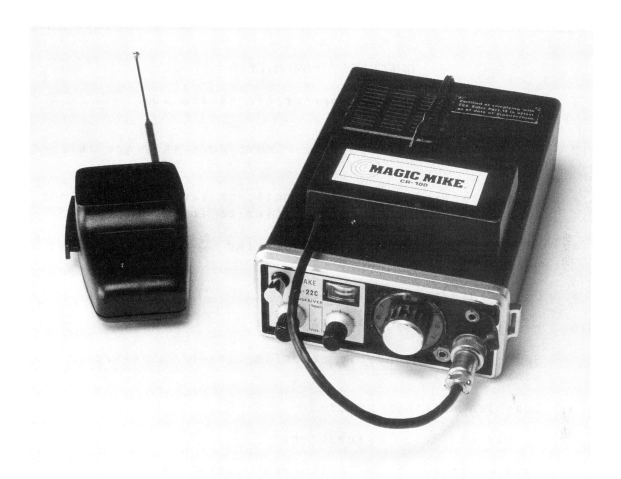

Fig. 14-21. A remote mike can be used to remotely talk through your T-hunt radio.

Cordless Microphones

The easiest way to talk remotely through the hidden T is with a commercial cordless microphone link. The wireless broadcasters which are sold as children's toys, and even most professional cordless mikes, are not suitable. They aren't designed to control the push-to-talk circuit of a transmitter. Shop carefully, however, and you can probably locate a cordless mike system made for the Citizens Band market which can easily be used with a vhf-FM rig.

These wireless mikes are intended to allow use of the dash mounted radio from anywhere inside a camper or from the driver's seat without getting the cord tangled in the steering wheel. The link is crystal controlled and uses frequency modulation. The transmitter is inside the hand mike. Figure 14-21 shows the system, sometimes sold at hamfests for about $30. Range in the clear is 100 feet or more with fresh mike batteries, which last for many hunts. There's hardly any telltale squelch tail.

Note that in this discussion the term "control link" has been carefully avoided. FCC rules 97.3(1), 97.86, and 97.88 deal with control links and auxiliary operation, calling out some special regulations for such uses, such as restricted link frequencies. As long as you're at the same "address" as your hidden transmitter and can easily walk over and shut it off if something goes awry, you're still in local control and these stringent rules don't apply to your cordless mikes and R/C control lash-ups.

SOME FINAL THOUGHTS ON HIDING

Once you've found the spot, and decided what equipment is needed, the hardest part is over and the fun begins. Get everything together and head out early for the hiding location. The checklist of Fig. 14-22, while not

```
┌─────────────────────────────────────────────────────────────────┐
│                     CHECKLIST FOR HIDERS                          │
│                                                                   │
│  1.  Transmitters(s) with appropriate cables and                 │
│      microphone.                                                  │
│                                                                   │
│  2.  Batteries and/or regulator.  Make sure they are fully        │
│      charged.                                                     │
│                                                                   │
│  3.  Noisemaker and/or timer/ID unit, or tape recording and       │
│      tape.  Don't forget fresh batteries for these items.         │
│                                                                   │
│  4.  Antenna and supports (pipe, string, C-clamps, baling         │
│      wire, duct tape, etc.).                                      │
│                                                                   │
│  5.  Coax cable and SWR meter.                                   │
│                                                                   │
│  6.  Basic antenna assembly tools.                               │
│                                                                   │
│  7.  Special items (shovel, camouflage paint, traffic            │
│      cones, etc.).                                               │
│                                                                   │
│  8.  Trophies or other prizes.                                   │
│                                                                   │
│  9.  Liquid and solid refreshments.                              │
│                                                                   │
│  10. Remote control gear, with fresh batteries.                  │
│                                                                   │
│  11. Paper, pencil, calculator.                                  │
│                                                                   │
└─────────────────────────────────────────────────────────────────┘
```

Fig. 14-22. Checklist for hiders.

guaranteed to be complete for every hunt, may help prevent you from forgetting an important item. Don't forget to maintain radio silence on the way to avoid giving a free triangulation bearing to any hunter.

Plan for any contingency when you hide. If you have extra rigs, antennas, or batteries, bring 'em along. You might find they'll be needed for one reason or another. Before starting the hunt, give any special instructions. Must hunters find the transmitter or the antenna? Will there be any special instructions at the transmitter? What frequency will be used to communicate with the fox during the hunt? If time is a factor, be sure to declare when the hunt begins, so all can start simultaneously.

Many hiders do not acknowledge on the air when or if they have been found until the hunt officially ends. This keeps the level of interest up. Certainly the callsigns of those who have finished should not be announced during the hunt, as their cars might have been spotted by other hunters, giving them a clue as to the approximate location.

Chapter 15

The Bunny Box: A Cigar Box Sized Rig for Hiding

A self-contained low power transmitter is worth its weight in GaAs-FETs to the active amateur DFer. Besides being quick and easy to conceal for sport hunts, it can serve as an rf source for antenna pattern testing. It also can be used in demonstrations of RDF units and T-hunting to radio clubs and prospective hams.

The synthesized 2 meter transmitter of Fig. 15-1 measures 7-1/2 × 5-1/4 × 1-5/8 inches and is complete with batteries. It has been used for a number of creative hunts where high power was not needed. It has been buried on a hill, with the AWG 30 antenna wire running up the side of a survey marker, looking just like a crack in the wooden stake. It has been concealed in a portable outhouse and hung from a freeway overpass. It was even hooked to a rubber duckie and strapped onto a carnival ferris wheel. (No wonder the signal went up and down!)

With a transmitter like this one, you can always be ready to hide when called upon. Since individual needs and desires vary, we won't describe this transmitter as a build-like-the-book construction project. Instead, we'll go over the considerations and some ideas for you to use in making up a portable bunny for your own use, and give you the schematics as a starting point.

PHYSICAL LAYOUT

Twenty-five years ago, it was popular to build small construction projects in cigar boxes. Good cigar boxes, particularly wooden ones, are hard to come by nowadays. A surplus pack which used to house D lead acid battery cells was used in this model. Something waterproof might be a better choice, but it's just as easy to seal the entire transmitter in a plastic bag for burial or wet weather use.

The size is determined in large part by the batteries, which should be chosen in accordance with your needs and the formulas of Chapter 14. Ten 1/2-sub-C size Ni-Cads are shown here, rated at 0.7 Ah. They power the rig for about four hours continuously, depending on the output power setting. Considerably more time is available in the intermittent mode.

AUDIO AND TIMING

A very simple 16-step tone generator was used with this unit, to conserve space in the box. Figure 15-2 shows the schematic of the circuit. A TLC556 CMOS dual timer (RS 276-1704) is recommended for minimum battery drain. For the same space-saving reason, no microphone

Fig. 15-1. When closed, it looks like an ordinary black box from most angles, but the miniature transmitter is complete and ready to hide.

or ID circuit was included. Instructions and IDs can easily be given with a separate handheld transceiver. To give CW identification, kill the tones by grounding the junction of R13 and R14, then key the battery line to the unit.

Signals from the Q6 and Q7 outputs of U2 are used to key the transmitter at a 25 percent duty cycle for intermittent signal hunts, as explained in the section on power and keying. With the values shown, the transmitter is off for 45 seconds and then on for 15 seconds. R1 and R2 can be changed to speed up or slow down the tone sequence, but the on and off times in the intermittent mode change accordingly. Many other different sequences can be selected by exchanging resistors R5-R10. Divider R13/R14 steps down the tone output to about a half volt peak-to-peak to drive the synthesizer VCO. A separate phase modulator is not needed.

A 2-3/8 × 1-5/8 inch piece of perf-board contains the circuit, in the lower left corner of the box in the photo. A wiring pencil is a convenient way to wire it up. Wire wrap is also fine if you can spare the room under the board. Use sockets for the ICs and do not install the chips until wiring is complete, to prevent static damage.

THE RF SYNTHESIZER

With so many different hunt groups in the area, it was important to us to be ready to transmit on any 2 meter frequency. This synthesizer circuit allows your Bunny Box to do just that. It uses readily available CMOS ICs to produce 400 frequencies or more while drawing only about 25 mA.

The synthesizer was originally designed by Dale Heatherington, WA4DSY, for use with the Motorola

HT-220 handheld transceiver, and was later adapted by K0OV for the Drake TR-33C. It has been modified slightly here to produce an output at one-ninth of the 2 meter transmitting frequency. It uses only two crystals, three ICs, seven transistors, and a few miscellaneous small parts. Any 5 KHz-spaced channel in a 2 megahertz range of the 2 meter band (146.000 to 147.995 described) can be selected. Tune-up is simple, since there is only one tuned circuit. If your junk box is well stocked, you'll probably only need to buy the FETs, ICs and crystals, which makes the cost surprisingly low.

The block diagram in Fig. 15-3 shows the simplicity of the unit. Q1 is the reference oscillator, whose output is divided by 16384 at U3. The resulting frequency (555.555 Hz) is one ninth of 5 kHz, the interchannel spac-

ing. The output of Q3, the voltage-controlled oscillator (VCO) stage, is buffered by Q5/Q7, and a sample is mixed with the 15.5555 MHz offset frequency at Q6. The difference is applied to U1, a programmable divider which is set by frequency select switches S1-S11 to divide by a selected number between 800 and 1599, corresponding to the possible output frequencies between 144 and 148 MHz.

Divider U1's output, at 555.555 hertz, is compared to the reference frequency by the phase detector in phase locked loop IC U2. The phase detector output is filtered and controls the frequency of the VCO via tuning diode D2, closing the loop. Regulator transistor Q1 supplies stabilized + 10 V to all circuitry from the + 12 V battery source.

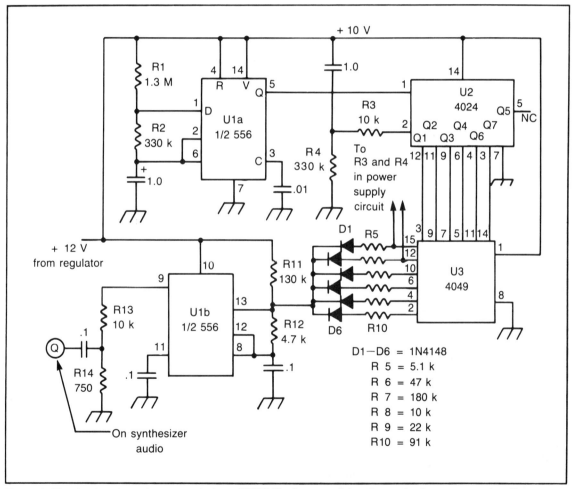

Fig. 15-2. The simple noisemaker uses only three CMOS ICs. It can modulate the synthesizer or a crystal oscillator. Interchanging resistors R5-R10 will provide different tone patterns.

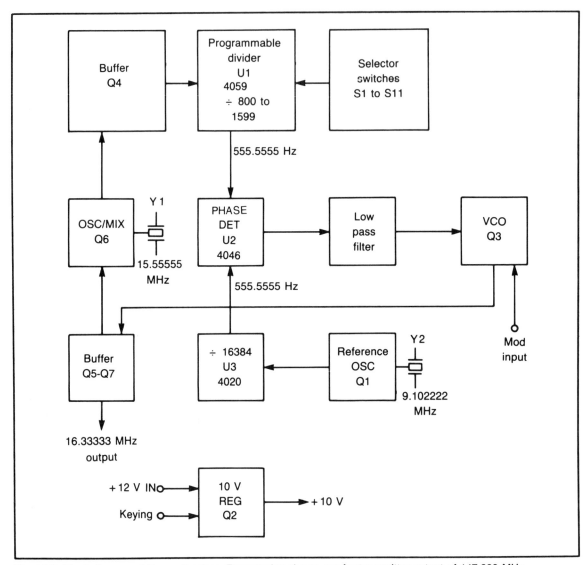

Fig. 15-3. Block diagram of the synthesizer. Frequencies shown are for transmitter output of 147.000 MHz.

The WA4DSY synthesizer, shown in the upper left corner of the box in the photo, was built on the original HT-220 double-sided PC board, which measured 2-1/4 × 4-1/4 inches. This board is no longer available commercially, but a suitable single-sided board can be readily designed around the parts placement layout shown in Fig. 15-4. Standard 0.1 inch spaced perf-board could also be used. In either case, the component side of the board should be a ground plane. Keep leads short and don't use wire wrap on the synthesizer portion of the project.

The schematic diagram of the synthesizer is given in Fig. 15-5. Erie 518 series ceramic trimmers work well for C1, 5, and 7, but surplus piston types are even better if you can find them. The buffered output of Q5/Q7 is low impedance and can be hooked to the following stages with small RG-174 coax. Crystals are in HC-18/U holders, 0.002 percent tolerance, operating into 20 picofarad load capacitance. Best results have been obtained with High Accuracy types from International Crystal (address in Appendix A).

L1, the VCO coil, is wound on a 1/2-watt carbon resistor of 1 megohm or greater. Wind 40 turns of #36

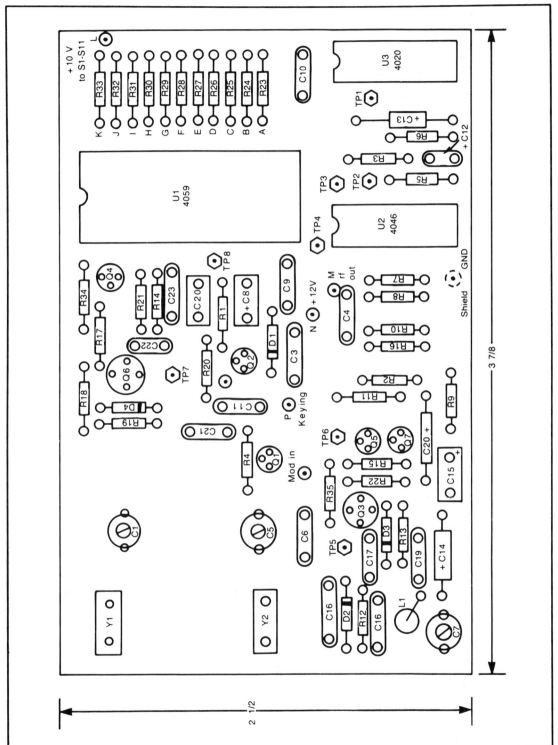

Fig. 15-4. Suggested parts layout for the synthesizer.

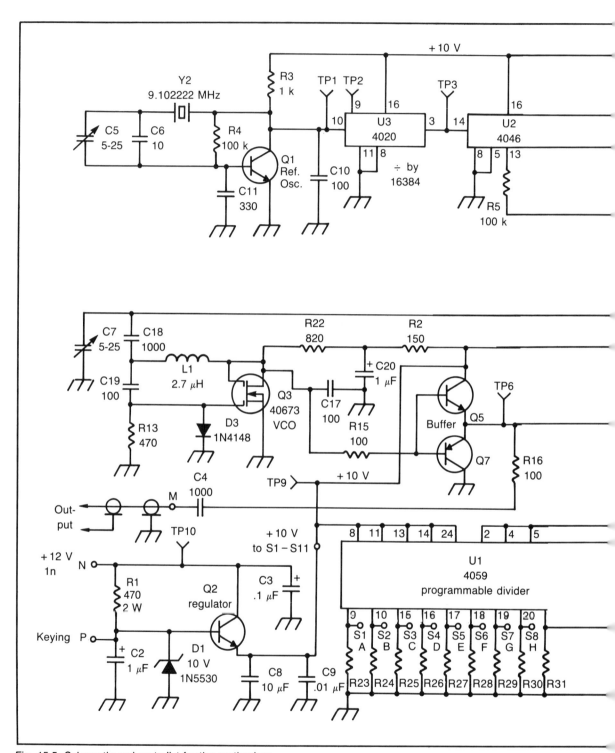

Fig. 15-5. Schematic and parts list for the synthesizer.

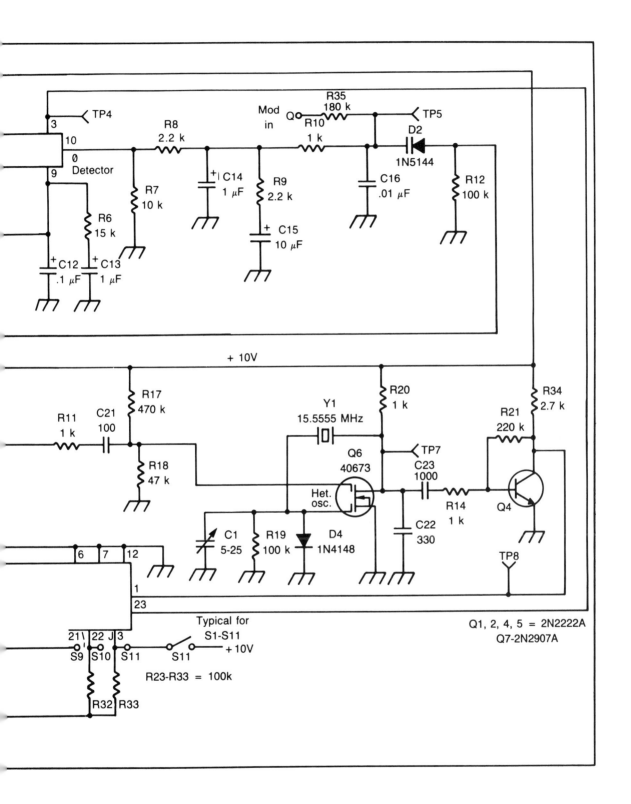

225

magnet wire close spaced on the resistor and solder the ends to the resistor leads. Coat the coil with lacquer, mount it vertically through a hole, and glue it down to keep vibration from modulating the VCO. Add some wire loops at the test point locations for ease of alignment and testing.

Npn bipolar transistors are low cost switching types, such as 2N2222, 2N4401, and MPS3704. For Q7, use a 2N2907, 2N4403, or MPS3703. A 22 picofarad at 4 volt varactor tuning diode (1N5144) is used at D2.

Use low-profile sockets on the ICs. Troubleshooting is far easier if you can break the loop by pulling out an IC. You may also have to try more than one 4020 at U3 to find one that works with 9 megahertz input at 10 volts VCC. This chip has been made by Motorola (MC14020CP), Fairchild (34020PC), and RCA (CD4020AE), and some units from each have worked at that frequency, while others have not.

The easiest way to select the transmit frequency is with an eleven section dipswitch, S1-S11. The desired output frequency is set using binary coded decimal (BCD) notation. Figure 15-6 shows how to do it. Although some may consider it cumbersome, this method is entirely adequate because frequency is set only once for each hunt. For something fancier, BCD coded thumbwheel switches could be mounted on the box. Two-pole rotary switches are available surplus and can also be used. Figure 15-7 shows how the rotary switches are connected up to the dipswitch lines.

A frequency counter and VTVM or FETVM are needed to align the unit. An oscilloscope is not needed unless troubleshooting is necessary. Connect the counter to TP2 through a 20 picofarad capacitor. Apply 12 V to TP10 and verify +10 V on TP9. Now adjust C5 for 4.551111 megahertz on the counter. Verify that the 555.555 hertz reference frequency is present at TP3. Set the frequency select switches for 147.000. Connect the VTVM to TP5, the VCO control voltage, and adjust C7 for a +5 V reading. Remove the VTVM probe and connect the counter to TP6, the VCO output. Adjust C1 for 16.33333 MHz.

If the board does not seem to work, check the test points with a good low capacitance probe and high bandwidth scope. If there is no counter indication at TP2, check for a clean square wave there. If not present, look for the oscillator output at TP1. It should be a rounded sawtooth wave of about 10 V peak to peak. If the signal is present at TP1 but not TP2, try another 4020 IC at U3.

The 555 hertz square wave at TP3 is easy to see, but the divider output at TP4 is difficult to catch on the scope,

as it is a 2 microsecond pulse at a 555 hertz rate. The mixer output at TP7 is a 1 volt peak to peak sine wave at about 700 kilohertz with the 15 megahertz oscillator signal superimposed on it at less than a half volt peak to peak. At TP8, the 15 MHz signal should be filtered out and the 700 KHz signal somewhat clipped or squared off.

This synthesizer circuit is fairly foolproof, and several units have been successfully constructed. It is ideal for the low power bunny box, but it is not recommended as is for a high power rig. The measured in-band spurious outputs are −53 dB with respect to the 16 MHz output signal. This is acceptable for QRP, but will not meet FCC requirements for a rig of 25 watts or greater output. The VCO is also somewhat microphonic. Advanced constructors with adequate test equipment may wish to experiment with VCO shielding, bypassing, potting, and other techniques to improve the spurious output suppression and reduce the vibration sensitivity of the unit.

Proper power supply voltage must be maintained for the duration of the hunt. The synthesizer may fall out of lock or the reference oscillator may stop if the regulated voltage falls below 10 volts. Check to be sure that the 10 V regulator doesn't oscillate at any input voltage, and add resistance or inductance between the regulator and battery if needed.

The original design of the circuit was done when only the upper two megahertz of 2 meters was used for FM. Although additional switching is shown in the schematic, the VCO range does not cover all of the band without retuning of C7. To cover other parts of the band, make the 5 volt VCO tune-up setting at the center of your intended range of use.

TRIPLERS AND FINAL AMPLIFIER

Two tripler stages are needed to get the 16 megahertz synthesizer output to two meters. The design of frequency multipliers and low power output stages is well covered in the general amateur radio literature, and we will not dwell on it here. This is a good opportunity for you to experiment with the latest rf devices, such as VMOS transistors. Or, to save money, design something around the inexpensive devices available at your local parts store or carried by the many mail order outlets.

Another low-cost way to step up the synthesizer output in frequency and power is to convert a ready-made module. Surplus vhf radio boards may have just the stages you need, with only slight modification of the tuned circuits necessary to hit the right frequency range. Small capacitors can be padded across the LC circuits to lower

BCD FREQUENCY SELECTION

There are eleven switches, S1 through S11. A "1" indicates a closed switch, while a "0" denotes an open switch.

S1 selects the upper or lower 2 MHz of range. Set to "0" for 144 to 145.995 MHz or "1" for 146 to 147.995 MHz.

S2 selects the MHz range. Set to "0" for 144 or 146 MHz. Set to "1" for 145 or 147 MHz.

S3 through S6 are the 100 KHz selectors, and are BCD coded to select .0 to .9 MHz.

#	S3	4	5	6
0	0	0	0	0
1	0	0	0	1
2	0	0	1	0
3	0	0	1	1
4	0	1	0	0
5	0	1	0	1
6	0	1	1	0
7	0	1	1	1
8	1	0	0	0
9	1	0	0	1

S7 through S10 are the 10 KHz selectors, and are BCD coded in the same manner to add .00 to .09 MHz.

S11 adds 0 or 5 KHz for "0" or "1" setting, respectively.

Here are three examples of popular T-hunting frequencies:

Switch	145.83	146.565	147.495
1	0	1	1
2	1	0	1
3	1	0	0
4	0	1	1
5	0	0	0
6	0	1	0
7	0	0	1
8	0	1	0
9	1	1	0
10	1	0	1
11	0	1	1

Fig. 15-6. Frequency selection using dipswitches.

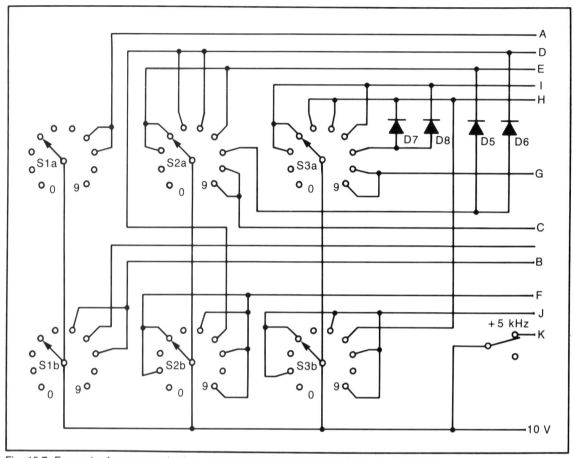

Fig. 15-7. For easier frequency selection, wire up rotary switches to replace the dipswitches.

their frequency, and turns can be removed from the inductors to raise frequency.

This modification technique was used in the prototype Bunny Box. The upper multiplier stages of a 100 milliwatt transmitter board were retuned. Originally the first stage tripled from 24 to 73 MHz and the second stage doubled to 146 MHz. Now both stages triple, with the output of the first stage at 48 MHz. The final stage transistor was replaced with a heftier device to upgrade reliability and increase output power.

The circuit board used was made in the early 1970's and thus doesn't use the latest devices, but it works fine. Though we don't recommend trying to duplicate it from scratch, the schematic diagram is shown in Fig. 15-8 for completeness, and to give builders an idea of what can be done with surplus boards. The board is in the lower center of the box in the photo.

It may be possible to damage the output stage transistor in some designs by operating the transmitter into a high mismatch. Consider adding some sort of SWR protection circuit. None was used in the model, so a 50-ohm termination is kept on the output when the antenna is disconnected, in case the unit is accidentally turned on.

POWER AND KEYING

Figure 15-9 shows a suggested power distribution and regulation scheme. The regulator for the first tripler prevents overdrive and provides a convenient method of keying the transmitter in the cycling mode.

The 2N4427 transistor used is capable of a watt output when properly heat sinked, so operation here is quite conservative. To conserve battery life when maximum power isn't needed, a 250-ohm potentiometer is incorporated in the B+ line to the final two stages, to allow easy reduction of power. Current in these two stages is adjustable from about 120 down to 35 mA.

This power reduction method is inefficient to be sure, and for higher power transmitters a far better way would be to use a simple voltage regulator stage. If this rheostat method is used, be careful not to exceed its power rating. Power can also be reduced by tapping down on the battery stack, with resultant unequal discharging of the cells.

There are occasions when it may be desirable to use the noisemaker to key and provide audio to a higher powered mobile rig without using the rf stages of the Bunny Box. The circuit of S2 in Fig. 15-9 does this. A three-position slide switch is used because it fit easily on the board, but a rotary switch could also be used by rewiring it appropriately. In the center position (terminals 2 and 3 tied together), the black box transmits a continuous signal. In the left position (continuity between terminals 1 and 2), the unit transmits intermittently. In the right position (terminals 3 and 4), the rf sections of the box are disabled and an external transmitter can be keyed at the 15 seconds per minute rate by connecting the push-to-talk output to the external rig. Wiring of the connector is for Drake and 1970's vintage Kenwood 2 meter gear, and should be changed as necessary to suit your rig.

The buffered output of the synthesizer can be heard directly at two meters within a few feet of this hidden T with the multiplier stages off. It was discovered in the first few hunts that during the 45 second off time, hunters could sometimes detect this signal when very close. To prevent this being a giveaway, diodes D1 and D2 were added to the switching circuit to shut down the synthesizer completely during the off time. Two diodes are needed because the two regulators being shut off aren't at the same voltages.

CRYSTAL CONTROL

Hams living in areas where activity is concentrated on a very few frequencies do not need to build a synthesizer into their miniature transmitter. A crystal oscillator to drive the triplers is easy to build. A voltage-variable capacitance diode will directly FM modulate the crystal oscillator. Speech amplifier stages should not be necessary, as the output of the tone generator circuit has plenty of voltage swing with changes to R13 and R14 as needed.

Recent editions of *FM and Repeaters for the Radio Amateur* by the American Radio Relay League describe small crystal controlled transmitters for several vhf bands. They are suitable as is for a rock-bound Bunny Box, or the multiplier and final stage designs could be modified for use with the synthesizer. Another option for a low cost crystalled rig is building a kit. Hamtronics, Incorporated, is one of several companies which sells such kits for many frequency ranges. See Appendix A for the address.

OTHER BANDS AND VARIATIONS

The information given so far serves as a general guideline if you want to build your miniature transmitter for another amateur radio band. Proper redesign of the synthesizer circuit should allow it to be used at either higher or lower ham frequencies. Building a Bunny Box for the aircraft band or CB use is unlawful, as it violates FCC requirements for the use of type-accepted equipment on these bands.

For the 1-1/4 meter band, try putting the synthesizer output at one twelfth of the transmit frequency, and using a doubler/tripler/doubler chain following. The 5 kHz switch can be eliminated, with frequency selection done in 10 or 20 kHz steps. Vibration sensitivity and temperature drift become greater problems as frequency is increased, and the synthesizer will probably be unusable on the 70 cm band. Use crystals there instead.

At lower frequencies the job becomes much easier. For the ten meter band, put the synthesizer output at one half the transmitter output frequency, and use a combination doubler and final amplifier stage, with an output low pass filter to prevent TVI. FM is used on ten meters from 29.5 to 29.7 MHz, and channel spacing can be 10 kHz. This results in only 20 channels, which simplifies switching. It may be difficult to get sufficient FM deviation without clipping at such a low multiplication factor. For hunts at the low end of ten meters, where FM is prohibited, replace the audio board with a keying circuit that alternates a continuous carrier (A0) with CW (A1) identification.

What about using the synthesizer circuit in a receiver? A low drain crystal replacement for the Little L-Per or other rock-bound units is very useful. If you take on this challenge, you'll want to be very careful to minimize spurious products. They cause far more trouble in a sensitive receiver than in the low power transmitter. The microphonic nature of the VCO will also cause aggravation. Tapping on the unit's case may cause pings in the speaker, and audible feedback is a possibility. The only cure may be potting the VCO section.

In the specific case of the L-Per, the second harmonic of the receiver's crystal frequency is used at vhf, and the fourth harmonic is used at uhf. The 16 MHz output of the board described here is not directly usable, as it cannot be multiplied into the 78 MHz region for two meters while keeping the frequency steps at 5 kHz. One

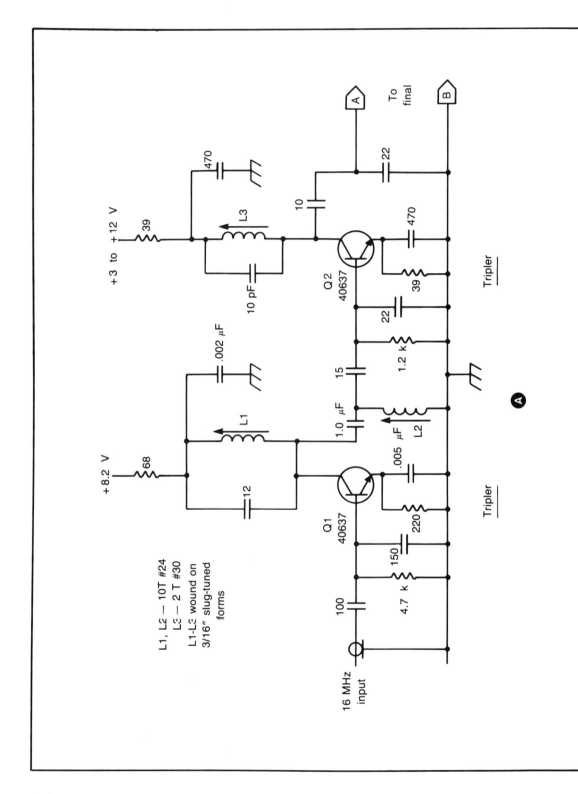

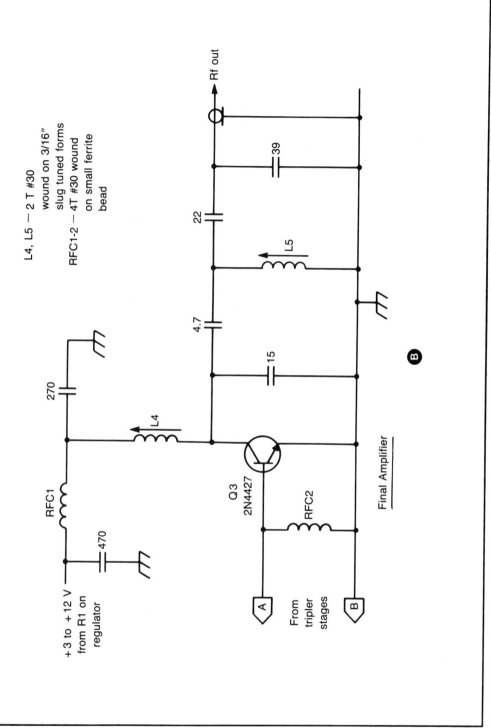

Fig. 15-8. Schematic of the triplers (A) and output stage (B) of the Bunny Box.

231

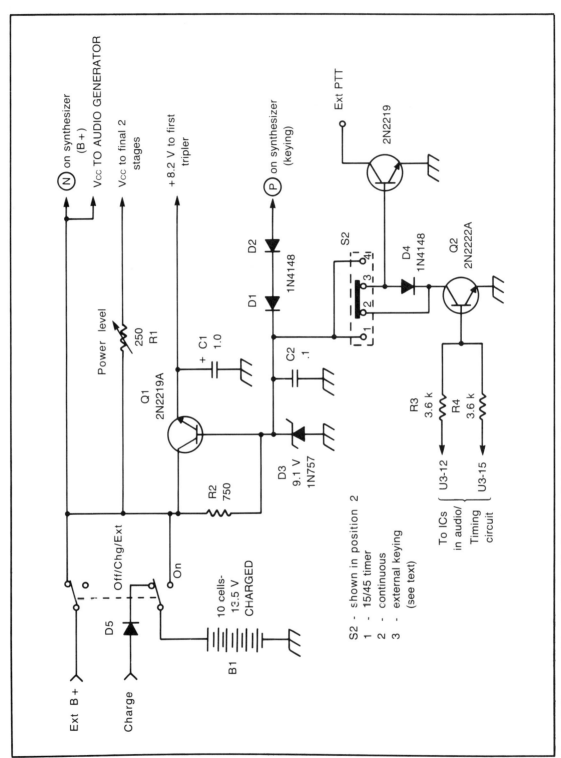

Fig. 15-9. Power regulators and intermittent mode switching circuits used in the Bunny Box.

solution would be a complete redesign of the synthesizer oscillator frequencies and division scheme. Another would be to put two triplers at the output of the board and inject the resulting 135 MHz local oscillator (LO) signal (receive frequency minus 10.7 MHz) into the L-Per at the first mixer.

For a receiver with 10.7 MHz i-f such as the L-Per, crystal Y1 must be changed to 14.36667 MHz. The purity and level of the LO signal is critical for optimum receiver sensitivity and freedom from birdies. Filter each stage and expect to spend a lot of time adjusting the LO level. Reception near 145.635 MHz may be impaired due to radiation of the 16th harmonic of the reference oscillator.

Be advised that the information in the last five paragraphs is based on an evening of cogitation with the schematic. None of the mods for higher and lower bands or DF receivers have been tried by us, so you're on your own. Think things through carefully before you start to build. You'll want to study the data sheet for the CD4059 to understand how division is programmed. You may want to get application note ICAN-6374, which explains how the CD4059 can be used in digital tuning applications. The note is published by the solid state division of RCA (see the appendix for the address).

Newer, more high tech synthesizer circuits are now available, and should be considered for high power transmitters. One very versatile design is marketed by A & A Engineering. Several bare circuit boards and complete kits are available, but you need to do some design work to apply them to your particular needs. Power requirements may make them impractical for some battery powered applications. Write or call the company for product information sheets on the synthesizer line.

Chapter 16

Hunting Without A Vehicle

T-hunting an Olympic event? It's possible! Ham radio operators have assisted the International Olympic Committee in the past with communications for participants and for the Olympic torch run. But how about using amateur radio in world class sports?

The idea isn't far-fetched, considering that fox hunting is already an international competition in some parts of the world. The fact that a hunter's equipment plays an important part in his success is no barrier to Olympic consideration. After all, bobsled and rowing competitors' success depends in part on the design and construction of their vehicles. Read on, and you too may want to become a world class hunter.

REAL RADIOSPORT

Call it a reflection on our culture if you will, but it is true that when hams in the USA think of T-hunts, they think of driving or riding in a car full of DF gear. Not so in Europe, where it's more like a track meet.

Leading the way in competitive on-foot DFing are the countries of eastern Europe. Region 1 of the International Amateur Radio Union (IARU) holds a championship meet every two years, in a different nation whenever possible. An international jury, with a member from each participating IARU national society, resolves any disputes and certifies the final results. The chairman of the jury is a member of the IARU Executive Committee.

As with any international competition, language is a potential barrier, but the host nation provides translators and few problems are encountered. Surprisingly, English is a common second language for many of the jury members, so it has become the official language of the jury.

To get to the IARU championships, entrants must be selected by a series of local and regional championship meets. Entrants need not be licensed amateur radio operators, since they do not transmit while competing. There are three categories of competitors:

- [ ] Seniors (male, 18 years or older)
- [ ] Juniors (male, under 18)
- [ ] Women (any age)

Just as in the Olympics, the competition is fierce, but a friendly spirit prevails at all times. A reception and banquet ends the meet, with participants exchanging QSL cards and small gifts. High government officials are the hosts, and television stations in eastern Europe cover the ceremony.

CHAMPIONSHIP RULES

One hidden fox is not enough for the intrepid hunters of Europe. There may be as many as five foxes, transmitting in sequence for two minutes each in Morse code at about eight words per minute. Power is five watts or less. CW is used on 80 meters, while tones on the AM carrier (MCW) are used on 2 meters. Championship races are held on each band, on separate days so that the same competition may enter both events.

The foxes may be visited in any order, but the rules often put a time limit on the search for each fox. The hunter carries a card, and punches it with a special punch at each fox to prove that he found it. His race time does not end until the visits all foxes and races to the end point. In some cases, the end point must be found from the map. In other cases, a homing transmitter is used at the end.

Forested, rural terrain is best suited for such hunts. The area chosen must be unfamiliar to all entrants. Each hunter is given a map of the area 15 minutes prior to starting. Typical maps are 1:50,000 topographical types. He also gets a list of frequencies, callsigns, and transmission schedules of the foxes. Each entrant provides his own radio equipment, watch, and compass.

Superregenerative or direct conversion receivers have the potential for causing interference to other hunters' gear. The rules prohibit such interference beyond 10 meters away, and equipment is checked for this before the hunt.

Some of the restrictions on the hiders tend to ease the job for hunters. For example, hiders are told to avoid railways, highways, fences, and power lines, which could throw off DF gear. The hiders must announce the polarization of their antennas, and not change it. The total course should not exceed six kilometers, and elevation change within the course should be less than 200 meters.

A staggered start helps to even out the competition and force hunters to work independently. Starts are timed to coincide with the transmitting pattern of the foxes. For example, assume the foxes are transmitting, one minute each, in sequence. Every five minutes, coinciding with the start of fox #1's transmission, a hunter from each category starts.

The hunters cannot listen to their radios before their start time. Successive starts by entrants from the same country and category are not allowed. Hunters are then timed individually until they finish. They may not assist one another in any way. Observers are kept behind the starting line to prevent any inadvertent help to the hunters.

A team competition is also part of the Region 1 championships, but it is by no means a relay event. Each country identifies two individuals to form a team, and the sum of the two individual scores is the country's team score. The countries can't just add up their two best individual scores—they must identify the team members before the contest begins.

For training purposes, manual transmitters are used, with the operators keying them in sequence. A much more sophisticated system is used for international meets. The foxes may be completely concealed, with only the special punch visible. Transmitters are turned on and off remotely by uhf signals from a timing and control point near the start. Transmissions are tape recorded off the air to provide documentation of any error or failure. A jury member hides near each fox to monitor the hunters.

Since the transmitters are on one at a time in sequence, there is no time to separately DF each transmitter. The hunters are forced to use their maps and compasses to plot their bearings and make multiple triangulations. There is always a mad dash from the starting point, because each hunter wants to take bearings from widely separated points on each fox.

How long does it take? The winning times for a five transmitter championship are in the order of 45 minutes!

Some interesting rules variations have been developed for local training hunts. In one special case, the hunters, upon finding the fox, switch it to a manual mode and send their call letters. Then the huntmaster and all hunters know that the fox has been found, by whom, and when. The hunters must have transmitting licenses for this, of course. For an extra measure of incentive to be first, the rules sometimes allow the first finder to turn the transmitter off and leave it!

ASIA, TOO

Fox-hunting on foot is not limited to Europe. It is a national sport in the Peoples' Republic of China as well, supported by the national sports organization. After being suspended for ten years during the Cultural Revolution, competition resumed in 1981. Traditionally it has been done on 80 meters, but recently 2 meter hunts have been added.

The All-China DF Competition draws about a hundred contestants, with teams from all over China and sometimes guests from nearby countries such as Japan. Each team has a manager, a coach, and three or four members. Categories are Individual Men, Individual Women, Pairs, and Boys. Prizes are awarded for each category on each band.

Five transmitters, running from 0.5 to 5 watts, are

deployed on a course that covers suburbs, forested terrain, public parks, and hills up to 200 meters in height. Men must find all five transmitters for a perfect score, while women and boys need only find four. The course is not easy, and the coaches stand by with salve to apply to the many scratches and abrasions that appear on the bare legs and arms of the most aggressive entrants.

The scoring system for the Chinese championships has an interesting twist: The results of a written exam are incorporated. The test, taken before the hunt, covers basic electronics, transistor circuits, and direction finding. For each five points scored on the exam, one minute is deducted from the hunter's elapsed time.

AMATEUR RADIO'S SPORTING GOODS

Championship competitors may use any kind of portable set, and creative ideas abound. Commercial DF gear is almost unknown in China; no L-Pers or BMGs here! The Chinese cooperate in home brewing their equipment, and the resulting units are quite sophisticated. One winner has used a dual-band rig with the receiver and batteries built into a long, slim, rectangular box that also served as the boom for the multi-element antenna. While hunting he held it up like it was a model airplane, ready to be sailed away. The idea was good—no feed line to get tangled or foul up the antenna pattern—but the element rods he used posed a danger to the hunter's eyes if he fell.

Most Europeans use a two element driven array antenna for hunting on 2 meters. For diving through heavy brush, elements are frequently made from strip metal similar to the type used in steel tape measures. (Sniffer builders take note!) Some Asian beginners simply use a whip antenna, performing "body fades" or using the null off the end.

Ferrite rod antennas are the norm on 80 meters. A pair of rods, about four inches long and a third of an inch in diameter, are taped together with double sided tape, and 22 turns of #28 wire is wound on the tape. The entire assembly is placed in a U-shaped aluminum channel, for shielding, and potted with resin.

The rod antenna is resonated with a variable capacitor and forms the input to a direct conversion receiver (see Fig. 16-1). Such a receiver has a local oscillator at the same frequency as the incoming signal, for AM reception, or is tuned a few hundred Hertz off frequency for a CW tone. An earphone is used to save battery power.

Because the local oscillator is on the received frequency, a shielded box is used to prevent leakage which might interfere with other receivers. Presence of the rf amplifier minimizes local oscillator radiation from the antenna.

Circuit enhancements include a built-in switched attenuator. One step of about 20 dB is generally sufficient. Varactor diode control of the oscillator may eliminate some mechanical problems. A potentiometer tunes the receiver over the narrow hunting range by varying the reverse bias voltage on the varactor diode.

Pattern of the rod antenna alone (Fig. 16-2A) is the

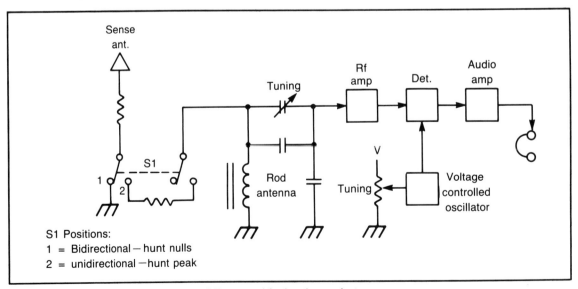

Fig. 16-1. Block diagram of the 80 meter DF set used for hunting on foot.

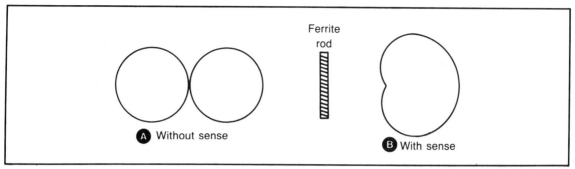

Fig. 16-2. Patterns of rod antenna with and without sense antenna.

same as a transistor radio loopstick. The peaks in the broadside directions are broad, and the nulls in the axis of the rods are very sharp and deep. The nulls work best for bearing accuracy.

Closing switch S1 to add the sense antenna gives a unidirectional pattern, as in (B). Signal from this antenna is 90 degrees out of phase with signal from the loop. Length of the sense antenna must be adjusted in a cut-and-try fashion, using a fox in a known location, to ensure enough front-to-back differentiation. About 6 dB differential is sufficient, and a flexible trailing wire fifteen inches long works fine. A stiff upright sense antenna is not required on 80 meters, and might result in an eye injury during a running hunt.

RADIOSPORT FOR AMERICANS

For European-style hunting to have a chance at becoming an Olympic event, it must take hold in more parts of the world. Already it is growing in Korea, Japan, and elsewhere in the Far East. Hams in North America must get involved next to make a World IARU Championship possible. One way might be through the Boy Scouts, who have staged fox hunts at their World Jamboree and had participants from over 90 countries.

Such a radiosport DFing event could be a great activity for your local Boy Scouts and Girl Scouts, or other youth groups. It combines technical achievement with good exercise and team spirit. It is also a fine way to introduce youngsters to amateur radio, which can be a life-long hobby and the start of a technical career for them. Only the hider need have a transmitting license, so any boy or girl can get involved. Don't let handicaps stop you either. How about a wheelchair DF competition?

Two Meters

The equipment you choose for a 2 meter on-foot hunt

depends on the transmitter power. For a reasonably strong signal, the sniffer system from Chapter 12 is excellent. If the signal is going to be only a few microvolts at the starting point, try the Shrunken Quad from that chapter, working into a portable receiver or handheld with a built-in or added S-meter. An attenuator may be needed when closing in on the fox.

The Shrunken Quad can be used with horizontally or vertically polarized signals by proper orientation of the driven element. Be sure to check both configurations as the hunt starts if there is doubt which polarization is being used by the fox.

Switched antenna DF units (L-Pers, BMG, and the like) are suitable for 2 meter radio orientation races (Fig. 16-3) with a few precautions:

☐ Use tip protection and great care to avoid injury from the antenna while running.

☐ Keep moving and take more frequent bearings, as there is greater bearing error on these units in the presence of multipath and reflections.

☐ Use a receiver S-meter if possible to help prevent running past the fox. An attenuation system may be needed to prevent pinning. The dual modes of the L-Per give it an advantage over the BMG unit in this regard.

☐ Expect great difficulty if the signal is predominately horizontally polarized.

160 Meters

European-style home brew DF units are not difficult to build, but there are even easier ways to hunt on 160 meters. AM transmitters can be hunted with modified transistor radios, available inexpensively by the sackful. Tune the local oscillator trimmer so that the top end of the receiver range is extended to about 2000 kHz. Then repeak the rf input circuits for 160 meters. No change to the i-f alignment is needed.

Fig. 16-3. Oh, no! Not more cactus! Clarke Harris, WB6ADC, hunting on foot with a SuperDF.

The ferrite antenna of a transistor radio gives very good nulls, making the radio into a sensitive handheld sniffer for 160. An S-meter isn't necessary for the AM signal—just listen to the modulation go up and down with the signal strength.

Making a hidden transmitter for 160 AM is just a matter of using some ingenuity. Circuits for QRP CW transmitters using a crystal and logic ICs are common in ham magazines and books, and are easy to throw together. A wireless broadcaster hobby kit for the AM broadcast band, retuned to above 1900 kHz, can also do the job and provide microphone input.

One hundred sixty meters has a lot of potential for challenging short range hunts during the daytime when there are no skywave signals coming in. Don't forget that the hidden T operator must have an amateur radio license, General Class or higher.

Chapter 17

Hunting Below 50 MHz

The different characteristics of radio wave behavior and equipment design make T-hunting at medium and high frequencies quite different than at vhf and uhf. Here is a look at what to expect from the signals and what to use to hunt. There's also a practical method for speeding up and automating the process with a rotating antenna unit and cathode ray tube (CRT) display.

LOCATING DX SIGNALS

It might appear at first that long distance DFing is easy. All we have to do is to have two or more stations point their beams or use loops to get bearings on a distant station, triangulate them, and we have a close fix on the source, right? Not necessarily. Hf signals propagate along the ground for only about 50 miles before being attenuated to undetectable levels. For distances over 125 miles or so, hf signals are propagated by the ionosphere.

The ionosphere isn't a nice smooth mirror-like reflector. It's a multi-layer non-homogeneous time-varying refractor and ductor. In other words, it's hard to be sure how high it is, and it's lumpy. The paths a signal takes through it are many and varied. They change with time of day, time of year, and with the sunspot cycle.

Scientists have studied and characterized several ionospheric mechanisms which result in errors in DF bearings. They include:

☐ Lateral tilts of the ionosphere (LATs). The various layers are constantly evolving as the hours and days pass. The effective height of a layer is not the same throughout its extent. As a result, a signal may be given some "English" when it is refracted, with a lateral deflection being the outcome. These are sometimes also called Systematic Ionospheric Tilts (SITs).

☐ Traveling ionospheric disturbances (TIDs). Irregularities in the layers can cause uneven refractive characteristics. They range from the size of a city block to hundreds of miles in diameter. They move around at rates that can reach 2000 miles per hour. The source of many of these TIDs is thought to be internal gravity waves. The uneven contours associated with TIDs cause changes in the refraction characteristics, and shifts in apparent wave direction occur. TIDs are the most common source of bearing errors.

☐ Wave interference (WI). When one transmitted signal arrives at the direction finder via different paths or modes of propagation at the same time, the DF cannot exactly resolve the direction. The effect is quite similar to multipath on vhf/uhf. The multiple signals can come from refractions from different ionospheric layers, from a different number of hops through the layers, or from TIDs. The resultant phase shifts between the sources is usually time varying and gives a fluttery azimuth movement. The name comes from the fact that the wave, in effect, interferes with itself.

Professional hf DFers can compensate for these phenomena, to some extent. Ionospheric soundings and measurements of the Doppler effects of the moving disturbances can help predict which readings are in error. Much remains to be learned, however.

As experienced amateur radio DXers will tell you, it's not uncommon to have an indicated beam heading 20 degrees or more away from the station you're working. The more distant the DX, the more likely an azimuth error becomes. If the signal is near the antipode (the exact opposite point on earth from the observer), the error might be 90 degrees or more. It doesn't take much azimuth shift at each DF station to cause hundreds or even thousands of miles of error in the triangulated fix when the target is far away.

Add to that the wide lobes of most hams' hf beams and quads and the inaccuracies of direction indicators and most rotors. There are site errors peculiar to the surroundings of each DF station. The patterns of identical antennas will vary considerably with their height and the nature of nearby structures and terrain features. Then there's the requirement for wide separation between DF stations so that triangulation can be accurate. If conditions are bad, the target signal may not be propagated to all DF stations in the network.

The end result is that a couple of hams simply aren't going to be able to use their station antennas to get a super accurate fix on a station a half a continent or more away. Accurate fixed site DFing of sky-wave propagated signals is very difficult unless complex equipment, beyond the means of amateurs, is used. The topic of DX DF is of great interest to the military, and engineers have studied it in depth for many years. With all this research and with the very latest equipment, their practical bearing accuracies on skywave signals are still seldom better than 3 or 4 degrees uncertainty.

Triangulation of DX signals must be done with care. Ordinary flat maps won't work at long distances. Such a map indicates, for example, that both Atlanta and Baghdad are directly east of Los Angeles (90 degree bearing). The correct great circle bearings for these cities are ac-

tually 80.5 degrees for Atlanta and 17 degrees for Baghdad. The best way to do long distance triangulation and take the shape of the earth into account is to use a computer.

For radio amateurs, the best method for locating sky-wave propagated signals, such as those of jammers on the hf bands, is to use our biggest advantage—our numbers. There are hams everywhere in the country and in the world. Chances are that one or more is close to the target signal. Triangulation from as many stations as possible gives a general indication of the area. Hams in the target area can provide signal strength information and ground wave bearings to narrow down the uncertainty even more. When the locality is deduced, it's time to send out the mobile teams.

Pinpointing the exact location of hf signals by amateurs, either for sport or for real, eventually ends up being done by ground wave. The remainder of this chapter concentrates on these methods for closing in.

LOOPS FOR 15 TO 50 MHz

For both mobile and on-foot hunting, the loop techniques used by hams in the 1950's still have a lot to offer today. To ensure that currents are the same magnitude and phase throughout the loop, it must be small compared to a wavelength, preferably less than 0.08 wavelength in circumference. This means that for 10 meter and Citizens Band hunting, a circular loop should be ten inches or less in diameter. Such a loop is small enough and light enough to be held out of the car on the end of a rod when bearings are needed. If it is much smaller than that, signal strength is drastically reduced.

If you begin your bunny chasing career on 2 meters and then try hf DFing, you're in for some surprises. Hf loops are far more sensitive to surrounding metal than vhf quads. Power lines are suddenly something to watch out for, as are trees, buildings, and just about anything else in the near field. It is helpful (perhaps even mandatory) to remove any other whip antennas from the car when null hunting with a loop. Here's your excuse to buy that motorized retractable antenna for the AM/FM set!

Some hunters claim that they get bearings through the windshield with their loops inside the car. That's one way to hunt on a rainy night, but we don't recommend it. Get the loop outside, in the clear, and on a good rotating system for best results. Nulls are most pronounced when the hidden T is toward the side of the car where the loop is mounted. The null is flattened somewhat if the car top forms a reflective surface in between.

Balanced and Unbalanced Loops

The basic single-turn unshielded loop for 10 meters or the Citizens Band is shown in Fig. 17-1. Made from copper or aluminum tubing for rigidity, it is resonated by a combination of fixed and variable capacitance. The center conductor of the coax to the receiver is tapped onto the loop about 3 inches from the mounting point.

The loop is tuned by orienting the lobe (not the null) toward a signal source, such as a dip meter or bench signal generator driving a short vertical whip antenna. The loop should be mounted on the vehicle as it will be in use.

Tune carefully with a non-metallic tool to avoid hand capacitance effects. Peak the resonating capacitor slowly, then adjust the coax center conductor tap point for maximum signal into the receiver. Repeat the peaking and tap point adjustments several times, as they interact. Then rotate the loop and check for proper nulls, which should be 180 degrees apart.

It could be argued that a loop of this size is more analogous to a coil than to an antenna, since it couples to the magnetic field of the incoming wave. In fact, we want to eliminate any antenna effect from the loop act-

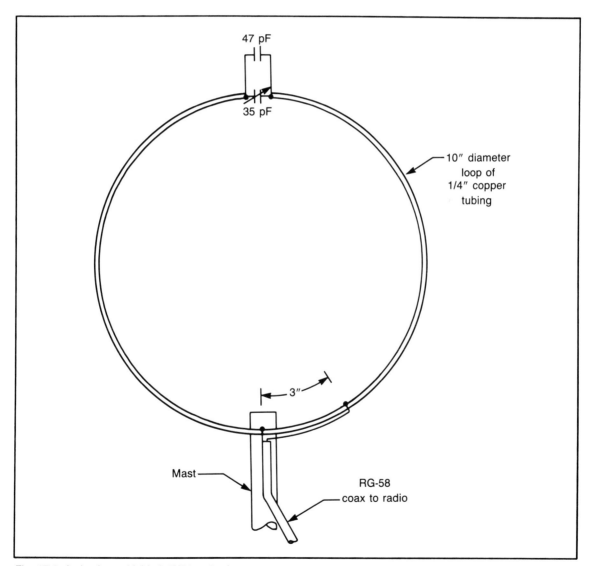

47 pF

35 pF

10" diameter loop of 1/4" copper tubing

3"

Mast

RG-58 coax to radio

Fig. 17-1. A simple unshielded 10/11 meter loop.

241

ing like a probe in space connected to the receiver input, which might reduce bearing accuracy. Such non-directional electrostatic field signal voltage adds to or subtracts from the magnetic component, degrading the nulls. This makes it very important to balance the loop electrostatically with respect to ground, so this antenna effect is cancelled.

A better balance can be achieved by making the loop from 300 ohm ribbon line (Fig. 17-2). For a balanced feed, the primary of the matching transformer is kept balanced to ground, and the transformer is mounted in a shield can. Some sort of wooden frame (not shown in the figure)

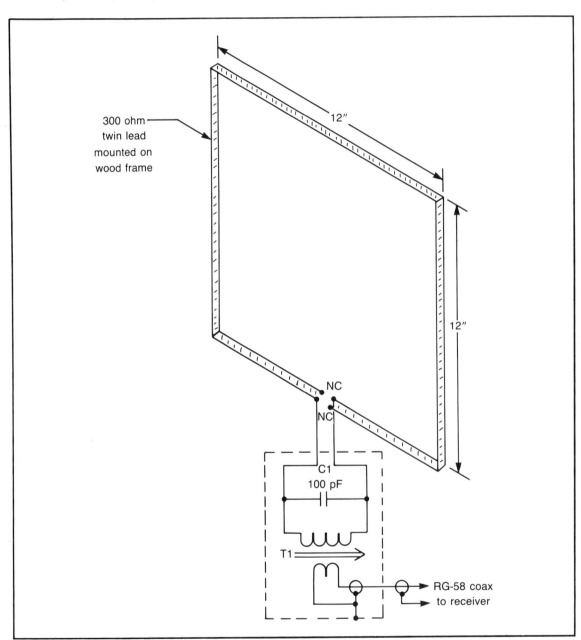

Fig. 17-2. 300-ohm TV twin lead can quickly be made into a loop for 10 meter hunting.

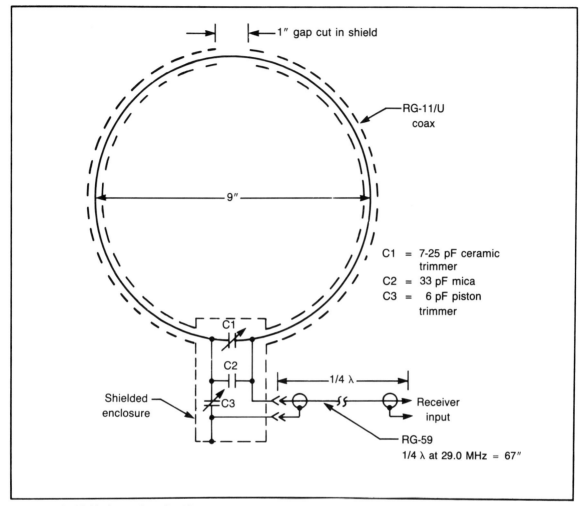

Fig. 17-3. A shielded coax loop for 10 meters.

should be used to keep the loop rigid. T1 is a surplus slug-tuned coil chosen to resonate with C1 at the desired frequency in the 26 to 29 MHz range. The secondary is two or three turns wound over the primary.

Shielded Loops

The most popular way to eliminate the antenna effect is to enclose the loop in a grounded electrostatic shield. The shield must be broken in one place to prevent it from acting as a closed loop itself.

Mechanical construction of a shielded loop can be done in many ways. If you expect to do a lot of hf hunting and want a rugged antenna system, make the outer shield of copper or aluminum tubing with the interior loop wire suspended on spacers inside, to keep all points on the inner loop equidistant from the shield. For a more simple starter antenna, just use a piece of RG-11/U coax with the shield continuity broken in the exact center, as shown in Fig. 17-3. Wrap tape over the point of discontinuity to avoid water ingress. C1 and C2 across the open end resonate the loop.

No transformers are used in this model. Matching from the high loop impedance to the low receiver input impedance is done with a quarter wavelength piece of RG-59/U or other 75 ohm cable. The matching line is 67 inches long at 29.0 MHz. (More information on the calculation of quarter wave matching lines is given in Chapter 14.)

Balance is kept by using C3 to compensate for the connection to the feed line. Tuning should be done on the

243

receiver to be used for the hunt. After peaking C1, check to see that the nulls are symmetrical, exactly 180 degrees apart. If not, adjust C3 as required for best symmetry, verifying the tuning of C1 after each adjustment of C3.

The inductance of the loop and its capacitance to shield and ground must be tuned to form a high Q resonant circuit.

PREAMPS

The capture area of a small loop is minuscule compared to a full size antenna, so signal pickup is much less. A low noise preamp right at the antenna can make up for much of this lack of sensitivity. A preamp is even more desirable with a sense antenna arrangement, because the addition of the sense antenna may cause a sensitivity reduction. Figure 17-4 shows one scheme, where a Field Effect Transistor (FET) is connected directly to the tuned circuit formed by the loop and its resonating capacitor. The high impedance of the loop at resonance is not loaded by the high input resistance of the FET.

The preamplifier circuit uses experimenter J-FETs such as the Motorola MPF-102 or HEP-802. Try to get both transistors from the same batch (date code) for best current matching. The input is tuned by the loop and its resonating capacitor. The output is untuned, making the preamp usable with a suitable resonant loop anywhere from 3 to 30 MHz.

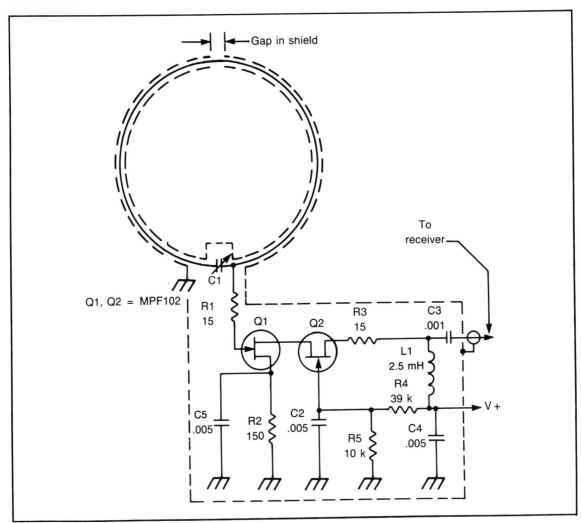

Fig. 17-4. A preamplifier can be combined with a loop to give greater sensitivity.

The cascode circuit gives high input impedance with excellent input/output isolation. Oscillation is unlikely, but it is still good practice to separate the input and output circuits and keep the device leads short. The preamp should be constructed on single-sided copper clad perfboard to provide a ground plane.

The preamp output is coupled to the low impedance coax line of any length to the receiver. Maintain good shield continuity from the preamp to the receiver by using a shielded connector (an RCA-type plug will do) at the preamp output. Battery power is easiest, but a separate line to vehicle power could be run. Power can also be supplied up the coax. Where strong out-of-band signals such as broadcast stations ride through the preamp, use a tuned circuit as the preamp output, with transformer coupling to the coax.

If you don't want to build a preamp from scratch, consider a kit. Hamtronics has tuned pramps for the 28 MHz region. A broadband preamp may be a better idea if you have individual loops for more than one band. The loop provides the tuning. A & A engineering supplies a bipolar broadband preamp kit based on a design by Wayne Cooper, AG4R.

UNIDIRECTIONAL LOOP SYSTEMS

Many hunters loathe the bidirectional ambiguity of a simple loop. Bearings must be taken and plotted constantly to prevent driving away from instead of toward the fox. Hf hunters soon want some sort of unidirectional indication. It turns out that by combining the loop output with rf from a non-directional antenna in the proper phase relationship, the pattern changes from a figure 8 (two nulls) to a cardioid (heart-shaped, with one null). (The principle of injecting signal from a vertical antenna 90 degrees out of phase with the loop was introduced in the last chapter, and patterns were shown.)

Warren Amfahr, W0WLR, put the antenna effect to work by using it to inject the electrostatic component, making a 10 meter loop unidirectional. His two-turn loop is detailed in Fig. 17-5. The stray electrostatic pick up has the correct amplitude and phase to produce a single null and single broad peak.

The loop consists of two turns of 1/4-inch copper tubing, with the ends flattened out and screwed to a one inch diameter dowel. Use spacers to hold the two turns 5/8 inch apart at all points. The capacitor is in the exact center of the two turns at the base. No matching transformer is used. The direct connection of the coax is in part responsible for the controlled antenna effect.

The first step in checking out the unidirectional loop

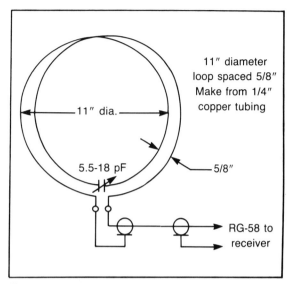

Fig. 17-5. The W0WLR unidirectional loop.

is to adjust the capacitor for resonance (maximum noise pick up in the receiver). Then use an on-frequency source to provide a signal while reducing the capacitance for unidirectional response. Tune carefully until there is a single null and a single peak, exactly 180 degrees apart.

The null on this antenna will be less sharp than the nulls on a bi-directional antenna. Many hf fox hunters prefer to use their bi-directional loops with a separate switchable sense antenna (pardon the tongue-twister) to get a single null only when needed to avoid ambiguity. Figure 17-6 is a schematic of how to accomplish it at 26 to 30 MHz. The sense signal is injected at the high impedance preamp input described earlier.

Use a collapsible vertical antenna about 40 inches long, mounted on the preamp box for sensing. It can be braced against the loop, but must be insulated from it. For sharpest and most accurate nulls during hunting, the sense antenna is disconnected by opening S1, and the bidirectional nulls are used. For the ultimate in convenience, use a small relay in place of S1, with a control switch for it at the operator's position.

Initial tuning is done for resonance with S1 open. Then the system is tuned for a single null by alternately shortening the whip and adjusting Rs. Changing the length of the antenna changes the phase of the injected signal. The single null is at 90 degrees from the two bidirectional nulls. If you can't get a single null, try adding a small amount of inductance in series with Rs. Be sure to mark the antenna system for peak and null direction after tuning is complete.

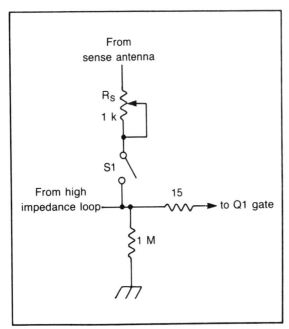

Fig. 17-6. A sense antenna signal can be inserted into the preamplified loop of Fig. 17-4.

LOOPS FOR 2 TO 15 MHz

Loops for the lower bands are built using the same basic principles. The single turn loop works well on 12 and 15 meters, but is not practical below 20 meters or so. Multi-turn loops are used on lower frequencies to increase inductance and achieve resonance.

A good example of a multiturn loop for 75 meters is a design by Frank Marshall, VE4CX, of Winnipeg (Fig. 17-7). He wound three turns of 14 AWG wire on a 12 × 16 inch wooden frame. Grooves are routed in the frame with a table saw to keep the turns spaced and in place. The loop is resonated with a 140 picofarad trimmer in series with the loop. It is mounted in a shield can at the bottom of the loop frame.

Rather than mount the sense antenna alongside the loop and have it in the way of rotation, Frank chose to use his regular mobile whip for sensing. The whip coax goes through a transmit/receive relay (for use with the transceiver), then to a small enclosure with the mixing circuit. The box can be located right at the receiver to make S1 accessible, because the series resonant loop matches the coax well.

L1 and L2 are each four turns of hookup wire, about one inch diameter, with windings taped together. R1 is a noninductive carbon pot, used for adjusting for a proper cardioid pattern as before. S1 is used in the opposite manner from the 10 meter system—it is opened for unidirectional response.

The electrostatic shield is made of aluminum foil. After winding the loop, carefully glue an eight inch strip of foil around the outer perimeter of the form and then slowly wrap it around the loop, making small cuts as necessary so everything fits flat. Leave a half inch discontinuity at the top of the loop. Wrap the completed loop carefully with plastic tape to hold everything together.

A more compact unit (Fig. 17-8) can be built for 75 meters if the loop is replaced with a coil on a ferrite rod, similar to the loopstick in a transistor radio. Inductive coupling matches the receiver input. A short whip is used for the sense antenna. The coupling link is two turns at the center of the winding, for balance.

Tune-up for both 75 meter loops is essentially the same. Use an on-frequency signal source. Set S1 for the bidirectional pattern (closed for the VE4CX loop, open for a rod loop). Tune C1 for maximum signal with the loop end-on to the transmitter. Note that this maximum signal direction is in the plane of the open loop and at right angles to the axis of the ferrite rod. Then switch S1 and tune R1 and L1 (on the rod system only) for a null. If null cannot be achieved, turn the antenna 180 degrees and repeat the whole procedure. L1 tuning may be very critical.

This method of achieving sense antenna operation is quite narrow band, and returning should be done if the hunt frequency changes more than a percent or so. This is most noticeable on the lowest bands, such as 75 meters. More information on hf rod antennas is in the earlier description of European fox hunting. Remember that you must never transmit into your loop if you have an active preamp.

SETTING UP FOR LOOP HUNTING

A battery-powered receiver (Fig. 17-9) is excellent for hf hunting. It rapidly adapts from mobile to foot use. It is less likely to be plagued by ignition interference. Nulls and peaks on CW, AM and SSB signals can be determined by ear, but an S-meter will result in better accuracy. (Chapter 5 will help you add one if the set lacks it.) Time spent getting rid of any ignition or other vehicle noise is well spent. It is much easier to locate signal nulls in a quiet system.

If you have hunted on vhf with a two meter quad, a short loop may take some getting used to. With a quad, maximum signal is obtained looking through the loops, but this broadside orientation gives nulls instead of peaks with a bidirectional small hf loop. Null hunting is trick-

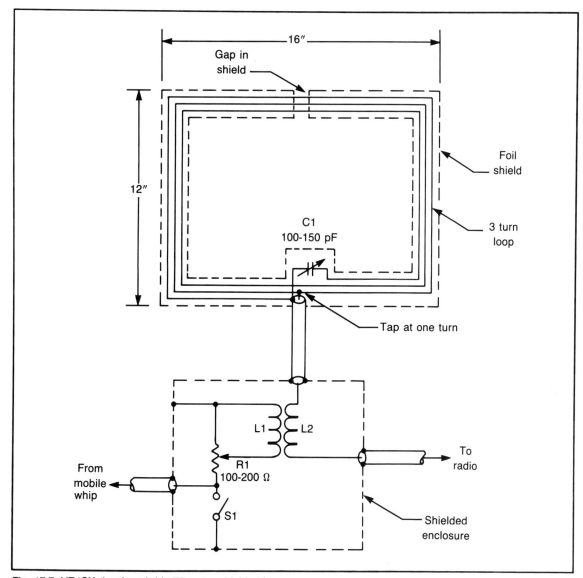

Fig. 17-7. VE4CX developed this 75 meter shielded loop.

ier than peak hunting, also.

As with any other hunt set-up, it's necessary to gain some experience with known sources before setting out on the first hunt. With the sense antenna disconnected, carefully rotate the loop and observe the location and deepness of the nulls. The nulls should be exactly 180 degrees apart and right through the loop. If they're not, your loop is not properly balanced to ground, or nearby objects may be affecting the pattern. When balance has been achieved, add your sense antenna, if any, and ad-

just amplitude and phase of the injected signal for a good cardioid pattern.

For serious hunting, the loop should be mounted to rotate freely on a non-metallic mast, using the techniques of Chapter 7. Get up high above the top of the vehicle. If you'll be hunting in the rain, insulate any exposed tuning capacitors to keep them from getting wet and becoming detuned. Be very careful where you take bearings. Try to find a spot that's clear for 100 yards in all directions from your car. T-hunt some known stations at first

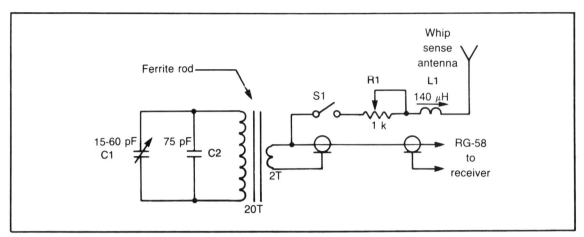

Fig. 17-8. Using a ferrite rod will reduce the size of the antenna, allowing a very small, possibly hand held DF unit.

Fig. 17-9. Hf hunting is much easier with the modern compact radios. Here a Sony 2002 and a 10 meter loop make a small easily-installed hf DF installation.

to get used to the characteristics of the setup and determine where good and bad locations are.

The combined effect of the receiver rf gain control and any preamp gain controls should give your system enough dynamic range to see nulls with both weak and very strong signals. However, rf overloading on very strong signals is likely with receivers such as the Sony ICF-2002 shown in the photo. The resultant pulling of the local oscillator on strong signals may impair direction finding performance, and certainly affects signal intelligibility. If you plan to hunt a lot of very strong hf signals, an attenuator is a good addition to the setup.

The attenuator can be put in the loop preamp box for convenience, or right at the receiver input, to reduce signal pickup by the coax. Requirements for attenuators at hf are less stringent than at vhf and uhf, particularly if you don't transmit through them. A simple pot attenuator (Chapter 6) will probably suffice. An external attenuator may be needed even with receivers having an rf gain control, since using the control may cause the internal S-meter to stop working adequately. In extreme overload cases, the receiver may have to be put into a shielded box (Chapter 12) to eliminate direct pick up.

It's impossible to get a good meter null when there are other signals on the frequency. This is what makes hunting on the Citizens Band with a loop arduous. It also points out the need to keep the system well shielded and grounded. To prevent signal pickup through the case and from the coax shield, the receiver may have to be in a shielded box, grounded to the vehicle frame with a very short strap.

Be sure you're nulling the signal of interest. If the signal is weak, it's easy to null the channel din instead of the signal being hunted. Channel chatter is probably coming from some concentrated population area, which may or may not be in the direction of the signal you seek. Because of user congestion, you'll probably have success hunting only the very strong signals on 11 meters.

Don't expect accurate triangulation help from local fixed stations with their beam antennas. Unfortunately, the beamwidth of low band trap multiband yagis and quads at typical heights isn't all that sharp. Unless you have a long boom monobander a wavelength high, don't expect anything approaching pinpoint accuracy. And how accurate is your rotor indicator? All in all, a fix from beams at fixed stations on hf gives the mobile hunters a good starting point, but that's about the most to hope for.

ADCOCKS FOR BASE STATIONS

Hf loops work best on the vertically polarized ground-wave signals. Ionospheric propagation causes the received signal to be of unknown and varying polarity and elevation. This can make loops totally useless for skywave signals due to their null-destroying horizontal component.

The greater accuracy of an Adcock is due in part to its sensitivity to only the vertical component of the incoming wave. In a properly built Adcock, the feeders are balanced to cancel the horizontal component. In the finest installations the horizontal members and the balanced feed line are also shielded to further reduce horizontal component pickup. Still, the Adcock is not truly precise on skywave signals, though it can be very good for ground wave DFing.

A well-built Adcock will give far sharper indications at a base station than a beam or quad. With care to keep everything balanced, an amateur can build a successful hf Adcock array. The classic Balanced H type (Fig. 17-10), despite all the problems that were detailed in Chapter 1, is still the easiest to get going. The elements should be as large as practical for good sensitivity at the null. Spacing should be a quarter wavelength or less.

Make every effort to build a symmetrical antenna. The supports and framework should be wood or other non-metallic material. Use ladder line between the elements and for the feeder. The coupling network matches to the unbalanced coax line, and should be near the center feed point. C1, a dual section capacitor, and L1 are resonant at the band of interest. C2 and C3 are used to help compensate for any lack of symmetry. They are adjusted for the best null depth on a nearby signal from a dip oscillator or other low powered source. Put the whole matching assembly in a shielded box.

Some users add a bazooka sleeve balun to help keep the coax from affecting the nulls. It is shown in the diagram. The Adcock should be in as clear an area as possible when in use. An efficient Adcock of this type becomes very large on the lowest bands. It turns out that other antennas such as loops can be substituted for the dipoles at these frequencies. The elements are spaced apart from each other on a boom and connected out of phase with balanced line as before.

AN INEXPENSIVE OSCILLOSCOPE DISPLAY

We have frequently pointed out that a problem with most amateur mobile DF equipment is that the real-time information displayed is limited. A gain antenna system with S-meter gives distinct amplitude information for each apparent signal direction (real or reflection), but only one at a time. Hf loops show the same thing, and it's necessary to stop often to check the null depth. Homing and

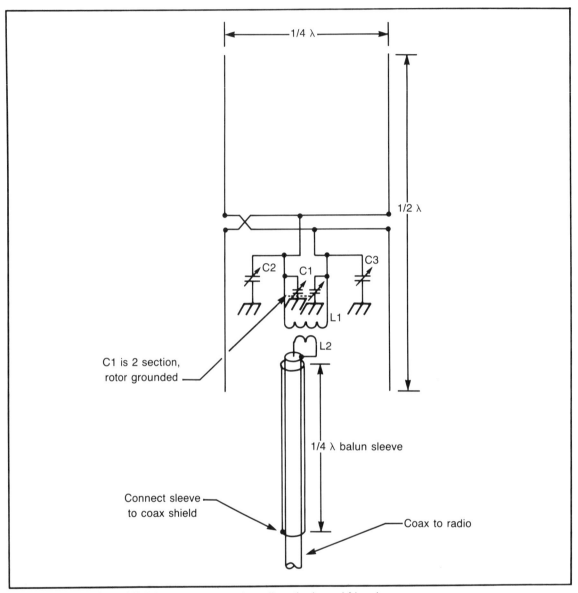

Fig. 17-10. The Balanced H Adcock antenna works well on the lower hf bands.

Doppler units display only a single "best" direction, often a compromise if multipath is present. Any amplitude information from the S-meter with them is non-directional. It is next to impossible to plot by hand all of the possible signal distractions in a multipath situation while in a moving vehicle.

It was with these problems in mind that the late Jim Davis, W6DTR, built the following DF system in the 1960s. It was so successful at the time that Jim was banned from hunting competitively at the club's hunts. He became the perpetual hider since he won nearly every hunt he entered and nobody wanted to hunt against him! Granted, this was before the advent of L-Pers, Dopplers, and BMGs. Yet none of these newfangled devices give as much instant information on possible signal paths and their relative amplitudes as the W6DTR system.

It traces out on a polar coordinate screen the receiver signal strength output versus azimuth as seen by what-

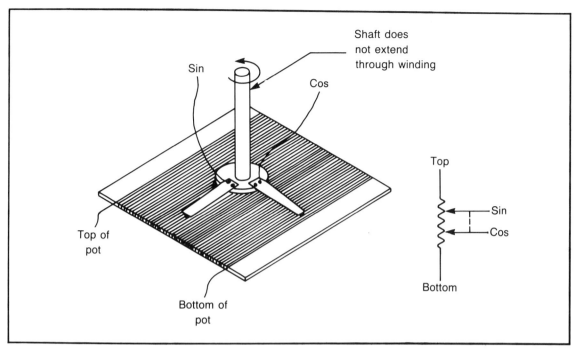

Fig. 17-11. The construction of the SIN-COS pot and the electrical circuit equivalent.

ever antenna system is used. It can be used on any frequency with almost any directional antenna. The method is described here not necessarily for duplication, though that certainly is possible. The purpose in presenting it is to give you food for thought. Perhaps a modernized version could be your next secret weapon.

How It Works

The secret of the display system is the method of continuously generating X and Y oscilloscope tube deflections from the rotating antenna position information. Attached to the antenna mast is a sine-cosine (SIN-COS) potentiometer. A SIN-COS pot has a specially tapered flat winding and two wipers that are 90 degrees apart as shown in Fig. 17-11.

With regulated positive and negative dc voltages impressed across the pot winding as in Fig. 17-12, the two tap outputs each have a sinusoidally varying voltage moving between $-V$ and $+V$ as the shaft is rotated. If we were to plot these two outputs we'd find that they are the same waveform except for a 90 degree shift relative to each other. When these SIN and COS outputs are connected, with proper polarity, to the horizontal and vertical inputs of a dc-response oscilloscope, the spot traces a perfect circle on the screen as the pot shaft is turned,

as in Fig. 7-13A.

At this point, all we have is another sophisticated antenna direction indicator which can be calibrated very accurately by marking the face of the scope tube. Now imagine what happens if the input dc voltage to the pot is replaced by positive and negative voltages corresponding to the received signal strength. This voltage for one side comes from the receiver AGC or S meter circuit. The

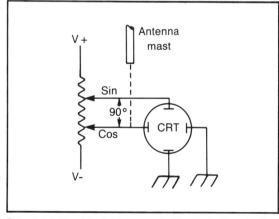

Fig. 17-12. The SIN-COS connected to CRT with only plus and minus dc voltages.

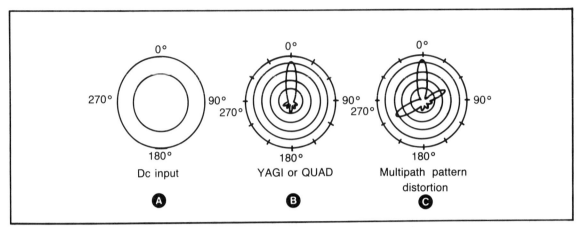

Fig. 17-13. The W6DTR DF CRT displays (a) dc input (b) a single signal without reflections or multipath, and (c) multipath.

other side of the pot is connected to the output of a negative unity gain amplifier fed by the same source, as Fig. 17-14 illustrates. Surprise! The pattern we get as the antenna is spun shows a representation of the signal amplitude in each direction.

The examples of Fig. 17-13B and C assume the use of a gain antenna such as a yagi or quad. In the simplest case of a single incoming signal (B), the scope pattern looks just like the response pattern of the directional antenna, including any side or back lobes. Multipath or additional received signals show up as additional lobes (C). By making a graticule for the scope face as shown with concentric circles as well as azimuth markers, it's easy to pick out the strongest of the incoming signal sources.

Of course sometimes the strongest signal source is a reflection, not the direct signal. This unit excels at iden-

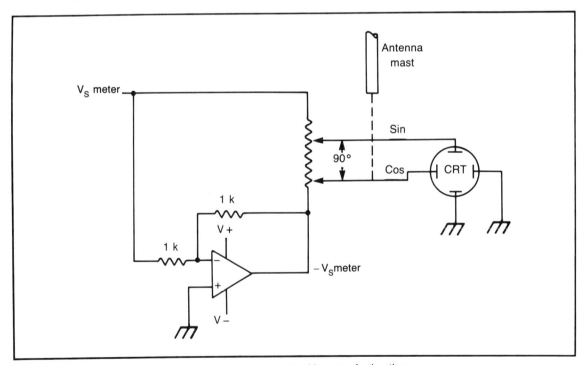

Fig. 17-14. The SIN-COS pot connected to +Vs-meter and −Vs-meter for hunting.

tifying these reflections in a moving vehicle. As the car travels along, the pattern can be continuously updated by rotating the antenna and watching the movement of the display. Figure 17-15 shows several rotations superimposed as the vehicle moves on a straight road. Notice that the reflections vary widely in intensity and azimuth compared to the direct signal.

While a Doppler unit gives an instantaneous display of direction, it is always a single azimuth reading which may be a compromise in strong multipath. The W6DTR system shows all received signal directions, and tells the exact signal level in each direction over time. Best results are achieved when the scope tube has several seconds of persistance, to allow the traces to pile up.

For enough persistance to see several spins at a time, a surplus radar scope tube is used. One such tube is the 5CP7. The 5 indicates the screen diameter and P7 indicates the long persistance yellow phosphor. Voltages for the deflection and gun electronics came from dynamotors then. Today, small inverters would be prefect for mobile use.

The original W6DTR antenna was hung out of the window and rotated by hand with a small right angle gear assembly. Watching the system in action was very interesting, because the apparent direction of the incoming signal often changed dramatically as the vehicle was driven down the street. Sometimes the best apparent signal direction would change almost 180 degrees when passing buildings and power lines. Not only did the direction change but the signal strength varied with terrain and surroundings.

W6DTR's sine-cosine system was used on both 75 and 2 meters. He used a loop on 75 and an Adcock on

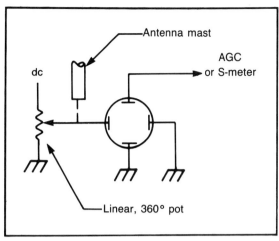

Fig. 17-16. A linear pot can be used with a CRT readout if a linear scale is used.

2 meters. Hunting the nulls is just about as easy with this system as hunting peaks, though the display does not look like the figures. W6DTR didn't put a motor on his unit. He just spun it by hand, but motorizing is certainly possible.

A motorized system may be the answer to a nearly instantaneous full-circle readout. A system could be constructed to locate jammers even when they are making very short transmissions. An antenna for motorized rotation should have low mass, low wind resistance, and have a center of gravity low and in the middle of the array. On 75 or 40 meters, we suggest using a ferrite rod loop spun by a motor, with slip rings to connect to the coil around the rod. A rotating transformer can be used for coupling.

For 2 meters, a phased pair of verticals fits the physical requirements nicely. Such an antenna was described in Chapter 4, and would be constructed for this application as two quarter wave whips. They would be a quarter wavelength apart, on a circular rotating ground plane,

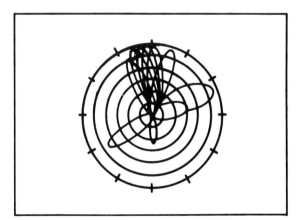

Fig. 17-15. When using a long-persistence phosphor, the CRT shows multipath and reflections as superimposed traces.

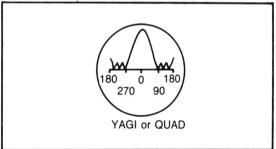

Fig. 17-17. An example of a linear display.

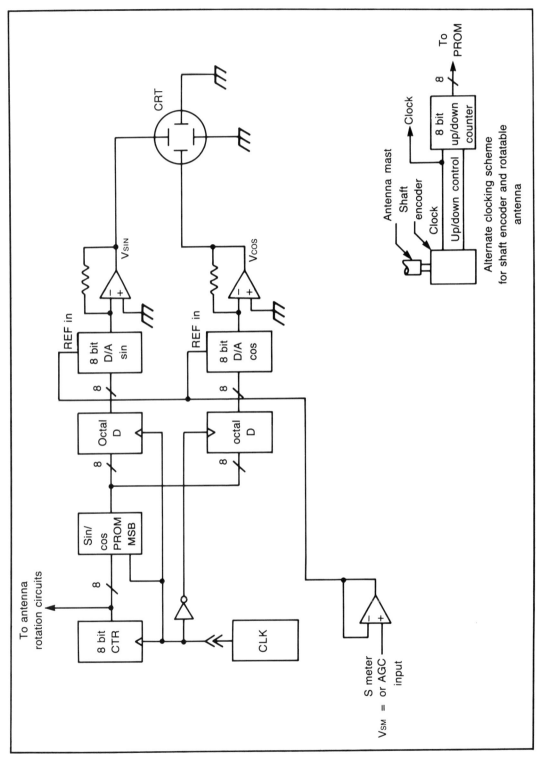

Fig. 17-18. An all-digital SIN-COS CRT display system. The counter driving the PROM could also drive an antenna with a digitally rotated lobe. Note that a Doppler will not work since the antenna pattern is not rotated. An incremental shaft encoder will allow this system to display a manually rotated quad, yagi, loop, etc.

with a three quarter wavelength coax phasing line. A good way of making a coaxial rotary joint at vhf would be to use BNC fittings with the bayonet portion filed down.

A SIN-COS pot is hard to find. Try some of the better surplus stores and mail order houses. It can be eliminated in a "poor man's" W6DTR unit as shown in Fig. 17-16. A linear 360 degree pot, as described in the chapter on antenna mounts and indicators, produces a display that is spread out linearly on the scope face. The peaks and valleys (Fig. 17-17) are interpreted to give the proper bearings.

An All-Electronic Method

A more modern, albeit more complex, approach to deflection generation is shown in block form in Fig. 17-18. The SIN and COS outputs are digitally generated, with the SIN and COS values stored in a look-up table in a Programmable Read Only Memory (PROM). Two 8-bit digital to analog (D/A) converters convert these digital values to sinusoidal analog values used to drive the CRT.

By toggling the MSB address bit with the clock, both the SIN and the COS byte can be loaded into the storage D registers in each clock cycle, with the SIN value stored in the bottom 256 bytes and the COS stored in the upper 256 bytes. Since the SIN/COS value is stored digitally, the -1 to $+1$ value must be converted to 0 to 255 respectively, with 128 being equal to zero. This digital number is then converted to an analog voltage by one of the multiplying D/A converters.

Instead of feeding the S-meter voltage to the top of a pot, it is applied to the reference voltage input of the D/A. With bipolar D/A converters, the X and Y signals to the plates of the CRT will swing both above and below ground. A mechanically rotated gain antenna can be coupled to this digital circuit by means of a shaft encoder as shown in the block diagram.

For a truly state-of-the-art system, you can design an electronically rotated antenna using multiple elements and diode switches. Connect the antenna directional control signals to the PROM address circuitry. It is vital to remember that the electronically-rotated antenna must produce a signal output corresponding to its pointing azimuth. A Doppler unit antenna is not suitable since the signal output is not directional; that is, the Doppler antenna is not actually pointed for maximum or minimum signal like a loop, quad, or phased array is.

Chapter 18

Direction Finding
from Fixed Sites

While most of this book has dealt with mobile hunting, many problems can be dealt with adequately or assisted by having one or more accurate fixed DF setups. It's easy to see how putting the antenna up in the clear on a tower should solve many of the multipath and reflection problems encountered in a car. If the DF system could be installed on a hilltop, perhaps at the repeater site, this would seem to be the ideal DF situation.

Unfortunately, things are not so simple. Most of the same causes of bearing error in a mobile system also happen at a fixed site. They may not be as severe, but can cause large bearing errors when least expected. The nasty thing about these errors is they may not be there all the time. They may change with weather, time of day, whether or not the neighbors are having company, etc. The following examples give you some idea what you may face while trying to achieve accuracy with a fixed site DF system.

VHF VAGARIES

The Coast Guard Auxiliary runs a net of fixed DF sites on the Southern California coast, as discussed in an earlier chapter. These sites are used to locate small craft in distress—either lost, out of fuel, or sinking. They oper-

ate in the vhf marine band, 155 to 163 MHz.

One of these stations is located in Newport Beach, California, about 35 miles southeast of Los Angeles. The station is installed on the roof of a 16-story office building that is itself located on the top of a hill. The view from the top of the building, both for the eye and radio waves, is outstanding. You can see up and down the coast unobstructed for many miles in both directions. When it's clear you can see Santa Catalina Island, 27 miles off the coast. The path to the island is totally and completely unobstructed.

The operators of this station use the marine radiotelephone station on Catalina to calibrate the DF unit, which is made by Regency and is specifically designed for the marine band. Since the location of the radiotelephone station is a known point, one would think that once the DF unit is initially calibrated the system would read correctly all the time.

In actual operation the bearing to the radiotelephone station tends to slowly drift. A 15-degree shift is commonly seen through the course of the day. This drift might be expected if the signal was propagated by troposcatter over a 100 mile or more path, but not when it's 27 mile direct line of sight. The drift appears to coincide with the inversion layer that develops over the area, apparently causing signal refraction. When there is a general cloud cover at about 1000 feet, bearings tend to be more accurate than when the inversion is present. The USCGA works around this problem by recalibrating off of the radiotelephone station regularly through the day.

Another effect on propagation over a water path is shown in Fig. 18-1. At the boundary between the land and water, refraction of the wavefront can occur, just as it does with light waves to cause an optical illusion about the depth of a pool or lake. From the figure we would expect this effect to be more severe when the wavefront strikes the coastline at a shallow angle as shown, rather than nearly perpendicular.

You may hear this phenomenon called either the "coast effect," the "land effect," or "coast refraction." Mobile hunters who search in areas near a river frequently report that the signal often seems to "follow the river," making it difficult to determine where on the shoreline the fox is located. They're also victims of the coast effect. Part of it may be the result of a boundary condition created by different air temperatures and humidities above water and land.

Still other sources of error can come from signal reflection by aircraft, from waveguide-like ducting, and from lacunae, which are islands of atmosphere with discontinuities in the refractive index due to moisture. Sometimes these phenomena cause the bearing to move rapidly and randomly through wide variations. This wandering of the bearing is called galloping.

These errors are not unusual and may easily be missed if your system is calibrated once and assumed to be correct. Check and recheck your DF setup at all times of the day and in all weather before you rely on it without recalibration. A good idea is to run a routine calibration on 3 or 4 known stations, if available, just before any serious DFing. If time doesn't allow a calibration check before taking a DF bearing, then make one immediately after and recompute your initial bearing, if necessary.

HF HEADACHES

Fixed site DF stations for the hf bands are not as affected by local atmospheric conditions, but they are subject to the changing state of the ionosphere. They are also more likely to be upset by the surroundings of the site. During World War II a portable DF system was installed on the East Coast on sandy ground overlooking the ocean. This usually accurate system gave wildly inaccurate

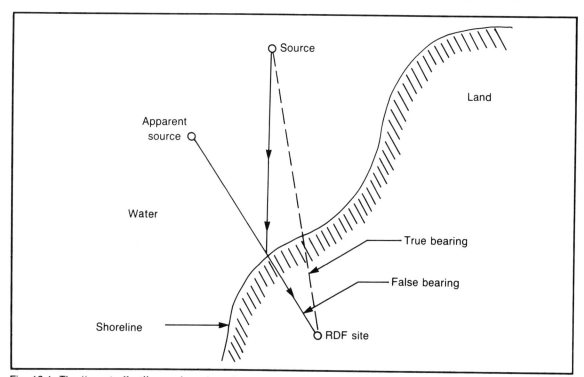

Fig. 18-1. The "coast effect" can often skew the apparent direction of the signal.

results in spite of the efforts of the Army technicians. An engineer from the manufacturer was called out for troubleshooting.

After much effort the problem was uncovered. The site chosen to install the equipment had been a real estate development in the 1920's. The streets and utilities had been put in, but the developer had gone bankrupt before any building had taken place. The site picked for the DF gear happened to be right in the center of what was to be an intersection. The metal pipes under the dry sandy soil provided a ground in four specific directions and no ground the rest of the way around the compass. The buried metal totally negated the high accuracy the system was designed to have.

Site-specific errors can be caused by many different objects and conditions peculiar to your location. Power lines, trees, a creek running by the house, all can contribute to deviations from the true bearing of the signal, assuming that it's coming from the direction of the transmitter in the first place. Local hills can cause reflections that can permanently degrade bearing accuracy. Or it could be something as transient as a car parked in front of your neighbor's house. You don't believe it? One group's tests have shown that a car parked at the base of the vhf DF antenna tower caused incorrect bearings.

Hf, vhf, and uhf setups each have their own types of errors. For example, while problems of reflections may happen at either hf or vhf/uhf, they are more predominant at the higher frequencies. When DFing lower frequency hf and lf/mf, the daytime ground wave may give accurate results while the nighttime skywave gives errors. It's best to assume that you have these problems until proven otherwise.

Now don't let all this frighten you away from fixed site DFing. With all of the problems, it isn't any worse than mobile DFing, except that you can't drive down the road to find a better spot. The secret is to be aware of the limitations and learn to work with them. A few people in an area with good fixed DF setups can do wonders to keep jammers and bootleggers under control. Fixed site triangulation gives the mobile hunters a head start on where to start looking.

SETTING UP A FIXED SITE DF STATION

Now let's turn attention to the how and what—how to do it and what kind of equipment is used. This depends first on whether the setup is locally or remotely controlled. The basic DF equipment can be the same but interfacing it to a readout will be different.

Local and remote DFs share three requirements: a means to rotate the DF antenna, a readout of the bearing, and a way to determine when the antenna is pointing toward the signal. Some DF systems, like the Doppler, don't need to be rotated and give the bearing readout directly while others—quad, Happy Flyers, L-Per, etc.—need to be rotated to the proper direction and this direction determined.

Related to these requirements is the problem of how to actually implement the necessary control and readouts. For example, with a Doppler some means must be found to either run antenna unit control signals up the tower or run the digital direction information back down to a readout next to the radio. Rf signal runs of more than 100 feet mean excessive signal attenuation. If the entire DF system is remotely located, some method must be used to run the signals up and down a data link.

If you are using a Doppler on a nearby tower, putting line driver/receiver pairs on all control and signal lines may be a solution, depending on the actual distance needed. If the system is remotely located, a modem can be used to convert the digital signals to audio tones. The Doppler Systems company units can be ordered with standard RS-232 output that can be directly interfaced to a modem. Doppler Systems also has a version that gives the bearing with a synthesized voice. This voice output can be directly connected to your repeater audio system or control talk-back transmitter.

A simple method of remotely controlling and reading out a mechanically rotating DF antenna has been developed for Happy Flyers DF units. The unit is installed at a remote repeater site on a standard rotator. The signal output from the DF unit that drives the meter is used to control the audio frequency of a voltage controlled oscillator (VCO). The VCO frequency at the zero crossing point of the meter is set to be 900 Hz. The audio signals go to the talk-back transmitter.

If the DF indicates that the antenna is to be turned to the right for crossover, the frequency is higher than 900 Hz. If the DF says the antenna is to be turned to the left, the frequency is lower. A frequency counter is connected to the audio output of the control link receiver and the antenna turned until the counter reads 900 Hz.

A possible improvement to the system is to use the LED outputs on the Happy Flyers DF to cause the tone to make a slight "step" in frequency at crossover. This would be in addition to the sweeping tone change that follows the meter movement. The step makes it much easier to determine the exact point of crossover.

This VCO frequency tells the operator when the antenna is pointed toward the transmitter but doesn't tell what direction that is. The rotator has an internal pot wired as a voltage divider. As the rotator is turned the voltage output changes, normally driving a meter to indicate direction. In this case this voltage is used to drive a second VCO, calibrated to read in degrees.

In use, the antenna is turned until the first VCO reads 900 Hz. Then the control link is used to switch the downlink audio to the second VCO. The frequency of this second VCO is read on the counter and the actual bearing determined by looking at the look-up table calibration chart. The VCO can be calibrated to be direct reading. Set the VCO so that at 0 degrees the output is exactly 1000 Hz and at 360 degrees it is 1360 Hz. When reading the counter, ignore the first digit and read the bearing directly.

In a more sophisticated system, the rotator indication could be transmitted down via a bit stream, ASCII, or RTTY. Even more high-tech would be an automated system that would lock onto the signal and turn the rotor automatically for crossover, then stop and transmit the rotator position data.

Due to temperature and component variations at a hilltop site, VCOs can be expected to drift. Use thermally stable components as much as possible. This problem is best solved by calibrating on known sources when DFing is necessary. With a map having pre-marked fixed sites with correct bearing from the DF location, correction factors can be generated very quickly when needed. Remember to recalibrate often, preferably just before or after you take a bearing on the signal to be located.

These are only a few ideas of how to remote a DF system. Many other ways using digital or analog techniques are possible. Other functions can be remoted in addition to the DF unit. Tuning anywhere in the band of interest can be done remotely with some receivers (the IC-22U is one easy radio to remote). You may also want to have multiple receivers. An additional AM receiver on 121.5 MHz can provide a very valuable public service if the site is very high and in a wilderness area.

CALIBRATING THE STATION

The first step in calibrating your DF system is to ac-
quire a map of the area of interest, preferably with your site in the middle. If possible, don't have it cover more area than necessary. Topographical maps are great. With them you may be able to see what terrain features are causing wild bearings.

Draw lines from your location to as many known stations as possible. For vhf/uhf the input of the local repeater is good, especially if there is a weekly ARES, traffic, or other net on the air. For each of these lines on the map use a protractor to measure the true bearing, relative to north, from your station to each of the other stations. These can be written on the map or made into a table. Assign a number to each of these bearings.

At different times of the day take bearings with your DF setup on these known stations and record the bearings on a chart by the appropriate number. You may find that all the bearings are offset by some amount, or that the bearings in some directions are more accurate than others. Continue to take these bearings for a few weeks as regularly as possible. If there is an obvious offset on all the bearings, recalibrate the system and start over.

If accuracy seems worse on certain days or times, note anything that might be different—for example, weather, time of day, or extra vehicles parked nearby. If after a few weeks the bearings seem to be generally consistent, at least in some directions, then generate a correction table. For example, if the bearings from 0 to 105 degrees seem to be accurate, then note this on a table. If bearings from 106 to 150 are shifted west by varying amounts, then try to determine from the data you have been taking for the last few weeks what the correction should be for each bearing. Obviously, the ideal is to have a correction for each degree around the circle, but you are limited to only the stations that you have been checking.

If all goes well the errors, when plotted, form a smooth deviation curve that fills in the holes between your calibration points. If there is too large a hole in your calibration chart where there are no stations to plot, try having a friend drive along roads in that area and transmit from announced points. While this is only a one-time test, if all the other data looks good you can at least get an idea what is happening in those directions.

Chapter 19

Commercial and Military Direction Finding Systems

Military RDF has come a long way since the battle of Jutland in 1916. It's now an important part of electronic warfare (EW). Sophisticated systems are in use on land, sea, and air. Unlike radar, which requires high power transmissions that can give away the position of forces, DF systems can be completely passive. On the battlefield, RDF is used to:

☐ Determine the position and movement of the enemy
☐ Determine what the enemy has (such as radar, EW, and/or communications equipment)
☐ Provide target data for weapons, jammers, and electronic countermeasures

The FCC and other federal agencies need the most up-to-date RDF equipment for rapid location and tracking of transmitters in all frequency bands. Single vehicle mobile drive-to-it or fly-to-it hunting is still being done, of course. But using computers, it is now possible to link widely separated fixed, portable, and mobile DF stations at high (9.6 kilobaud) rates to obtain multiple bearings, triangulate, and display the results of triangulation on CRT screens in full color.

In this chapter we'll look at some of the techniques

and equipment being used by the armed services and other governmental agencies. Perhaps there are some ideas you can use in your own RDF development effort.

ROTATING ANTENNAS

The W6DTR DF of Chapter 17 is an example of a mechanically rotating directional antenna system. When the antenna does not have to be physically large, mechanical rotation is quite practical. Rates from 1800 rpm for small cylinders down to about 300 rpm for larger reflector antennas are used.

In general, the narrower the antenna beamwidth, the greater the bearing accuracy. Antennas suitable for rotation include microwave horns, which have a fan-shaped beam, and conical spirals, which can cover from uhf through microwaves. Many systems use pairs of antennas with combiners to produce sum and difference outputs. The known characteristics of the sum and difference amplitude and phase patterns of the antenna at each frequency is often used to determine the direction to a higher accuracy than is possible with a single antenna. The method used is an extension of the two-antenna interferometer technique discussed in Chapter 10.

Rotating antennas work well for continuous signals,

but not as well for DFing pulsed signals such as radar emitters. The on/off radar signal characteristic foils attempts to obtain accuracy by utilizing the antenna pattern characteristics. Another disadvantage in a tactical environment is the moving equipment's susceptibility to detection by airborne motion detecting sensors.

LENS ANTENNAS

At microwave frequencies it is possible to use dielectric lenses to focus the rf beam to get directivity. R. K. Luneberg invented a lens, usually in the form of a sphere, which can focus the output of multiple feeds spaced around the circumference of the lens. The refraction index of the lens material varies with the distance from the center of the lens. Full 360 degree coverage and a narrow pattern are features of a Luneberg Lens system.

It turns out that if the many feeders are properly polarized at about 45 degrees, they can all operate simul-

taneously and DFing can be instantaneous. This feature is necessary for proper location of signals with very short pulses such as radar or spread spectrum.

STATE OF THE ART ADCOCK SYSTEMS

Ocean Applied Research (OAR) of San Diego, California, makes a broad line of DF equipment based on the Adcock principle. Models are available to cover 200 kilohertz to 520 megahertz in various ranges with appropriate antenna systems for fixed and mobile use. The design is proprietary to OAR, and is fully automatic with no moving parts.

The DF system used by OAR fits the "instantaneous amplitude comparison" category. Its primary advantage is near instantaneous response, which is very important for DFing short duration signals. A greatly simplified sample is shown in Fig. 19-1.

The example is a two-channel Watson-Watt DF. Two

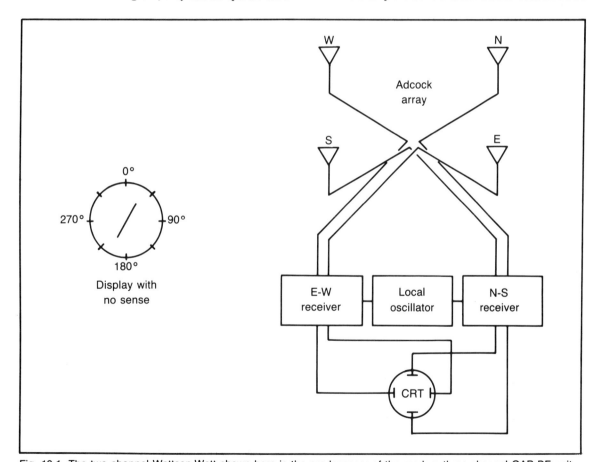

Fig. 19-1. The two channel Wattson-Watt shown here is the predecessor of the modern three channel OAR DF unit.

elevated H Adcock arrays are oriented at right angles. The outputs are fed to identical receiver channels, which must maintain gain tracking over the entire frequency range. The outputs drive the horizontal and vertical channels of an oscilloscope display.

Instead of only two channels, OAR uses a much more sophisticated three-channel system to provide a direct azimuth bearing on the display. The signals from the DF antennas are compared against a non-directional sense antenna. To avoid having to use three closely matched individual receivers, the signal from the sense antennas is modulated by the information from the two DF antennas. After detection, this modulation is stripped off and processed. The level of modulation, which is a function of the signals in the DF antenna channels, is used along with the sense information to give the direction indication.

OAR has done everything possible to bring the state of the art to Adcock DFing, resulting in a lengthy list of extra-cost options. Therefore, OAR's DF gear is usually sold on a "made to order" basis, with specifications determined by the using agency. Yet there are a few standard models:

☐ Model ADFS-320-405, shown in Fig. 19-2, covers 100 to 175 MHz high band vhf, with variations available for 140 to 165 MHz, 118-136 aircraft band, and the 156-162 MHz marine band. Antennas are monopoles or dipoles. AM or FM signals can be DFed. Tuning is by thumbwheel switch.

☐ Model ADFS-335 is similar but covers 70 MHz in the 400 to 520 MHz range.

☐ Model ADFS-928 is a synthesized VFO-tuned unit

covering the FM broadcast band, for use in finding pirate FM radio stations. Camouflaged antenna units ae available for vehicles, using monopole elements.

Using a CRT display of this type makes single vehicle mobile DFing about as easy and trouble-free as possible. All incoming signal azimuths are displayed, and the length of each vector tells the strength of the signal in that direction. Sudden obstructions and reflections are easier to distinguish from the main signal with this display system than with any other. However, on the down side, most OAR antenna units are designed to track vertically polarized signals only.

Equipment from OAR and similar government/military suppliers is truly "Rolls Royce" mobile hunting equipment by ham standards, costing several thousand dollars for a basic system. The sensitivity, accuracy, near instantaneous response, and CRT display make it worth the cost when the application demands this level of sophistication.

WIDE APERTURE WULLENWEBER SYSTEMS

The small aperture of most DF systems in use can be a major disadvantage. The signal can, and probably will, be distorted to some extent by objects and terrain in the near field, giving erroneous bearings. It is very difficult to determine the sources of these terrain feature errors to compensate for them.

For a fixed site installation, one solution is to use a wide aperture system, sometimes shortened to WADF. It samples the incoming wavefront over an area greater than one wavelength, so that disturbances in the wavefront are averaged out. This is done by using a num-

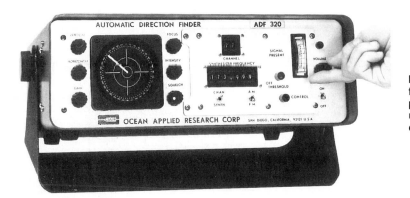

Fig. 19-2. The OAR DF, showing front panel controls and CRT display. A variety of antennas can be used with this unit. (Photo courtesy of Ocean Applied Research.)

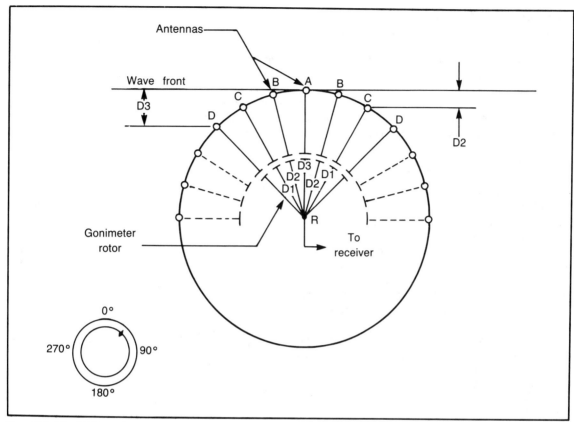

Fig. 19-3. The Wullenweber wide aperture hf DF system.

ber of antennas spread out over a large area. One wide aperture system is the Wullenweber.

The Wullenweber, or Circularly Disposed Antenna Array (CDAA), has a ring (or rings) of vertical monopole antennas 500 to 1000 feet in diameter at hf. Inside of this ring is another ring, which is a vertical shielding screen. An underground screen extends hundreds of feet out under from the monopoles to give a good, consistent ground. Physically smaller arrays can be used in vhf systems at up to 800 MHz.

In operation, the outputs of adjacent antennas covering 100 to 120 degrees of the circumference of the antenna ring are combined together, forming a narrow directional lobe. While electronic beam forming networks are used on current designs, a mechanically spun capacitive goniometer was used for beam forming on the original German design on up through the mid 1960's. We will use the goniometer design for this description since it is easier to visualize.

Since the antennas are in a circle, an approaching pla-

nar wavefront from the transmitter arrives at one antenna first, then perhaps two others, and so on (illustrated in Fig. 19-3). For a signal coming from a specific direction, the output of each of the antennas must arrive at point R in the figure at the same time, and therefore in phase, to obtain directivity. If the total line lengths from each antenna to point R were equal, the signals would arrive out of phase and the antenna would thus not be directive. Hence the need for the delay lines connected to the goniometer rotor plates.

The goniometer rotor is spun by an electric motor, capacitively coupling to about one third of the stators at one time. The outputs of all of the center rotors go through delay lines that bring the center antennas into phase with the two outer antennas. The time the wavefront takes to travel from A to D is delay D3.

Without the delay, the signal from antenna A would arrive at point R before antenna D by the amount D3. The delay line D3 brings the two antenna outputs into phase. Similarly, the delay between antenna A and C is the amount D2, and is corrected by delay line D2. The other antennas and associated delay lines work in the same manner.

Without rotation, the array forms a sharp unidirectional antenna. Beamwidth is a function of the size of the array and the number of antennas being scanned at a time. If forty percent of the antennas are active at a time, a twelve wavelength diameter array will give five degree beamwidth. Less than two degree beamwidth is obtained from a 30 wavelength circle.

As the rotor is spun, the antennas are selected in sequence around the circle, giving a moving, highly directional antenna pattern. A cathode ray tube (CRT) display is synchronized with the rotation of the rotor. As the antenna pattern rotates around, a circle is generated on the CRT display. A received signal from 45 degrees is indicated as a bump on this circle as shown in the detail of Fig. 19-3.

With the edge of the display calibrated in degrees, determining the apparent direction is easy. But in this basic system, other factors must be taken into account such as time of day, ionospheric conditions, and site errors. The Wullenweber equipment itself is capable of up to 0.1 degree bearing accuracy exclusive of propagation errors. In actual use the operational accuracy is around 3 to 6 degrees. Extensive use must be made of ionospheric models so that propagation errors can be predicted and corrected for on sky wave signals.

Wullenweber systems of various levels of sophistication have been manufactured by numerous companies and their characteristics have been studied extensively. Many technical journal articles are available giving detailed information about these systems.

NEW TACTICAL MOBILE DFs

The TC-5100 series by Tech-Comm of Sunrise, Florida, is a new mobile DF line for tactical use which is built around a small slot antenna unit. It's capable of DFing in all directions from 1.5 to 1000 MHz. The antenna looks a bit like a flying saucer, or perhaps one of those frisbee-shaped TV antennas for travel trailers, and is 22 to 34 inches in diameter, depending on the frequency range.

In battlefield action, the antenna is raised up on a mast about 25 feet, giving the greatest range. When needed for other applications, it can be operated when the mast is not extended, or even when in motion. Antennas for specific frequency ranges are available as well as a full frequency coverage antenna. The lack of switching in the antenna system overcomes two problems of

Doppler DFs: disruption of the audio and sensitivity to adjacent channel interference due to cross-modulation.

The antenna is a cavity-backed, top-loaded annular slot system. It has inputs for four or more directional antennas. The incoming signals are amplitude modulated at each input, and then a hybrid circuit combines the individual inputs. The receiver processes this combined signal normally, and then the DF processor correlates the DF modulation information, rejecting random rf energy. The directional information comes from the phase of the detected signal with respect to the phase of the modulation applied at the antenna.

This particular DF processor can also be used with other kinds of antennas, such as yagis or log periodics. This would be most suitable for a fixed site, and would allow greater sensitivity and coverage of any polarization. The processor will handle signals with almost any type of modulation, though AM detection is used for DF purposes.

TRIPLE CHANNEL INTERFEROMETERS

An extremely powerful RDF system is possible by extending the interferometer concepts described in Chapter 10. The Watkins Johnson Company of Gaithersburg, Maryland, makes one such computer controlled triple channel interferometer. Figure 19-4 is a block diagram of the system.

The antenna is simply three short vertical antennas arranged in an equilateral triangle formation. A little mathematics shows that if the relative phase of the incoming signal wave and its frequency are exactly known, the signal bearing can be determined, both in azimuth and elevation. This assumes a small aperture antenna and a plane wavefront signal.

The signals from the antennas go to three identical receivers which track each other in frequency and phase. The bandwidth-filtered i-f outputs of each receiver (at 500 kHz) go to 12-bit analog to digital converters, and then to a special purpose digital computer. An array processor in the computer performs the mathematics to extract the phase information for the signals of interest within the i-f bandwidth. Azimuth and elevation of their incoming wavefronts is determined rapidly by the use of Fast Fourier Transform (FFT) processing.

With the computer for signal processing, it is not necessary to have precisely matched antenna and receiver characteristics. Instead, a calibration is done using a reference signal fed into all three receiver inputs with equal amplitude and phase, and processed by all receivers as if it were any other signal. The computer uses the infor-

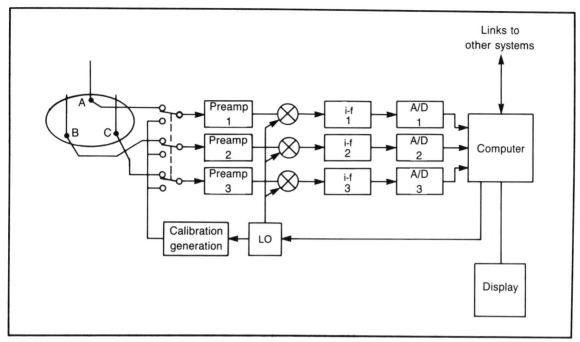

Fig. 19-4. Block diagram of the triple channel interferometer.

mation from this reference signal to generate correction factors for the phase and level information on incoming signals, putting this information into a look-up table for selected frequencies. Interpolation between these frequencies is done. The system must, of course, be calibrated for each of the frequencies in the table.

When systems such as this are installed at a fixed site, the computer can also be used to minimize site-specific errors. Correction factors are developed by the use of transmissions from distant stations at known locations, and are then put into the computer for real-time use.

Expected bearing accuracy of the system is a function of the incoming signal strength. Basic accuracy of 0.02 degrees is possible when the signal-to-noise ratio (SNR) is 50 dB. One degree accuracy is expected at 26 dB SNR, and the error approaches 10 degrees when the signal is only 6 dB above the noise. The SNR can be measured and used to compute the confidence level for any bearing. Of course these bearing accuracy figures do not include the possible effects of the ionosphere on skip signals.

TIME-DIFFERENCE-OF-ARRIVAL DFs

A similar concept, called Time-Difference-of-Arrival

interferometry (TDOA), can achieve bearing accuracies of one degree or better over a wide frequency range. Figure 19-5 is a diagram of an early TDOA system that demonstrates the basic principle. This method is sometimes called inverse LORAN. The incoming wave from 330 degrees azimuth is received by first antenna #1 and then antenna #2. The receiver and detector display a tick at a critical point on the received wave for each channel, producing two distinct blips on the scope screen. In this setup the two fixed delays are exactly equal and are used just to get the blips on the scope face after the undelayed signal produces a trigger.

One channel is inverted so that the operator can easily tell which receiver output is which. If the signal were coming from any direction from 0.1 to 179.9 degrees it would be received by antenna #2 first and the blips would be in opposite order. The distance d between the blips is a measure of direction. It is zero for 0 and 180 degrees and maximum for 90 and 270 degrees. There are ambiguities in this two-antenna setup which can be resolved by rotating the antenna assembly, varying the delay between the channels, or adding a third antenna.

By switching between antennas, a single receiver channel can be used in a TDOA approach. A form of this scheme is used in the BMG Super-DF, which does not have an azimuth readout. When groups of antennas and

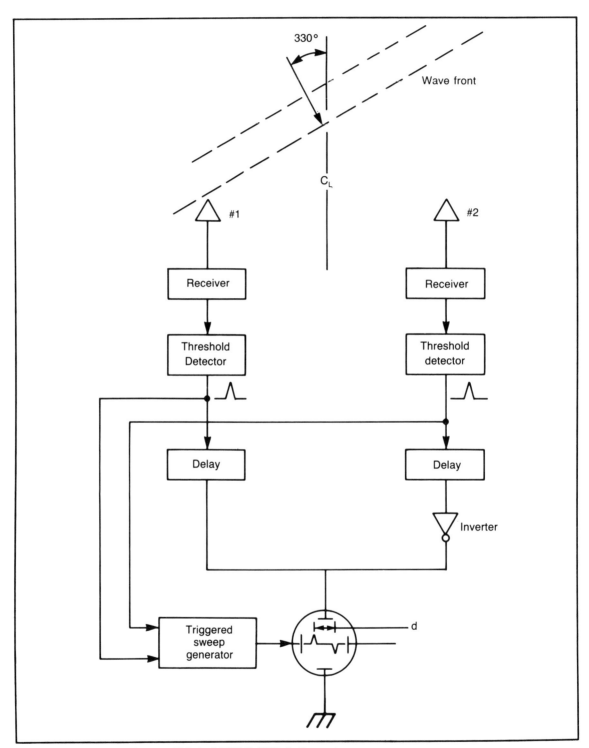

Fig. 19-5. Basic Time-Difference-Of-Arrival (TDOA) CRT display system.

digital processing are employed, a high performance RDF system results.

EM Systems of Sunnyvale, California, is using this technology in a much more advanced way. A complex switching matrix using power splitters, hybrids, and rf switches drives groups of antennas. The antennas can be in L, square, or other formations, but are all located at the facility. The matrix output drives the computer-tuned receiver. The audio and AGC outputs are selectively digitized and processed in a digital central processing unit (CPU). The CPU can be tied to remote displays and other smart DFs over its RS-232 link.

The operator can make a tradeoff between processing time and accuracy. He can choose from 0.5 to 9 seconds, with the longest time giving greatest accuracy. With the internal 12-bit analog to digital converter, display resolution to 0.1 degree is achieved. Overall accuracy is limited by the usual factors. Near-field multipath cannot be corrected for, but error due to the interaction between the antennas, and with the antenna structure, can be compensated for by calibration data stored in EPROMs. The uncalibrated system has an expected accuracy of 4 to 6 degrees. A typical calibrated system is accurate to one degree.

An extension of this scheme is the long baseline TDOA DF. Like the one just described, it uses the time difference of arrival at two antennas to determine the direction of the unknown signal, but the long baseline TDOA separates the antennas by tens or hundreds of miles. The incoming signal is sampled at the same time (within a few microseconds) at each receiving site and the position of the source is determined by the difference in arrival time. Three stations are necessary to get a fix.

With the speed of radio wave propagation on the order of one nanosecond (one thousandth of a microsecond) per foot, an error of one microsecond will equal a 0.2 mile error in locating the unknown transmitter. The receiving sites use atomic frequency standards for synchronization. At a given time all the sites simultaneously digitize the outputs of their receivers. The sites are tied together with data links to transfer the information, in digital form, back to a central processing center.

The digitized data is processed using FFT algorithms. Knowing the time of arrival differences between two stations places the unknown on a hyperbolic curve. In the example of Fig. 19-6, stations #1 and #3 place the unknown on the hyperbola ab. Stations #2 and #3 place the unknown on hyperbola cd. The place where the two curves cross is the location of the unknown signal.

Theoretically, a system of this type is capable of extremely high accuracy, limited only by the timing and measurement accuracy in the receiving systems, at least for signals between the stations. The curve intersection angles become much smaller when the unknown is far

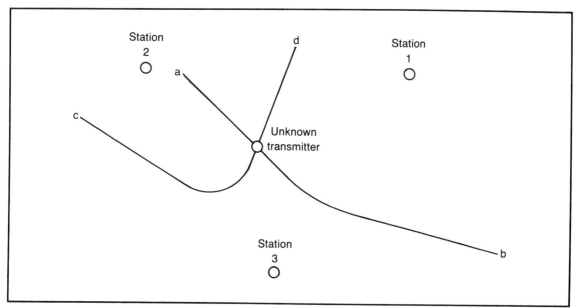

Fig. 19-6. Example of long baseline TDOA signal location. Two stations can place the signal somewhere on a hyperbola between the two stations. Using a third station locates the point on the hyperbola.

away from all stations. Of course in the real world all the problems of DFing at hf apply, and the height and other characteristics of the ionosphere must be measured accurately and modeled in the computer.

So how accurate is the system? Assuming an ionospheric height of 500 kilometers, a timing accuracy of one microsecond, a baseline (spacing between stations) of 1000 miles, a one-hop path, and absolutely optimum conditions, unknowns can be pinpointed within 10 miles.

COUNTERMEASURES AGAINST RDF

A military unit in the field must be constantly aware that enemy RDF is a real threat. The following techniques are used to minimize the risk:

☐ Place radio equipment such that hills or other terrain masks are between the antennas and the forward edge of the battle area (FEBA).

☐ Use directional antennas aimed away from the FEBA, and keep them in the rear part of the forces. Half-rhombics and terminated long wires can be used easily in the field at vhf.

☐ Keep transmissions as short as possible.

☐ Spread out communications and teleprinter transmitters instead of concentrating them in one area. Use wire line remoted radio sets.

☐ Deploy decoy transmitters in areas away from command posts and other critical locations.

☐ Use other means of communications whenever possible, such as field phones and messengers.

Chapter 20

A Mobile Computerized Triangulation System

Your own small microcomputer or programmable calculator can help you win transmitter hunts. The computer is far more accurate than triangulation by hand, particularly when you consider the difficulty of plotting bearings on a large map in a moving car at night. It's faster, too.

To demonstrate the possibilities for computer-aided triangulation and navigation, we developed a simple system. It is capable of being used with very inexpensive computers such as the Sinclair model ZX-81 (Timex 1000) and a battery operated monitor. It can also be used with lap-top computers such as the Tandy Model 100 or with a programmable hand calculator.

Two programs were written for the triangulation system. The first was made as simple to use as possible, to be suitable for hunts where time is a factor. The second has added features for versatility and accuracy, and is better for mileage-only hunts where bearings can be taken relative to reference transmitters from a stationary vehicle.

THE COORDINATE SYSTEM

The program accepts map coordinates in an X-Y system, beginning at the lower left corner of the map as shown in Fig. 20-1. The letter/number margin coordinates of street maps can be used, provided that the vertical sides of the map run exactly north/south. Simply change the letters to numbers, with zero at the bottom. Before using such a map, check it carefully to be sure all divisions are equal width. We found one map where one major vertical division was 25 percent wider than all others.

For Program #1 (Fig. 20-2), choose a map that includes the entire hunt area. Do not use one that covers so much extra territory that the hunt area is only a small portion of it; this reduces accuracy. Measure as accurately as you can the number of miles in each horizontal division, and put this in program line 80 as the horizontal correction factor variable (SH). Similarly, measure the miles in each vertical division and put it into line 90 as SV.

For accurate interpolation between the parallel lines, make up a scale for each map as shown in Fig. 20-1. Coordinates should be entered into the computer as decimal numbers. For example, the X on the map in the figure is at coordinates 2.43, 4.51. If the margin coordinates of the map aren't exactly linear or north/south, draw your own north/south and east/west lines on the map at one-mile or other appropriate intervals, and put in SH and SV accordingly.

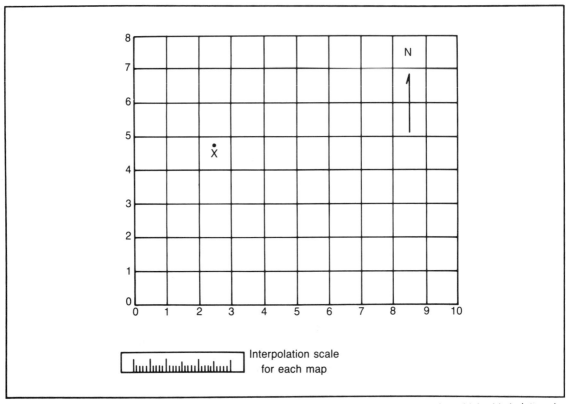

Fig. 20-1. Maps should be marked with coordinates in this manner. The small scale is an overlay which aids in interpolation between the lines.

Program #2 (Fig. 20-3) is more versatile, and allows for multiple maps. Figure 20-4 shows a set of four maps covering a single hunt's boundaries, with slight overlap.

If the initial bearing shows that the signal is from the southeast, for instance, map #3 is used by selecting it from the program menu. Maps may have different scales, with

```
    ENTER STARTING POINT CO-ORDINATES? 5.22,3.06
    (20)ENTER STARTING BEARING RELATIVE TO NORTH? 42
    (30)ENTER YOUR PRESENT CO-ORDINATES? 8.31,7.5
    (40)ENTER PRESENT BEARING RELATIVE TO NORTH? 222
    BEARINGS WHICH ARE EQUAL OR OPPOSITE CANNOT BE TRIANGULATED"
                    GET ANOTHER BEARING"
    (30)ENTER YOUR PRESENT CO-ORDINATES? 7.17,9.04
    (40)ENTER PRESENT BEARING RELATIVE TO NORTH? 181
    ---> THE HIDDEN T IS LOCATED AT  7.10, 5.15
    YOU ARE   7.78 MILES AWAY NOW
    GO   0.14 MILES WEST AND   7.78 MILES SOUTH

    READY FOR ANOTHER BEARING
    (30)ENTER YOUR PRESENT CO-ORDINATES?
```

Fig. 20-2. The screen display for Program #1.

```
        CHOOSE YOUR MAP FROM:
1)  F.R.C.  HUNT,  NORTHEAST  QUADRANT
2)  F.R.C.  HUNT,  SOUTHEAST  QUADRANT
3)  F.R.C.  HUNT,  SOUTHWEST  QUADRANT
4)  F.R.C.  HUNT,  NORTHWEST  QUADRANT
5)  GUIDE TO L.A.  AND ORG.  COUNTY,  START @ BASTANCHURY AND 5T  COL.
6)  AAA  LOS  ANGELES  AND  VICINITY,  NOHL  RANCH  HUNT  START
7)  AAA  ORANGE  COUNTY  AND  CORONA,  START  AT  HOME
8)  CALIFORNIA,  SOUTHERN  HALF,  ALL-DAY  HUNT  START
ENTER MAP NUMBER? 3
        CHOOSE REFERENCE TRANSMITTER FROM:
1)  CODE  PRACTICE  STATION,  147.24  MHZ
2)  LA  HABRA  HEIGHTS  REPEATER,  147.435  MHZ
3)  K6QEH/R,  146.97  MHZ
4)  KC6K/R,  146.79  MHZ
(610)ENTER  REFERENCE  TX  NUMBER?  3
(390)ENTER  STARTING  CO-ORDINATES  OR  0,0  TO  USE  PROGRAMMED  ONES?  0,0
---> REFERENCE T IS AT 257.5 DEGREES RELATIVE TO NORTH
(640)ENTER  REFERENCE  BEARING  RELATIVE  TO  VEHICLE?  5
---> YOUR VEHICLE HEADING SHOULD BE 252.5 DEGREES
(520)ENTER  STARTING  HIDDEN  T  BEARING  RELATIVE  TO  VEHICLE?  134

     * * * * READY TO BEGIN THE HUNT.  GOOD LUCK! * * * *"

IF YOU NEED TO ENTER A NEW REFERENCE TRANSMITTER, ANSWER 0,0"
                TO NEXT QUESTION
(600)ENTER  YOUR  PRESENT  CO-ORDINATES?  10.2,6.3
---> REFERENCE T IS AT 44.4 DEGREES RELATIVE TO NORTH
(640)ENTER  REFERENCE  BEARING  RELATIVE  TO  VEHICLE?  66
---> YOUR VEHICLE HEADING SHOULD BE 338.4 DEGREES
(650)ENTER  PRESENT  HIDDEN  T  BEARING  RELATIVE  TO  VEHICLE?  32
YOUR BEARINGS DO NOT CONVERGE!!  GET ANOTHER BEARING.
(600)ENTER  YOUR  PRESENT  CO-ORDINATES?  10.6,6.5
---> REFERENCE T IS AT  43.7 DEGREES RELATIVE TO NORTH
(640)ENTER  REFERENCE  BEARING  RELATIVE  TO  VEHICLE?  90
---> YOUR VEHICLE HEADING SHOULD BE 313.7 DEGREES
(650)ENTER  PRESENT  HIDDEN  T  BEARING  RELATIVE  TO  VEHICLE?  165
---> THE HIDDEN T IS LOCATED AT 18.15, 2.37
YOU ARE   6.46 MILES AWAY NOW
GO   5.66 MILES EAST AND   3.10 MILES SOUTH

READY FOR ANOTHER BEARING
IF YOU NEED TO ENTER A NEW REFERENCE TRANSMITTER, ANSWER 0,0"
                TO NEXT QUESTION
(600)ENTER  YOUR  PRESENT  CO-ORDINATES?
```

Fig. 20-3. The screen display for Program #2.

separate correction factors for each.

HUNTING WITH A COMPUTER

To be a timesaver in a competitive T-hunt, the system must be easy to use and anticipate the needs of the hunter. The first program was written with this in mind. It is loaded from tape before the hunt begins, and the starting coordinates are entered. (They can be kept as data on tape if you always start from the same point.) Since the starting bearing is taken from a high and reflection-free point, it is used as the reference for all future bearings. After entering this bearing, the computer asks for a bearing from another location. This can be gotten from a distant base station, if such is permitted, or taken later in the hunt. The user enters the second coordinates and bearing with respect to north, and the computer displays the distance to the transmitter.

As an aid to navigation in urban areas, the computer also displays the distance to the transmitter in terms of north/south and east/west directions. In addition, the expected map coordinates of the hidden T are displayed. If the triangulated location is off the map, directions and distances to the hidden T are still computed properly. The coordinates are displayed as if the map were expanded, with higher numbers at top and right, and negative numbers at left and bottom.

The computer then waits for another bearing, which is triangulated again against the initial bearing. This is done as many times as desired while the hidden T is approached. If bearings are entered which diverge instead of converge, a warning message is printed and the computer awaits another bearing.

It is important to note that two bearings with the same azimuth, or azimuths 180 degrees apart, cannot be trian-

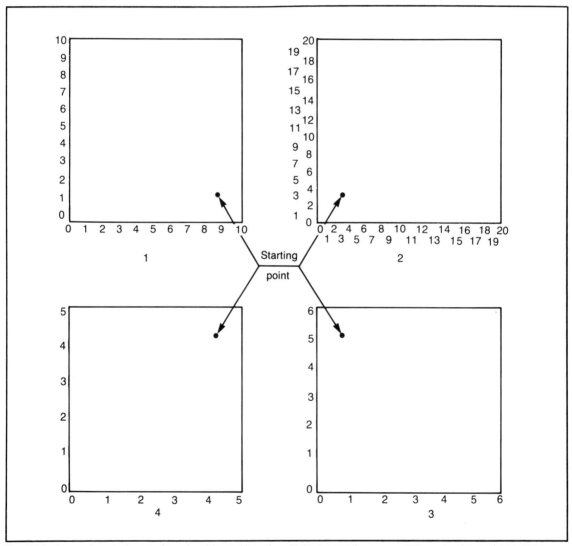

Fig. 20-4. A sample set of maps for a local hunt. Use of multiple maps allows versatility and enhanced accuracy.

gulated either manually or with the computer. Normally, roads and terrain layout will prevent following an exact straight line course to the transmitter, and this problem won't arise. If it does, the program warns the user to get a new bearing.

This program may also be used in another way when the rules allow the use of a helper at another location to give you a bearing. Enter the helper's bearing as the starting bearing and your own bearings thereafter. That way, each time you take a bearing along the way, you get a triangulated fix.

Program #2 includes some other timesaving conveniences. Upon loading, the program asks which map will be used for the hunt. This is necessary because the correction factors and coordinates of the reference transmitters are different on each map. The user is then asked which pre-programmed reference transmitter will be used. The chosen reference should be powerful enough for bearings over the entire area of the hunt. Although the program does allow the use of a different pre-programmed reference later in the hunt, it is better to stick to only one to reduce the possibility of error.

It is possible to use reference transmitter locations not actually on the map, provided the location of the reference T can be extrapolated in terms of the coordinates of that map. The coordinates are negative for X-west-of-map and Y-south-of-map references. Figure 20-5 gives an example of an off-map reference. If the selected reference transmitter isn't programmed for the selected map, an error message will be displayed. Each time a triangulation is desired, the hunter's coordinates are entered, then the bearing of the reference transmitter, followed by the bearing of the hidden T, both relative to the vehicle.

For each computation, the program determines a correction factor for the hidden T bearing based on the reference bearing. It then performs the triangulation using this corrected bearing and the starting bearing. The screen displays triangulation information as before. It also displays what the bearing (relative to north) to the reference transmitter should be, based on the coordinates entered, and what the car heading (relative to north) should be, based on the reference bearing. This is done as a check on the user and the system. If the predicted vehicle head-

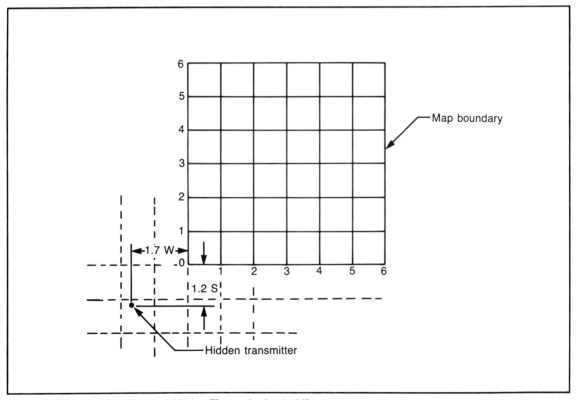

Fig. 20-5. Off-map references and hidden T's can be located if necessary.

ing is significantly different from what your car compass or instincts tell you, find out why. Such an error can be caused by:

☐ Inaccurate coordinates. Are you really where you think you are? Is your car pointing where you think it is?
☐ Inaccurate bearings. Did that last low tree mash the quad?
☐ Reflections. Inaccurate reference bearings can occur if you're hunting in a multipath area with the quad set for horizontal polarization and the reference repeater is vertically polarized.

If you're short on time and don't want to take a reference T bearing using Program #2, you can enter a compass bearing. This should only be done if you are confident of the direction your vehicle is headed and are sure of the on-axis performance of your antenna. Here's how to do it: After entering the coordinates, the computer displays the expected direction of the reference T. Enter this expected direction as your reference bearing. The computer tells you that you are headed 0 or 360 degrees. Now enter your bearing with respect to north.

While it is possible to write the program to allow entry of scale factors and reference transmitter coordinates at run time, it is far easier in the case of small computers like the TS-1000 to put them directly into the program and then SAVE the program on tape. When the tape is loaded prior to the hunt, all data is ready.

EXPLANATION OF THE PROGRAMS

To be sure these programs are easily adaptable to today's myriad of computers, we wrote them as simply as possible, with a minimum of programming tricks. The listing is in BASIC-80, better known as Microsoft™ BASIC. BASIC-80 is widely known and easily converted to other forms of BASIC such as that used in the ZX-81 or the lap top Radio Shack computers.

Figure 20-6 is the listing of Program #1 and Fig. 20-7 is the listing of Program #2. To aid you in modifying the programs to the particular BASIC dialect you need, or enhancing them, here's a run-through of the program of Program #2. (Almost all of the statements of Program #1 are also in Program #2. A list of variables common to both programs is provided in Fig. 20-8.)

Statements with numbers below 100 are for entry of constants and remarks. After entry of map number at statement 190, a loop (200-220) reads the scale factors and starting coordinates from that map from data statements 1000 to 1003.

Subroutine 1230 to 1380 enters a reference transmitter number and reads its coordinates in a similar manner in loop 1330-1350. The data statement containing the coordinates (1010-1080) is chosen by a RESTORE command (1301-1308), which in turn is selected by the map number in line 1290. Lines 1360-1380 are a trap to verify that the reference T coordinates are programmed for that map.

Revised starting coordinates are entered at 390-420 if desired, and the program branches to subroutine 1440-1510, which determines the expected bearing to the reference transmitter, first in polar representation at 1440-1470, then converting to compass representation at 1471-1475. Statements 1456-1457 handle the special case of the reference being directly north or south of the hunter, to prevent a divide-by-zero error. Statements 1472-1473 resolve the 180 degree ambiguity of the arctangent function.

At this point the program prints the expected direction of the reference T (1476) and asks for the hunter's reference bearing (1480). From this, the vehicle heading H is determined and printed at 1490-1500. The starting point bearing is entered (520), it is added to the heading to determine a compass bearing (530-540), and the hunt is begun (550-575).

If the hunter wants a new reference transmitter for his next bearing, he inputs 0 to the prompt in statement 600 and the program jumps to subroutine 1230-1380 to select it as before. Otherwise, latest coordinates are accepted and subroutine 1440-1510 again determines the exact vehicle heading, which is added to the second bearing at 660-665.

Triangulation is performed and location of the hidden T is printed at 670-910. The formulas for triangulation are:

$$x_t = \frac{x_2 \tan \theta_2 - x_1 \tan \theta_1 + y_1 - y_2}{\theta_2 - \theta_1}$$

$$y_t = (x_t - x_1) \tan \theta_1 + y_1$$

θ_1 and θ_2 are bearings 1 and 2 converted to polar form

Statements 680-685 form a trap to handle the special case of equal or opposite bearings, which cannot be triangulated. In such case, the program jumps back to the entry point for another field bearing.

If the bearings diverge, the formula's predicted location will be in the exact opposite direction of the field bearing. Lines 743 and 745 check for this condition and

```
10 REM             DF2.BAS 1-2-84
11 REM             DIRECTION FINDING PROGRAM #1 BY J. D. MOELL
12 REM             FOR "TRANSMITTER HUNTING: The Art and Science of Radio
13 REM             Direction Finding."         Microsoft [TM] Basic
70 LET DR=0.0174532
80 LET SH=2
90 LET SV=2
390 INPUT "ENTER STARTING POINT CO-ORDINATES";X1,Y1
520 INPUT "(20)ENTER STARTING BEARING RELATIVE TO NORTH";B1
525 LET P1=TAN((450-B1)*DR)
630 INPUT "(30)ENTER YOUR PRESENT CO-ORDINATES";X2,Y2
640 INPUT "(40)ENTER PRESENT BEARING RELATIVE TO NORTH";B2
670 LET P2=TAN((450-B2)*DR)
680 IF ABS(B2-B1)<>0 AND ABS(B2-B1)-180<>0 THEN GOTO 690
684 PRINT "BEARINGS WHICH ARE EQUAL OR OPPOSITE CANNOT BE TRIANGULATED"
685 PRINT "                    GET ANOTHER BEARING"
686 GOTO 630
690 LET XT=((P2*X2)-(P1*X1)+Y1-Y2)/(P2-P1)
700 LET YT=P1*(XT-X1)+Y1
730 LET HD=(XT-X2)*SH
740 LET VD=(YT-Y2)*SV
743 IF SGN(COS((450-B2)*DR))=SGN(HD) THEN GOTO 750
745 IF SGN(SIN((450-B2)*DR))=SGN(VD) THEN GOTO 750
747 PRINT "YOUR BEARINGS DO NOT CONVERGE!!  GET ANOTHER BEARING."
748 GOTO 630
750 LET DT=SQR(HD^2+VD^2)
753 PRINT USING "---> THE HIDDEN T IS LOCATED AT ##.##,##.##";XT;YT
755 LET EW=ABS(HD)
756 LET NS=ABS(VD)
760 PRINT USING "YOU ARE ###.## MILES AWAY NOW";DT
830 IF HD=>0 AND VD=>0 THEN GOTO 870
840 IF HD=>0 AND VD<0 THEN GOTO 890
850 IF HD<0 AND VD>0 THEN GOTO 910
860 PRINT USING "GO ###.## MILES WEST AND ###.## MILES SOUTH";EW;NS
865 GOTO 920
870 PRINT USING "GO ###.## MILES EAST AND ###.## MILES NORTH";EW;NS
880 GOTO 920
890 PRINT USING "GO ###.## MILES EAST AND ###.## MILES SOUTH";EW;NS
900 GOTO 920
910 PRINT USING "GO ###.## MILES WEST AND ###.## MILES NORTH";EW;NS
920 PRINT
930 PRINT "READY FOR ANOTHER BEARING"
940 GOTO 630
2000 END
```

Fig. 20-6. This BASIC triangulation program is written in BASIC-80, but is readily adaptable to most other dialects.

print a warning message at line 747 when bearings diverge.

THE MOBILE SET-UP

The Timex computer and RAM pack require 9 Vdc at less than one half ampere, readily available in the car with the use of a three-terminal regulator, as shown in Fig. 20-9. It might appear that the car battery is an uninterruptable power supply for the computer, but beware. What if the engine must be restarted during the hunt?

275

```
10 REM             DF3.BAS 1-15-84
11 REM        DIRECTION FINDING PROGRAM #2 BY J. D. MOELL
12 REM        FOR "TRANSMITTER HUNTING: The Art and Science of Radio
13 REM        Direction Finding."        Microsoft [TM] Basic
20 LET DR=0.0174532
100 PRINT "   CHOOSE YOUR MAP FROM:"
110 PRINT "1) F.R.C. HUNT, NORTHEAST QUADRANT"
120 PRINT "2) F.R.C. HUNT, SOUTHEAST QUADRANT"
130 PRINT "3) F.R.C. HUNT, SOUTHWEST QUADRANT"
140 PRINT "4) F.R.C. HUNT, NORTHWEST QUADRANT"
150 PRINT "5) GUIDE TO L.A. AND ORG. COUNTY, START @ BASTANCHURY AND
    ST. COL."
160 PRINT "6) AAA LOS ANGELES AND VICINITY, NOHL RANCH HUNT START"
170 PRINT "7) AAA ORANGE COUNTY AND CORONA, START AT HOME"
180 PRINT "8) CALIFORNIA, SOUTHERN HALF, ALL-DAY HUNT START"
190 INPUT "ENTER MAP NUMBER";M
200 FOR X=1 TO M
210 READ SH,SV,X1,Y1
220 NEXT X
230 GOSUB 1230
390 INPUT "(390)ENTER STARTING CO-ORDINATES OR 0,0 TO USE PROGRAMMED
    ONES";X,Y
400 IF X<>0 THEN GOTO 430
410 LET X=X1
420 LET Y=Y1
430 GOSUB 1440
520 INPUT "(520)ENTER STARTING HIDDEN T BEARING RELATIVE TO VEHICLE";B
530 LET B1=B+H
535 IF B1=>360 THEN LET B1=B1-360
540 LET P1=TAN((450-B1)*DR)
545 PRINT
550 PRINT "   * * * * READY TO BEGIN THE HUNT.  GOOD LUCK! * * * *"
555 PRINT
560 PRINT "IF YOU NEED TO ENTER A NEW REFERENCE TRANSMITTER,ANSWER 0,0"
570 PRINT "                TO NEXT QUESTION"
600 INPUT "(600)ENTER YOUR PRESENT CO-ORDINATES";X2,Y2
610 IF X2=0 THEN GOSUB 1230
620 LET X=X2
630 LET Y=Y2
640 GOSUB 1440
650 INPUT "(650)ENTER PRESENT HIDDEN T BEARING RELATIVE TO VEHICLE";B
660 LET B2=B+H
665 IF B2=>360 THEN LET B2=B2-360
670 LET P2=TAN((450-B2)*DR)
680 IF ABS(B2-B1)<>0 AND ABS(B2-B1)-180<>0 THEN GOTO 690
684 PRINT "BEARINGS WHICH ARE EQUAL OR OPPOSITE CANNOT BE TRIANGULATED"
685 PRINT "                 GET ANOTHER BEARING"
686 GOTO 600
690 LET XT=((P2*X2)-(P1*X1)+Y1-Y2)/(P2-P1)
700 LET YT=P1*(XT-X1)+Y1
730 LET HD=(XT-X2)*SH
740 LET VD=(YT-Y2)*SV
743 IF SGN(COS((450-B2)*DR))=SGN(HD) THEN GOTO 750
```

Fig. 20-7. This program is more complex, but allows for multiple maps and reference bearings from fixed transmitters.

```
745 IF SGN(SIN((450-B2)*DR))=SGN(VD) THEN GOTO 750
747 PRINT "YOUR BEARINGS DO NOT CONVERGE!!  GET ANOTHER BEARING."
748 GOTO 600
750 LET DT=SQR(HD^2+VD^2)
755 PRINT USING "---) THE HIDDEN T IS LOCATED AT ##.##,##.##";XT;YT
760 LET EW=ABS(HD)
770 LET NS=ABS(VD)
780 PRINT USING "YOU ARE ###.## MILES AWAY NOW";DT

830 IF HD=)0 AND VD=)0 THEN GOTO 870
840 IF HD=)0 AND VD<0 THEN GOTO 890
850 IF HD<0 AND VD>0 THEN GOTO 910
860 PRINT USING "GO ###.## MILES WEST AND ###.## MILES SOUTH";EW;NS
865 GOTO 920
870 PRINT USING "GO ###.## MILES EAST AND ###.## MILES NORTH";EW;NS
880 GOTO 920
890 PRINT USING "GO ###.## MILES EAST AND ###.## MILES SOUTH";EW;NS
900 GOTO 920
910 PRINT USING "GO ###.## MILES WEST AND ###.## MILES NORTH";EW;NS
920 PRINT
930 PRINT "READY FOR ANOTHER BEARING"
950 GOTO 560
990 REM DATA FORMAT FOR LINE 1000-1002 IS SH(1),SV(1),X1(1),Y1(1),ETC.
1000 DATA .66,.66,3.74,3.95,.75,.75,5.86,16.35,.75,.75,25.12,16.35
1001 DATA .75,.75,13.96,6.22,3.5,3.5,10.9,6.45,7.2,8.8,9.5,3.4
1002 DATA 3.6,3.8,5.85,15.48,11.3,11.3,10.8,2.4
1009 REM DATA FORMAT FOR LINES 1010-80 IS XR(1),YR(1),XR(2),YR(2),ETC
1010 DATA 0,0,0,0,-1.95,2.55,8.28,7.76
1020 DATA 0,0,0,0,-0.51,14.91,0,0
1030 DATA 0,0,0,0,18.63,14.91,0,0
1040 DATA 0,0,10.91,13.74,9.03,5.03,0,0
1050 DATA 0,0,10.25,8.3,9.75,6.2,11.6,7.15
1060 DATA 5.6,6,8.5,4.7,8.3,3.9,9.1,4.3
1070 DATA 0,0,5.4,19.2,4.33,15.59,6.95,7.21
1080 DATA 9.7,3.2,10.3,3.7,10.5,3.4,10.7,3.6
1230 PRINT "   CHOOSE REFERENCE TRANSMITTER FROM:"
1240 PRINT "1) CODE PRACTICE STATION, 147.24 MHZ"
1250 PRINT "2) LA HABRA HEIGHTS REPEATER, 147.435 MHZ"
1260 PRINT "3) K6QEH/R, 146.97 MHZ"
1270 PRINT "4) KC6K/R, 146.79 MHZ"
1280 INPUT "(610)ENTER REFERENCE TX NUMBER";R
1290 ON M GOTO 1301,1302,1303,1304,1305,1306,1307,1308
1301 RESTORE 1010:GOTO 1330
1302 RESTORE 1020:GOTO 1330
1303 RESTORE 1030:GOTO 1330
1304 RESTORE 1040:GOTO 1330
1305 RESTORE 1050:GOTO 1330
1306 RESTORE 1060:GOTO 1330
1307 RESTORE 1070:GOTO 1330
1308 RESTORE 1080
1330 FOR X=1 TO R
1340 READ XR,YR
1350 NEXT X
1360 IF XR<>0 THEN RETURN
1370 PRINT "CHOSEN REFERENCE NOT ON CHOSEN MAP!! CHOOSE AGAIN"
1380 GOTO 1230
```

```
1440 LET HR=XR-X
1450 LET VR=YR-Y
1454 IF HR<>0 THEN GOTO 1460
1456 IF YR=>Y THEN LET Z=90
1457 IF YR<Y THEN LET Z=270
1458 GOTO 1472
1460 LET PR=VR/HR
1470 LET Z=ATN(PR)/DR
1471 IF Z<0 THEN Z=Z+360
1472 IF HR<0 THEN LET Z=Z+180
1473 IF Z=>360 THEN LET Z=Z-360
1474 LET BR=90-Z
1475 IF BR<0 THEN LET BR=BR+360
1476 PRINT USING "---> REFERENCE T IS AT ###.# DEGREES RELATIVE TO
     NORTH";BR
1480 INPUT "(640)ENTER REFERENCE BEARING RELATIVE TO VEHICLE";B
1490 LET H=BR-B
1495 IF H<0 THEN LET H=H+360
1500 PRINT USING "---> YOUR VEHICLE HEADING SHOULD BE ###.# DEGREES";H
1510 RETURN
2000 END
```

VARIABLES FOR BASIC TRIANGULATION PROGRAMS

B	Bearing relative to vehicle
B1	Initial hidden T bearing with respect to north
B2	Latest hidden T bearing with respect to north
BR	Expected reference T bearing with respect to north
DR	Degrees to radians conversion factor
DT	Air line distance to transmitter from hunter in miles
EW	East/west distance to transmitter from hunter in miles
H	Heading of vehicle with respect to north
HD	Horizontal grid distance to transmitter from hunter
HR	Horizontal grid distance to reference T from hunter
M	Chosen map number
NS	North/south distance to transmitter from hunter in miles
P1	Tangent of polar representation of initial bearing
P2	Tangent of polar representation of latest bearing
PR	Tangent of polar representation of expected reference T bearing
R	Chosen reference transmitter number
SH	Horizontal map scale factor in miles per division
SV	Vertical map scale factor in miles per division
VD	Vertical grid distance to transmitter from hunter
VR	Vertical grid distance to reference T from hunter
XR	Horizontal grid co-ordinate of reference T
XT	Horizontal grid co-ordinate of hidden T
X1	Horizontal grid co-ordinate of initial bearing
X2	Horizontal grid co-ordinate of latest bearing
YR	Vertical grid co-ordinate of reference T
YT	Vertical grid co-ordinate of hidden T
Y1	Vertical grid co-ordinate of initial bearing
Y2	Vertical grid co-ordinate of latest bearing

Fig. 20-8. List of variables used in both programs.

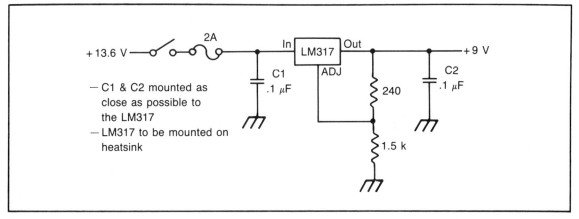

Fig. 20-9. A simple regulator circuit will power the ZX-81 and similar computers from the car battery.

The program will crash if the starter drops the car battery voltage below about 9 volts. Consider a set of Sealed Lead Acid batteries (see Chapter 14) or other independent power source for the computer.

A battery-operated tape recorder is required for program loading and a portable TV set is used as a monitor. Car battery glitches are not a problem with these peripherals. Some states prohibit operation of a TV in a vehicle if it can be seen by the driver. Take no chances. Make it visible only to the beam-turner and navigator.

The TV set we used for experimentation has its circuit ground connected to the positive side of the supply. To use such a set, the antenna and any other grounded external metal parts should be taped up or removed to prevent contact with vehicle ground. For dc isolation between the computer and the set, put 100 picofarad capacitors in series with each lead of the twin lead from the 75-to-300 ohm transformer or game switch where it connects to the TV antenna input. This is necessary because the internal balun has no dc isolation. A self-powered computer such as the Tandy Model 100 won't have these problems and won't need to be loaded with a cassette recorder. It has a constant battery backed up memory.

Before use on a hunt, the computer system should be checked carefully for compatibility with the radio gear. If the computer puts hash into DFing gear, check both for poor case shielding. There should be a good connection between the circuit ground and case, or the metallization inside a plastic case. Computers having plug-in RAM packs and other accessories may have to be enclosed in an overall shielded box in stubborn cases.

The monitor may also be a cause of QRM if it or its cables are poorly shielded. Good shielding and grounding practices should be used with lines for power, data,

tape recorder, and monitor. Keep them as short as possible. With proper care, there will be no degradation in DF sensitivity due to the presence of the computer, and transmitting will cause no adverse effects on the computer system.

CORRECTING ERRORS

In the heat of a competitive hunt, it's easy to make an error in entering a location or bearing. Sometimes these errors cause a program break and at other times wrong data results. It is time-consuming to restart the program, because all starting variables must be put in again.

With an interpreter BASIC system such as used in the ZX-81 or Model 100, restarting is unnecessary in most cases. It is generally possible to GOTO the appropriate input statement and enter the data again. To aid the user in recovering from mistakes, screen prompts have a line number in parentheses, representing the line number to restart the program.

For example, in the second program, let's say that you're nearing the T and inadvertently enter 9 when asked for a new reference T number. You realize your mistake, press the BREAK key, and enter GOTO 610. This gets you back to the new reference T entry point without destroying your starting bearing data.

These numbers are for use only during the hunt after the initial selection of map and taking the starting point reference bearing is done. If errors occur before this, restart the program.

IMPROVING THE SYSTEM

There are many ways to improve both the hardware and software of this system. Most require changes very specific to the computer and hunting equipment used. Computer technology is changing very rapidly, so we'll just give some general descriptions here, to help you imagine what you can do with the computers available to you now.

☐ More preprogrammed locations. The coordinates of certain key intersections or high points could be programmed into the computer to save time during the hunt.

☐ North reference. If you have a very accurate car compass, use it to determine your vehicle's heading for accurate bearings relative to north. Modify the program so that for each input the car's heading is entered, followed by the hidden T's bearing with respect to vehicle heading. An even better way is to digitize the compass information and tie it directly into the computer.

☐ Direct bearing input. Some Doppler-type DF units provide bearing data in digital form on a rear-panel connector. This can be tied to the computer and entered automatically on command.

☐ Computerized beam/quad bearings. A shaft encoder can digitize the azimuth information from the antenna mast. This data can be read automatically by the computer to within 0.1 degree, instead of being entered manually. Since a hunter cannot usually peak a quad to that accuracy, consider fully automating the bearing-taking function by also feeding S-meter data into the computer through an analog-to-digital converter. Then the computer can take the bearing automatically. As the hunter twirls the mast, the computer determines the point of maximum signal and displays its azimuth while performing the triangulation.

☐ Bearing correction. A directional antenna on a vehicle may have the peak of its frontal lobe off-axis in some directions due to proximity effects of the vehicle body. A correction factor can be applied within the computations if these effects are documented on a pattern range, via a look-up table incorporated into the program.

☐ Graphics. Depending on the capabilities of the computer and programmer, added graphics can cover a wide range of complexity. A simple schematic representation of the triangulation would show the starting point, reference transmitter, vehicle location, and predicted hiding point. At the other end of the scale, the entire map could be put into the memory of some larger machines. This was not done in our system because the graphics resolution of the ZX-81 is inadequate for even a simple display.

In the two programs described, no averaging of multiple bearing was done. In most hunt situations when the same equipment is used for each bearing, the predicted locations tend to become more and more accurate with each successive reading. Averaging under these circumstances is counterproductive.

There are cases where averaging several bearings is useful and contributes to overall accuracy. For example, the bearings of several fixed stations on a jammer can be averaged to help cancel the effects of individual equipment variations. This can be done in either program by taking the bearings in pairs, storing the results of triangulation in an array, and averaging the X and Y coordinates to find the average location. Bearings which differ markedly from the average be flagged and discarded, and the results re-computed for greater accuracy.

OTHER USES

The programs described here were designed with the needs of the mobile hunter in mind. Their triangulation routines are easily adaptable to DFing of jammers by bearings from multiple base stations. Several active hunters, including Edward Crandall, WD8CBE, and Paul Wirt, W6AOP, have done work to set up jammer locating nets using computers.

Give computer DFing a try. You won't be competitive with the big government DF installations, but you may win more hunts.

Chapter 21

Dealing with
Mischief and Malice

Perhaps the main reason you're reading this book now is that a jammer is making you miserable. You can't have a pleasant QSO with your buddy without a carrier, some phone tones, or a bunch of four-letter words covering him up or timing out the repeater. You're ashamed and embarrassed to have the radio on where your family and friends can hear these goings-on.

Maybe you got so upset that you called the engineers at your district FCC office. Were you surprised when they politely told you that they couldn't just drop everything, hop in their fancy van, and chase down all the culprits? You shouldn't be, considering that they must also solve problems with almost every other user group in the radio spectrum.

Are you now ready to do something about the problem yourself? That's good, if you're willing to do the right things. One reason jamming gets out of hand is that it's easy for the "good guys" to be apathetic and say, "Let George do it." But the more operators who are willing to do the DFing and write the letters, the sooner the bands can get cleaned up.

ORGANIZING THE HUNTING TEAM

In Orange County, California, there are dozens of repeaters and even more simplex groups. Yet jamming on the local machines there has been an unusual occurrence, compared to some other parts of the metropolitan Los Angeles area. One reason is that several clubs and non-club groups sponsor regular sport hunts. Seldom does a weekend go by without at least one hunting opportunity, and often there are more.

Just the knowledge that plenty of hunters are ready and listening is a powerful deterrent. But results are even better when a club frequency or repeater has an extra measure of concern in its user population. The true dedication of the hunters—their willingness to stop whatever they are doing and go find jammers—will best convince the turkeys (as ARRL calls them) to go to other systems.

Don't think that you can bluff your way out of a jamming situation by talking a lot about direction finding but not actually fielding some teams. The jammers won't be impressed. We've heard lots of talk on some repeaters about the mountaintop remote control DF systems on their drawing boards. But a paper "Ultimate RDF" system never put a jammer off the air. The same amount of energy expended in putting quads on a few cars and doing some searching pays off much sooner.

This is not meant to be a put-down of remote RDFs, computer bearing plotters, and other high-tech ideas. On the contrary, they should be encouraged. We even have a few such ideas of our own. But they only make the job easier and faster, and that only when they are up and working. In the meantime, hunt with what you have.

Just having an occasional fun hunt isn't necessarily enough, either. The hunters must be dedicated, ready to hunt an intruder at any hour, without fanfare. Actually putting an offender off the air (legally!) is a great deterrent, while talking about what you're going to do someday is no deterrent at all.

If you believe that coordinating your hunters on the jammer's frequency will deter him, think again. You'll just give him the opportunity to play cat-and-mouse with you by letting you get close and then going off the air. Find another frequency on which to coordinate, and keep it secret. One group in Omaha put up a repeater on the 220 MHz band primarily for coordinating its 2 meter jammer hunts. Use the telephone whenever possible, for the highest security.

IS IT REALLY JAMMING?

The term jammer originally implied the use of a high power rig to cover up or capture out a station. Today it has come to mean anyone who maliciously disrupts a QSO or net. The classic jammer, one who puts on a strong carrier, modulated or not, probably knows he can be readily traced. Chances are he is running a considerable amount of power, and is in a populated area. If he has the signal on long enough to be truly annoying, it's easy to triangulate bearings, get vehicles on the scene, and sniff him out.

Once caught, the jammer can be shown to be guilty of transmitting without identification (97.123), and possibly malicious interference (97.125). The second charge is harder, because if licensed, he could claim the rig was left on by accident. If the same station or stations are jammed regularly and selectively, however, that alibi falls apart. And if he brags about preventing a particular person from getting through, you've nailed him for sure.

In a few areas, the misfits of amateur radio have discovered that they can cause just as much trouble by doing some things that appear perfectly legal, or can be justified by twists and apparent loopholes of the FCC rules. When a club in southern California put up a repeater in the then-new 144.5 to 145.5 MHz sub-band, which happened to have an input frequency 5 kHz away from an established simplex group, the simplex users didn't immediately vacate. They kept using their high powered base stations, usually identifying as required. Often they timed out the repeater, or covered up mobiles and handhelds trying to use it.

Repeater users accused the simplex operators of being jammers. The simplex types said they didn't hear the weak stations, who were not on the same frequency anyway. "Besides," said the simplexers, "We were here first, long before the repeater was built. You repeater jockeys are the ones bothering us." Who was right? Were any of them jammers?

The FCC finally had to take a stand on the problem of repeater versus simplex operations, and chose to back up the local coordination councils. Band users are now expected to abide by generally accepted band plans. Simplex operators are not allowed to interfere with established coordinated repeaters in the repeater sub-bands. New repeaters coming on the air must not interfere with established coordinated systems, simplex operations, or weak signal activities.

Cases of interference documented by repeater councils may result in FCC citations. If you transmit on an established repeater input frequency, it is assumed by the authorities that you intend to be repeated. Several FCC enforcement actions have resulted from this clarification of policy.

Not as easily solvable is the problem of obscenity. What words are obscene? Does a ham's so-called freedom of speech give him the right to say anything, no matter how filthy? Or is ham radio not comparable to the press or a park soapbox, making the freedom of speech argument not applicable? The courts have not given the last word on this issue, so despite the existence of 97.119, hams should not expect vigorous enforcement of this rule while the controversy remains.

BAIT-ERS, BAIT-EES, AND REVERSE JAMMING

Some hams just delight in provoking arguments. They have developed the technique of harping on a point, or on a person, in a manner that makes listeners' blood boil. They sometimes break into a QSO legally and then take it over with long harangues. What many listeners may not realize is that they (the listeners) are simply being baited. The baiter does it because he gets kicks from the reactions.

While many listeners call such a person a jammer because they feel forced off the air, a close examination of the situation frequently shows that the baiters have broken no rules. They haven't jammed another's transmissions. Station IDs have been regular and complete. Obscenities haven't been used. Where's the violation? There are no FCC rules against epithets or noxious conduct.

"Ah," you say, "the guy that carries on for hours with

his personal attacks is really doing broadcasting. That's a violation of 97.113, isn't it?"

Sorry. Typical ham conduct will seldom allow that charge to stick. He'd be guilty of broadcasting if he went on for hours with no one else in contact with him. But unfortunately there's always a ready supply of well-meaning hams ready to come back to him when he stops his tirade, just to tell him what a jerk he is. The minute someone does that, the haranguer is in a QSO and can't be said to be broadcasting.

The FCC finds that so-called "good guys" are often one of the biggest roadblocks to successful solution of these problems. While investigating the famous West-CARS net interference problem, FCC DFers discovered that besides the few anti-net troublemakers there were many, many pro-net hams who got so irate that they committed violations just as serious. They engaged the interfering stations in QSOs, legitimizing their presence on the frequency. They made their own unidentified transmissions. They deliberately QRMed the net-breakers, making the DFing job much harder.

What was FCC to do, revoke the tickets of everyone involved? There certainly was justification. To quote a decision by an FCC Administrative Law Judge in a vhf jamming case: "The Review Board has previously emphasized that this Commission cannot tolerate the use of 'vigilante tactics,' noting that the one who uses such tactics becomes part of the problem and only aggravates the situation."

We are not for a moment condoning the behavior of the scoundrels. It has no place in Amateur Radio, or anywhere else on the air. The point to be made is that the problem of so-called jammers (perhaps "disrupters" or "misfits" are better appellations) often cannot be solved merely by DFing and prosecution.

SHUT OFF THE REPEATER?

When jamming and obscenity take place on the hf bands or vhf simplex, it's bad for those who are trying to listen or communicate. When the same thing happens over a repeater, it's even more of a problem. Not only is the smut heard over a wider area, but the licensee and control stations must be concerned with their responsibility for the emission of the repeater transmitter.

Fortunately, the FCC has not made a policy of automatically citing both the offender and the repeater controller when a violation occurs on the repeater input frequency. Were the Commission to do this, some licensees would feel forced to shut down their open machines immediately. They would not want to be made the 24 hour

a day judge of the content of users' transmissions. To do such conjures visions of control operators with fingers poised on the kill button like the host of a broadcast radio call-in show.

If troublemakers knew that repeater controllers shared the responsibility for their nefarious activities, they would revel in their power. With one four-letter word, one joker could bring to a halt the activities on a machine serving hundreds of users, no matter how important the public service taking place at the time. Controllers would be wondering, "How quickly must I turn it off? Must I turn it off during an emergency QSO? How long must I leave it off?"

FCC field personnel realize the value of keeping repeaters on the air, and understand that it's impossible for repeater owners to control their users. Yet repeater owners do have the responsibility to keep their machines under control. This was best explained by Larry Guy, Engineer-in-Charge of the Los Angeles FCC district office, when he addressed a southern California radio club.

When asked about the problem of what constitutes profanity, Mr. Guy stated, "I've never asked a repeater licensee to determine what is profane, obscene, or indecent. The U.S. Supreme Court can't even do that . . . but certainly every repeater owner knows the difference between music and communications. That's what I ask him to do."

That implies that the control operator has some responsibility for the content of transmissions. What should he be asked to do? Mr. Guy was eager to clarify his position, and said, "If somebody uses a 2 meter transmitter and works through a repeater to transmit a threat to do bodily harm to someone, that person is committing a criminal violation of Title 18, Section 875c. The repeater licensee is *not* in violation.

"In essence, the content of the communication determines who's going to be held responsible. If somebody plays music over a 2 meter station being repeated, that person playing the music is in violation of FCC rules, not the repeater operator.

"So what is the repeater operator responsible for? If the repeater operator permits someone to play music over his repeater *continuously*, then he is not controlling his repeater. He will be cited for failure to control his repeater . . . *not* for playing music, but for failure to control. If someone continuously transmits threats, the control operator will be cited for failure to control, not for transmitting threats. The repeater licensee does have responsibility for *his* transmitter."

This FCC official, who has been involved in many

southern California jamming cases, summed up his down-to-earth approach to amateur band policing by saying, "Believe me, I'm going to be very, very generous in my judgment because I don't want to lose a case . . . So if you get a violation from me, it's going to be for good cause."

PSYCHOLOGY MAY HELP

All too often there's no need for extensive DFing to find the local radio ruffians. They may give callsigns and locations regularly, and be proud of their ability to disrupt things while staying on the very edge of legality. In such situations it's logical to wonder what goes on in the heads of these people, and to try to figure out a way to "rehabilitate" them.

In 1980, during the height of the problems on the Mt. Lee repeater and others in the Los Angeles area, the Fullerton Radio Club program chairman decided to see how a clinical psychologist would analyze the situation. Dr Roberta Trieschmann heard tapes of obscenities being injected into hospitalized childrens' ham radio conversations with Santa Claus. She also witnessed live jamming on the machines.

In her subsequent talk to the club, she commented that interference problems on amateur radio seem to parallel our entire society's "me first" attitudes, where etiquette and consideration for others is on the wane. Also, being an anonymous jammer provides a convenient way for frustrated souls to react to the sense of powerlessness they may feel about other areas of their lives.

Dr. Trieschmann had the following suggestions for counteracting the activities of the disrupters:

☐ Don't respond verbally. Even if only one person responds in some way, reinforcement and encouragement are provided to the jammer. Remember, what he primarily wants is attention. You won't hear smut on un-busy repeaters. No audience, no incentive.

☐ Reduce anonymity. Unless he has a specific vendetta, a jammer usually doesn't jam people he knows personally. Your repeater club or other local ham group should reach out to encompass all users and active hams. Who knows, you may turn a misfit into a productive ARES member!

☐ Support the FCC. Take whatever political action is appropriate to help FCC get the funding it needs to combat interference successfully. Volunteer your services to the new volunteer enforcement assistance program.

☐ Set an example. Put articles in your club newsletter on proper procedure and common courtesy. Help newcomers and oldtimers improve their operating skills. Make sure your own operating procedures are beyond reproach. *Never* lower yourself to the level of the offenders.

RESISTING THE URGE

Victims of very severe jamming may consider the psychologist's suggestions simplistic. It is true that by the time a problem escalates to a long-lasting daily occurrence with several perpetrators, psychological methods alone won't solve it. The above recommendations help *prevent* the escalation. If the reaction of the "good guys" had been appropriate, perhaps the escalation would not have happened. But at any stage in the war against jamming, *do not respond* must still be the watchword.

Think about how the typical repeater user (yourself?) reacts to deliberate interference. There's an immediate urge to sternly tell the offender to knock it off, just as you would discipline a foul-mouthed child. But the jammer isn't your child. He doesn't respect your authority. He knows you aren't in a position to punish him. He came on the radio to get a reaction from someone, and he got it. Why should he stop now? Why shouldn't he say more and see what other reactions ensue?

So again we restate the most important rule in combatting jamming: DO NOT RESPOND! No matter how much anger wells up in you, don't express it on the air. Go to the phone to complete your QSO. Listen for your contact on the input to the repeater if there's no QRM there. Stand by and try again later. But *never* react verbally.

If this prohibition on response is to work, it must be absolute. Don't say, "Somebody threw a carrier on you on that last transmission, Ted." Don't come back with, "That jammer's back again, George, but you covered him up fine." Don't even say, "Let's go to five-two simplex to get away from those tones."

In short, don't acknowledge the clod's presence. Most hams who operate the DX bands are used to telling their contacts exactly how well they are making it through the QRM, so this kind of restraint, though vital, is against their natural instincts. So talk through, QSY, or QRT, but don't mention the malicious QRM. Let him go away wondering if his transmitter is working.

Resisting the urge yourself is hard. It's even harder to get a large group of hams to follow this advice, which goes contrary to their natural parental instincts. When it is followed, however, the results are dramatic, as this following true story shows.

About a week after an Orange County, California, repeater put up an auxiliary receiver in a high location, increasing the coverage considerably, a new mobile station appeared one morning. He gave a callsign (phony?) and blabbed for about three minutes some filth that would embarrass a hard-core sex offender. When he stopped, there was total silence, even though it was morning drive time. Perhaps everyone was just stunned, or immediately turned off the rig. He tried again, and again no one replied at all. He disappeared, and hasn't been heard on that system since. No performer likes to play to an empty hall.

The psychology of "good guys" is almost as interesting as that of jammers. Many are so threatened by jamming that they feel compelled to react, perhaps without realizing it. A mountaintop repeater's control operator once complained to us about a ham who picked on certain officials of his repeater organization. We suggested that the officials just ignore him, because by responding they were only egging him on. "Oh, we always ignore him," the controller replied.

Listening to that repeater a few days later we heard an unidentified station make an obscene remark to this same control operator. Instantly the controller replied, "I'll bet you don't have the guts to say that to my face!" Of course the interloper persisted in his remarks. The controller tried to resume his QSO, occasionally mentioning how the "turkey" wouldn't be able to jam him because he was talking on a secret control input frequency.

You can be sure that the jammer isn't through with that repeater. He might even be challenged enough now to get a scanner and try to find the special control frequency to cause trouble there. The control operators had convinced themselves that by not QSOing with him, they were ignoring him. Actually, they were taunting him. They were indirectly saying, "You got me mad, but you can't jam me now, ha ha!" Maybe not now, but look out later!

So we say again, resist the urge and ignore the transmissions of illegal operators. Don't talk to them, and don't talk about them. Treat them like the squelch tail at the end of your transmissions. It makes a noise, but you don't talk to it or about it.

If you say anything at all, you are playing the jammer's game with him, and on his terms. You're doing what he wants. Ignore, and the game must end. As somebody once said, "Never wrestle with a pig. You both get filthy, and the pig likes it."

KEEPING HIM ON THE AIR

Many groups justify baiting an illegal station for a noble purpose—keeping it on the air while the DFers close in. Is it a good idea?

First, consider whether it's legal. FCC rule 97.89 defines legal communications by amateur radio operators. With a few specific exceptions, hams are allowed to talk only to duly licensed amateur stations. If the station is not identifying legally, there's a good chance it may not be licensed. In that case you are technically in violation of 97.89 if you QSO him. If FCC is listening and cites him, might you not also be cited?

Second, consider the legal aspects of holding him this way. Suppose FCC is listening, drives up to his location, and catches him. If you have kept him going, couldn't he claim that he was a victim of entrapment? That argument may sound silly, but it could make it harder to get a conviction. It's hard enough already.

Third, consider if it's worthwhile. Why continue to subject the listeners to him if he is ready to go away on his own? The commotion may entice other troublemakers to join in. So just ignore him. If the DFers can't get all their bearings for triangulation taken before he leaves, they can do it the next time he shows up. If he doesn't come back again, all the better. Sure it's frustrating to have him disappear in mid-hunt, but you can start from that point the next time. If you locate him and can't get him to stop, the legal action will be far more frustrating.

So a jammer disappeared is better than a jammer caught. It's certainly better for listeners and those wanting to use the frequency normally. If you ignore the law-breaker and don't goad him on, there's a good chance he'll go away quietly and you can save your gas for the next fun hunt. If he doesn't give up, you'll have the transmissions you need to quietly start the DFing.

PROSECUTING THE OFFENDER

Far too often, attempts to resolve an interference situation eventually break down. The interferers may have a personality problem that cannot be overcome. It may be that some of the so-called "good guys" take out their anger by baiting the jammers, compounding and escalating the problem.

In the case of the Southern California DX Club repeater mentioned earlier, the ruckus eventually led to the suspension of at least one ham's license. But if you're hoping that a court conviction will end the problem in your area, be prepared for a long wait. It takes a long time to gather evidence and build a successful case. Delays and appeals can seem to take forever.

There are several strikes against a group wanting speedy justice. First, FCC and prosecuting attorneys give the case very low priority if it doesn't involve safety of life or property. It's not hard to see why bureaucrats may consider malicious QRM cases to be in the same category as family feuding, and not worthy of the expense of prosecuting. It also explains why most of the successful prosecutions on record have involved additional offenses such as threats to individuals or public officials.

Then there's the problem of the unfamiliarity of prosecuters with radio law. FCC's Larry Guy explains it this way: "When we take a case down to the U.S. Attorney, we have to have a great deal of evidence. The first reaction is, 'Why don't you send him a letter?' When we point out we've already sent fifteen, he says, 'I don't think the judge wants this kind of case.'"

Guy continues, "The D.A. is right, basically, because not too many of these lawyers and judges understand radio, and they know it's going to get them involved in something over their heads. So what we have to do is go down to the U.S. Attorney's office and make it as simple as possible. We have to show a great deal of hue and cry before they will take it, and we just sit there spoonfeeding them until they're comfortable with the case."

Given the difficulty of successful prosecution in the federal courts, it's often best for the FCC to handle matters itself and minimize court involvement. The FCC cannot by itself impose prison time, but it can deal out heavy fines and suspend or revoke licenses. The jamming problem is not viewed lightly, either. In revoking the station license of a La Crescenta, California, ham, the judge's decision stated:

> The Commission has stressed that malicious interference in any radio service is a serious matter, that it is 'the most serious violation' found in the Amateur Radio Service [and] that it warrants the 'most stringent penalty.'

Backing up the judge is a statement by FCC Private Radio Bureau Chief Robert Foosaner at the 1984 ARRL National Convention. He said:

> As little as five or six years ago there were no cases in the law books of license revocation for malicious interference. Today there are several. Today there is ample precedent for a judge to revoke a license, and it's happening more and more. All sorts of defenses have been raised and rejected. The point is simple: Jamming is jamming. It always occurs

for a reason. We don't care what your reason is. If we catch you (and if you do it often enough *we will catch you*) you can say good-bye to amateur radio.

WE'RE FOURTH ON THE LIST

FCC field personnel, many of whom are amateur radio or CB operators themselves, are as eager as anyone to resolve malicious interference problems in these services. They understand the importance of these services in emergencies, and the need for backing up the self-policing efforts. But even the most ardent personal radio enthusiasts must realize that FCC has a limited staff, and there must be priorities. Specifically they are, in order:

1. White House and presidential communications
2. Communications involving safety of life and property
3. Business and Public Safety Band communications
4. Other

Southern California amateurs may have more understanding of FCC's problem when they recall that many recent U.S. Presidents have been from the west or had Western White Houses, with their enormous communications needs. That FCC office is regularly called in to assist in searches for ELTs, because of its outstanding DF capabilities. Business and public safety band problems are immense there, with land mobile frequencies expanding into the uhf TV bands. How does FCC ever find time for priority number four?

The "other" category is dynamic, changing from time to time as needs change. It is different for rural areas than for urban centers. While the category includes ham and CB radio, it encompasses more than jamming. TVI and rfi complaints, helping settle repeater coordination wars, and monitoring for infractions such as harmonics and out-of-band operation all have to be done.

"Other" also includes making a thorough inspection of fifteen AM/FM/TV broadcast facilities in each district yearly, plus follow-up rechecks. Broadcasters have their problems, too—about 250 instances of pirate broadcasting across the nation must be DFed and investigated yearly. The southern California office must personally inspect 150 ship's radio rooms per year, checking the reliability of equipment and qualifications of the operators. It's small wonder that most district FCC offices welcome volunteer ham enforcement assistance, if it is done properly.

VOLUNTEERS TO THE RESCUE

In the early 1980's, the jamming problems of southern California were becoming intolerable. As Larry Guy puts it, "Southern California was getting a reputation worldwide, particularly on the 2 meter band, and occasionally on some of the lower bands. And of course that reflects on the (FCC) district office. So we decided that we'd work Saturdays and Sundays and try to get rid of some of these guys."

There had been considerable pressure from the ham community, even to the point of involving members of Congress. "If a congressman gets involved," says Guy dryly, "it has an elevating effect." When the legislators realized that FCC could not legally use the volunteer help of hams, they did their part by passing Public Law 97-259. This important revision to the Communications Act of 1934, signed into law September, 1982, added two important weapons to the war against jamming. First, it exempted the amateur radio service from the secrecy of communications provisions of the Act. Now hams can report the content of communications, and make recordings of them, for enforcement purposes.

Secondly, the revision gave any citizen the opportunity to use his eyes, ears, and equipment to help solve the problem of illegal activity on the ham and CB bands. The helper doesn't even need to be a licensed operator. Volunteers and evidence they obtain can now be used by FCC to track down and prosecute rules violators.

One of the first cases of non-FCC enforcement assistance occurred during the 1983 Empire State Games in Syracuse, New York. When the amateur radio support communications received interference from an unidentified station, Captain David Stevenson of the Onondaga County Sheriff's department used a DF set to find the offending station. The efforts of Capt. Stevenson, who was not a ham, led to a fine and license suspension for the guilty operator.

Volunteer enforcement assistance was a long time in coming, and holds great promise for helping deal with understaffing at FCC. The amateur radio service has been known for being self-policing, but a formal working relationship with FCC in enforcement matters was lacking until this law was passed. Now, through the American Radio Relay League, the national association of amateur radio operators, we can all work together. While FCC remains responsible for citations, apprehension, and prosecution, the volunteers can do much "leg work," establish the *modus operandi* of illegal operators, and help bring them to justice sooner and at less expense to the taxpayers.

Another important role that amateur radio volunteers can play is that of a screening committee. Many interference complaints of lesser significance have in the past been taken straight to the FCC, when they could have been ironed out by the operators themselves or by the ham community. The volunteer enforcement program can be an important vehicle for handling these minor tiffs. FCC help can then be called in to assist in solving the ones that don't respond to negotiation and peer pressure.

It is important that members of this auxiliary to the FCC Field Operations Bureau receive suitable training. They must be sensitive to the pitfalls to successful prosecution. The Freedom of Information Act and the Privacy Act put restraints on what can be done. ARRL's responsibility is to organize volunteers and help train them. To get yourself or your group involved, contact your ARRL Section Manager or Division Director to find out the status of the ARRL/FCC Amateur Auxiliary in your area.

SUBMITTING YOUR OWN REPORTS

When you successfully locate an illegal operation, and have shown that FCC rules have been violated, you can submit a statement to the FCC even if there isn't an organized volunteer enforcement group in your area. Your letter to the FCC district office helps establish patterns and helps show where the problem areas are. If your information is complete enough, an official warning may be issued to the offender. While not carrying the weight of an official Notice of Violation, it lets the offender know that he has been located and a record of his activities is on file.

Your letter to the FCC should be as complete as possible. Don't go on and on about your suppositions of the operator's motives. Stick to facts and tell them precisely. Give exact times, dates, frequencies, locations, durations, and signal strengths. Tell how you arrived at any conclusions. Figure 21-1 is an example, with pertinent facts changed, of an actual volunteer DFing report to the Los Angeles FCC office. If your information is useful, you will receive a reply similar to the letter shown in Fig. 21-2.

One way to help ensure accurate observations and reporting is to carry a portable tape recorder on the jammer hunt. Make a recording of both the sounds from the radio and your own running commentary of the time and your location. FCC may or may not be interested in it, but it will be a big help to you later. You'll easily be able to document:

☐ Who was being jammed? Everyone, or just selected stations?

```
P. O. Box 7388
Brea, CA  92621
23 May 1984

FCC Engineer in charge
3711 Long Beach Blvd., Suite 501
Long Beach, CA  90807

Dear Sir,

The following may be useful for your Amateur Radio interference files.

On May 22, 1984, we were preparing to participate in a regularly scheduled Amateur Radio
Emergency Service net on the KC6K/R two-meter repeater.  At 7:35 PM, about five minutes after
the start of the net, a strong interfering signal on the repeater input frequency (146.19
Mhz) captured the transmissions of the net control station and several other participants.
At first, the interfering signal had no modulation, and later sent random DTMF telephone
tones continuously.  There was no voice or CW identification.

After about 5 minutes of the interference, we began to DF the interfering signal.  Equipment
was a four element quad antenna mounted on the car, feeding an ICOM IC-255 FM receiver
through a step RF attenuator.  Our starting point was the intersection of Imperial Highway
and State College Boulevard in Brea.  The interference continued, with transmissions lasting
about one minute, and occurring about every three minutes.  There seemed to be no attempt to
single out any particular station for jamming.

When we were about one half block from the transmitter's location, the signal was so strong
that all available attenuation (112 dB) did not sufficiently reduce the signal into the
receiver for bearings.  We completed the search using an amplified field strength meter
connected to the quad.  A later check of the field strength meter setting showed that the
signal from the quad exceeded 5000 microvolts with the car parked directly across the street
in front of the property where the transmitter was located.

Bearings from several locations along the street showed that the interfering signal was
emanating from 1404 Warburton Way in Fullerton.  The property contains several amateur HF and
VHF antennas.  Of the two vehicles in the driveway, one had license plate WA6QRM.

The interference was in progress when we pulled up in front of the property at 8:05 PM.  It
then immediately ceased and did not recur.  We remained in the area for 15 minutes and heard
no further transmissions on 146.19 MHz from that property.  No attempt was made to make
contact with the residents.

Sincerely,

A. T. Hunter  WB6XZY
Ann Other Hunter  WB6XZZ
```

Fig. 21-1. A sample letter to the FCC reporting results of DFing.

☐ The nature of the jammer's modulation.

☐ Was any ID given, or were there any other identifying sounds?

☐ How long were the jamming transmissions? How often?

☐ When did they start? When did they stop?

☐ Was anything said that indicates malicious intent?

Proving malicious intent is a key factor to getting stern enforcement action. A review of some recent FCC "show cause" and suspension orders shows the sort of evidence that is needed. In the case of James W. Smith

(W6VCE) the notice said:

The following transmissions monitored on February 19, 1984, apparently show your intent to wilfully and maliciously cause interference: "KF6Z just came up and announced that my radio, my other radio that is jamming the 147.990 frequency, had burned up and was off the air . . . no, that's not true. It's still on, but the only problem is that it's not on 147.990. It's on another frequency which I am jamming."

In the Randy L. Ballinger (WB6MMJ) case, the tran-

FEDERAL COMMUNICATIONS COMMISSION

FIELD OPERATIONS BUREAU

July 18, 1984

ADDRESS REPLY TO:

3711 Long Beach Blvd.
Suite 501
Long Beach, California
90807

Refer to Case #LB-84-XXXX

Dear Mr. Hunter

Your letter (telephone call) of complaint has been received. The person that you mentioned has been asked to relate his license data and details of his operation to this office. In addition, he has been advised, if he happens not to possess a valid license for the radio, of the consequences regarding the operation of unlicensed transmitting equipment.

Any further information that you may have, or a list containing the dates and times of the operators activities would be helpful in this investigation. Also, in the event these activities cease completely or for any considerable length of time this information would be helpful to us.

Your continued patience and cooperation is appreciated.

Very truly yours,

L. D. Guy
Engineer in Charge
FCC-District #11

Fig. 21-2. Typical response letter from the FCC.

scripts showed he singled out individuals:

> The communications from the interfering signal included the following (while jamming the communications of amateur radio station W6LDG): "LDG ain't gonna hear nothing. I know he can't. Zed W ain't gonna hear nothing. I know he can't. Ah we'll take care of AEE and SSB. I don't think AEE can hear anything either . . . so you go ahead and jam the repeaters and we'll go ahead and jam you . . . I'll bet you can't hear anything. Ha, ha."

In the case of Calvin C. Plageman (WB6DSV), the statements of the defendant were less direct, but still resulted in FCC action:

WB6SUS: Yes sir, roger roger . . . (WB6DSV transmitted over the remainder of this communication as follows).

WD6DSV: WD6DSV calling N6DSD

NF6B: . . . give me a break, Cal. I have not relinquished it yet.

WD6DSV: I'm just not going to let SUS get involved here. Sorry about that, Oley. Can't be helped.

NF6B: What you are saying is you are going to jam him because you don't like what he talks about?

WD6DSV: That's affirmative.

Tape recording of statements such as this are of great help in putting together a malicious interference case. But to complete the case it must be shown conclusively that the statements came from the individual being accused. You must be willing and able to testify that you heard the words, and that they came from the alleged jammer's transmitter. Such testimony was made by FCC engineers in each of the above cases.

After you have sent in your letters and tapes, and the enforcement process has begun, don't sit back. Continue to gather evidence by DFing and making recordings when further interference happens. The additional material may be needed to bolster the case.

Should you make contact with the jammer once you find him? The decision is up to you. If it's someone you know, who probably left his transmit switch on by accident, by all means knock on the door and tell him. On the other hand, the FCC recommends you do not force an encounter with someone who is obviously engaged in obscene or malicious QRM. He may be on booze or drugs and unable to control his behavior. Some have threatened the lives of federal marshals. Is it worth the risk if you're alone? In such cases it's best to not only avoid an encounter, but to avoid being seen.

It's best to have the authorities with you if you must confront the jammer. John Moore, NJ7E, says that he contacts the police or sheriff and tells them why it is necessary to identify the source of radiation from the location in question. He explains that he has reason to believe that violence or personal danger is a possibility, and asks that an officer accompany him.

"They don't like to do it," he says, "but they will, just to avoid an incident. I tell them that of course I'm not going to do any violence, but the guy I want to talk to might. Please just walk to the door with me. They'll show up and say, 'You know you don't have any authority?' I tell them that I know that, and I'm not a vigilante. I just want some help because I need to identify this fellow so I can turn him in to the Feds. They go along with that."

In his area, this procedure is called a "civil standby." The officer is there only to prevent an assault—not to enforce any radio laws. Don't expect the police to do the FCC's job.

Keep in mind that formal documentation to the FCC should be reserved for severe, continuing, deliberate interference situations. Don't send a letter to the FCC every time someone gets a little hot under the collar and cusses or takes over a frequency. These one-time problems usually solve themselves, or can be patched up with a little diplomacy.

The ARRL recommends that a letter be sent to the offender before the FCC is informed, to give him a chance to clean up his act. The letter should be non-accusatory, but should state clearly the facts that lead to the conclusion that the addressee is the perpetrator of the problem. If he had thought he was anonymous, he may stop causing problems immediately. Lawyers are often particularly good at writing such letters, if such help is available to your group.

GENERATING THE HUE AND CRY

Once a major offender has been found and apprehended, don't rest on your laurels. Getting prosecutors to file charges and judges to dole out stiff penalties requires the active interest of the community. If no one from the affected radio service shows up at the hearing or trial, how can the judge be expected to believe that the matter is of great significance?

During the sentencing phase of one such trial in Los Angeles, hams were invited to write to the judge to express to him the seriousness of the violations as perceived by the users of the repeater involved. The result was a disappointing trickle of responses, resulting in a sentence lighter than most would have liked.

Sure it's hard to take off work to attend hearings and trials. It's also inconvenient to write letters and make calls. But that's the way the system works, and it's a small price to pay for the benefits to be gained.

TECHNICAL TRICKS

Occasionally it's possible to use technical methods to solve an illegal operation problem without resorting to a big DF effort, as the following story shows. It actually happened, but the name has been changed.

The Pirate Patch

In the summer of 1983, the users of the HFEA Amateur Radio Club repeater in southern California began to hear autopatch calls coming through the system once or twice a day. The technical committee thought it was a typical multiple-repeater intermodulation problem and looked for another repeater carrying the same audio, but none was found. Then it was noted that both sides of the conversation were present on the repeater input, but from different directions.

Clearly, someone was using the repeater as a range extender for his personal autopatch. He would transmit to it on the repeater input. His machine listened on the output and transmitted the telephone audio on the input back through the repeater. Sometimes the unit would transmit on its own, apparently signalling that his phone was ringing. To prove the theory, the repeater was turned off momentarily while the user was making a call, and it was obvious that he lost control of his patch. There was no CW or voice ID on the autopatch, a clear violation of amateur radio rules.

Club members were cautioned not to talk about the interloper on the air, but to try to get bearings and catch any ID given by the user. This was difficult. He was on for only short times, and the user's ID was always given so fast and slurred that it was impossible to understand.

It was decided to try a technical approach to identifying the unwelcome guest. First, the control operators

began to tape record all transmissions through the repeater, 24 hours a day. This was done with a COR circuit hooked to a receiver tuned to the output frequency. The recorder came on automatically whenever the repeater was keyed up.

Those hearing the tapes were never able to figure out a callsign, but the recordings made it possible to determine what phone numbers were being called, as well as the patch access codes. The numbers were gotten by playing the tape audio directly into the repeater's control input, and reading the numbers from the LEDs on the tone decoder outputs. Two residential numbers were obtained, and checked in the reverse telephone directory. This directory, available at the local public library, gave a name and street address for one of the called parties, but the other was unlisted.

At this point it was decided to change tactics. To prevent further abuse of the system, a filter was added in the audio line to the repeater transmitter. This narrow notch filter passed everything except tone H3, at 1477 Hz. Users didn't notice its presence and the repeater autopatch functioned normally. But the renegade patch could no longer hear digits "3," "6," "9," and "#" through the repeater. Since "6" was used in the access code, the illegal patch stopped working.

The control operators hoped the problem was over, but the mystery operator continued to try unsuccessfully to access his box. One of the controllers then got him into a short QSO. He gave his name as Mort, and slurred his callsign. Asked to give it phonetically, he gave a WB6 call which could not be found in recent call directories.

Obviously Mort was a bootlegger, and a brazen one. The decoded telephone numbers were called. The unlisted one was never answered, but the other was picked up by a middle aged man. The controller told him he was trying to locate Mort, the radio ham. The man said Mort was dating his daughter, who was not home. He did not have Mort's phone number.

Success finally came another evening when the girlfriend was reached and persuaded to pass along Mort's number. When it was dialed, the patch "rang" through the repeater. Mort answered, but denied being involved with radios, autopatches, or even knowing about the hobby, saying, "Is that like CB?" He was assured that it wasn't and thanked for his time. The patch immediately disappeared, and has not been heard on the repeater since.

The Simple Tone Notcher

The LC notch filter (Fig. 21-3) used in the above sit-

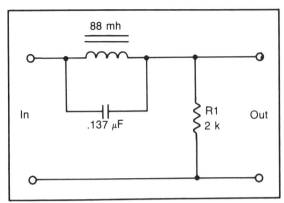

Fig. 21-3. Schematic of the LC notch filter.

uation was chosen over the popular active filter designs because its simplicity made it possible to put it into operation in just a few minutes. No matching of resistors or capacitors must be done. It requires no power and is readily installed.

The inductor is a standard 88 mH telephone toroid coil, available surplus. Tune it by paralleling capacitors to achieve resonance at the desired notch frequency. By selecting a variety of large and small capacitors, any tone frequency can be easily hit regardless of tolerance of the capacitors. For example, a 0.1, a .022, and a .015 microfarad capacitor from the junkbox were perfect for this particular filter.

For good Q, the filter should follow a low impedance stage such as an audio operational amplifier and drive a load of less than 1000 ohms. Add resistor R1 if needed to load it. If Q is too low because source or load impedance is too great, repeater audio response suffers and the notch is not deep enough. Be sure to put the filter in line to the transmitter after audio for any tone decoders or autopatch inputs is stripped off. Filtering only one frequency will be sufficient because each frequency is used in three or four digits.

Other High-tech Countermeasures

Hams are known for their ingenuity. Turn your club's technically inclined members loose and they may think of some other clever ways to help solve your repeater interference situation. If persistent kerchunkers are a problem, add a counter/timer circuit to the control logic that requires the user's carrier be on for at least two seconds before the repeater comes up when inactive. The two-second delay should be defeated when the repeater is inactive or in QSO, but activate itself when it detects more than one very short transmission. Other groups have solved this problem with a voice-operated circuit, requiring that the user talk to make the repeater transmitter come on.

Some groups try to use continuous sub-audible tone access (often called PL, a Motorola trademark) on their repeaters to eliminate undesired stations. Sub-audible tone systems were not designed for this purpose, and are not very secure. More and more new ham rigs come with all standard sub-audible frequencies built in. Going to a non-standard tone frequency does not guarantee success and is a big headache for legitimate users. A strong jamming signal will capture the repeater and prevent users from being heard whether the jammer has the correct tone or not.

Remotely controlled DF setups are well within today's technology. Your repeater's high location may be the perfect spot for a DF station, to give triangulation bearings with any user's base station. The digital output of a commercial Doppler unit or beam rotor can be placed as a bit stream on a sub-audible carrier, detectable by any user with a decoder and perhaps his home computer as a readout device.

A short burst of ASCII data can replace the courtesy beep, giving direction, relative signal strength, and perhaps frequency offset. Repeater owners who are experimenting with this idea report that just the presence of the burst is a deterrent to some troublemakers.

New developments in digital and rf technology are making possible even more exotic schemes to control repeater access. Here are a few of our "blue sky" ideas. We leave the development up to you.

☐ Directional access lockout. Once the DF unit at the repeater has found the direction of the offending signal, the signal could be locked out of the repeater by adding logic to keep the repeater transmitter off when the incoming signal is from that direction. Remote command circuitry must be provided to tell the control system which signal is to be DFed and then locked out.

☐ Null steering antenna. The directional access lockout above does nothing to reduce the level of a strong jammer relative to other signals. A special repeater antenna with a null adjustable in azimuth could be rotated electrically or mechanically to put high rejection in the direction of the jammer's signal. Null steering antennas are now used in state-of-the-art radar systems. Unfortunately this system and the directional access lockout are not practical on systems where most users (and jammers) are in the same direction from the repeater.

□ Multiple receiver sites. Some repeaters now use several receivers at scattered sites connected to a "voting" system to choose the receiver with best signal for retransmission. By changing the control logic and by monitoring the individual links by control operators, receivers being jammed could be locked out of the system to allow users to access receivers where the jamming signal is weak or not heard. With enough receivers a miniature cellular system results. A side benefit is that the strength in the individual receivers helps determine immediately the area to start the hunt in.

□ Signature recognition lockout. Traits of the offending signal such as carrier shift, key-up transients, or even voice pattern form a unique "signature" of every transmitter and user. These traits could be computer recognized and used to control a lockout circuit. A very sophisticated system would be required for rapid signature analysis.

As pointed out before, these tricks will not by themselves solve a malicious QRM problem, but they can speed up the solution process or make it more bearable.

ARE YOU ABLE TO TURN IT OFF?

In this chapter we've shown that there are technical, psychological, and legal ways to combat jamming and bootlegging on the bands. But there's one final thought to share: It's possible to work too hard, worry too much, and end up being more of a potential prisoner than the jammer, in a psychological sense.

The problem of illegal operation and malicious interference can easily become a very emotional one. Repeater owners and regular users usually have invested a great deal of time, money, and effort into their system, and this is all threatened when the trouble starts. At such times it's easy to forget that amateur radio, despite all of the fine public services it can provide, is primarily a hobby.

Are you a balanced ham? There's a time to DF, a time to write letters, and a time to make technical innovations. All are important. But when the effort starts affecting your health or your family life, that's the time to turn the switches off for a while. Sometimes that's the hardest thing to do. Are you compelled to keep punishing your ears and stomach listening to the problems and combatting them hour after hour, day after day? Or are you able to turn it off occasionally?

If you can't put it all in perspective, and walk away from your shack once in a while, the radio mavericks are controlling your life. By constantly reacting, you are giving them power over you. They don't deserve it.

Chapter 22

Other Uses for Your RDF Skills

The principles of radio direction finding have many uses beyond those we have discussed so far. In this chapter we'll explore some unusual RDF applications, which may be useful to you as a radio amateur right now.

STALKING THE WILD FOX

All through this book you have been urged to go out and find the fox, or bunny—figuratively speaking. There are some DFers who follow these critters quite literally. They're trying to understand and preserve the wildlife on our planet.

With RDF, naturalists can study the migrating and hibernating behavior of animals without disturbing their lifestyle. Fish can be tracked as they move up and down stream. Ornithologists can multiply their effectiveness by using radio to augment visual sighting. Automated tracking systems can determine the patterns of animal movement without continuous human presence, saving valuable funds.

Transmitters and antennas must be extremely rugged to survive on animals in the wild. Underwater species and a few land animals such as badgers cannot be successfully fitted with external transmitters. They must be implanted, severely reducing the monitoring range.

Fish transmitters are most easily placed in the stomach, with the antenna extending back to the throat. Large animals such as elk get external transmitting collars with a halo type antenna. The antenna for a bird is often placed in its tail feathers. A large bird such as an eagle easily tolerates a six-ounce transmitter on its back held in place with a harness of surgical tubing.

Receivers need to cover many channels. Oftentimes each animal is assigned to its own frequency to avoid the transmitters jamming each other and to simplify identification. Scanners are popular for this purpose, and are used by bird trackers with small budgets to discover when tagged birds enter their area. Field DFing is done with traditional methods, but when appropriate, aircraft or even satellites are used.

The transmitter is usually crystal controlled with very few stages. Lithium batteries give the longest life in a small size. (No periodic recharging possible here!) Output is often pulsed at low rate to conserve battery life. A typical pulsed transmitter is on the air for 30 milliseconds each second. Pulsed transmitters are needed when high peak power must be used to overcome the path loss. Pulsed signals are harder to DF, though, and aren't practical when scanning receivers are used to detect the presence of the animals. So when scanners are to be used for reception, transmitters either operate continuously or are keyed intermittently for several seconds at a time.

Frequencies ranging from 27 to 500 megahertz are used as appropriate for animal RDF. Lowest frequencies are used for fish and other underwater species to minimize attenuation of the water. Much higher frequencies are used for birds to shorten the antenna. By tracking birds air to air with aircraft, range of 100 miles is possible.

Short range telemetry of vital signs is usually done in the license-free biomedical bands at 38-41, 88-108, and 174-216 MHz. Biomedical information from inside a large animal may be received in the external collar and retransmitted at a higher power level. There are no true standards for modulation or encoding. Each group tends to work out its own schemes, with the level of sophistication determined by its needs and the size of its grants.

DF techniques for wildlife management are a combination of the old and the new. Old fashioned multi-element yagi antennas are still the basis of most long range locating systems. They provide high gain to pull in the weak signals of the tiny transmitters. Most monitoring is done from a distance, so large angular swings of the signal are not a problem. Pinpoint accuracy is seldom required.

Two or three receiving setups locate the animal's general whereabouts adequately by triangulation. Usually this is sufficient, but on some occasions it's necessary to actually track down the animal. An example of this is the use of a hypodermic dart to anesthetize an animal. By putting a transmitter in the dart, the animal can be found quickly if he runs away when first injected.

By using multiple receivers, antennas, and computers, automatic habitat monitoring systems have been developed to cover an area such as a meadow or a river segment. Multiple animals can be monitored, giving individual and group behavior information, 24 hours a day for weeks at a time.

NAVIGATING BY WIRE

We have been dealing mostly with RDF situations where the location of the rf source is unknown, and it is necessary to seek it out. Some applications of RDF such as VOR, TACAN, and ADF make use of a known fixed transmitter location for navigation by radio. The following technique is a cross between DF and navigation, and is novel because radio waves are not used.

The time is 1956. The place is Camp Tuto, Greenland, on the edge of a dense ice pack. It is neces-

sary to make regular trips 220 miles into the interior of the island from the camp, but it is feared that heavy snowfalls and "whiteouts" will result in some of the vehicles leaving the trail and becoming lost. There is great danger of a lost vehicle falling into one of the many deep crevasses. A navigation and communication system along the route is needed, but radio transmitters cannot be used for security reasons.

The answer was provided by the ingenuity of engineers at the Mechanical Division of the General Mills Company. (Yes, General Mills, the people who bring you Cheerios! The company's mechanical engineers occasionally took on government contracts.)

A 440-mile loop of wire was installed about a foot under the snow surface, encircling the trail, carrying 60 Hz ac current at about one half ampere. Because the outgoing and return conductors were parallel and only 75 feet apart, little outward radiation of the 60 Hz signal took place, but the strong magnetic field within the loop could be used to keep vehicles on course.

Figure 22-1 is a block diagram of the system. Chuck Lobb, KN6H, who was a member of the design and installation team, explains its operation:

"Each vehicle had a special sensor consisting of three magnetic loop detectors, each about three feet in diameter. One was in the horizontal plane, one was vertical in the direction of the trail, and the last was vertical and perpendicular to the trail. Each loop contained 4000 turns of 30 gauge wire. The amplitude and phase of each loop output was a function of the position of the loop with respect to the buried wires.

"By properly combining the outputs of the loops in a 13-tube receiver, the driver was provided with two zero-center indicator meters. One told him his vehicle's orientation (azimuth) relative to the axis of the trail. The other told him how well his vehicle was centered between the wires at the edges of the trail. Audible warnings sounded if the vehicle got outside of the loop, or if loop current failed. With this system, progress could be made even when visibility was zero. The drivers could learn to use the system in just a few minutes."

Another method of RDF was used when the trail wires broke, as they did several times a year due to shifts

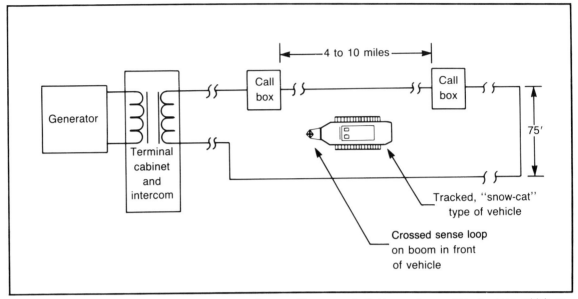

Fig. 22-1. Block diagram of the Arctic Trail Marking System. The magnetic field was strong within the loop, which extended 220 miles inland.

in the ice. At those times, the only loop current was due to dielectric effects in the ice, so magnetic detectors could not be used. Instead, a small handheld probe and amplifier was used to detect the electrostatic field around the buried trail wires. The field peaked when the probe was directly over the wire, but the peak disappeared when the wire break point was encountered. This told the repair crew where to dig to find the break.

Several way stations along the trail had junction boxes where intercom units could be plugged in series with the loop, allowing trail vehicles to communicate with the base camp. Due to the wire's weight, dark color, and heating from the continuous current, the loop was soon many feet underground. It was eventually abandoned in 1963 when breaks were very frequent and repairs became too difficult.

HUNTING CABLE TELEVISION LEAKAGE

When Community Antenna Television (CATV) first began, cable companies put signals on their lines only at the standard vhf television frequencies. Since the mid 70's, added stations and premium services have required the use of more of the spectrum to the point that today's modern systems carry programming to subscribers on channels ranging almost continuously from 54 MHz to 450 MHz or higher. In some systems, frequencies below 50 MHz are used to send programming or data "up-

stream" from originating sources or subscribers to the "head end."

With CATV's use of much of the spectrum not allocated to over-the-air television, it is vital that all cable signals stay within the company's coax lines and equipment, and that outside signals be kept out of the cable system. Any breaks or discontinuities in the shields of the cables or the enclosures around amplification and distribution equipment can result in interference to and from the system.

Most commonly, such problems result from corroded connectors or shield breaks along the cable, caused by age, stress, weather, or improper assembly. Signal levels are quite high along the main cable, and if the radiating coax is up on a power pole instead of buried, QRM over a wide area can result. Some CATV customers contribute to the problem by improperly adding lines to additional TV sets, game switches, recorders, and FM receivers. Cable technicians often find that ordinary unshielded twin-lead has been used for these runs.

Amateur radio operators and aircraft communicators, among others, must be concerned with the problem of cable radiation, because their vhf and uhf frequencies are also being used for TV signals within the supposedly "closed" cable systems. Table 22-1 shows the CATV channels that coincide with ham and ELT frequencies at vhf. In some areas these channels may not be used, and in others their use may be prohibited by contract. When leaks do occur on these channels, though, amateur repeaters can be locked up, and ingress of amateur vhf signals can cause TVI to CATV viewers.

The CATV Company's Hunting Methods

Before you can successfully locate a cable QRM problem, it's important to understand the technical requirements on CATV systems, and how companies go about keeping the system tight.

Some cable systems give letter designations to the mid-band (88 to 174 MHz) and super-band/hyper-band (above 216 MHz) channels. Other companies use numbered channels. Standard systems have 6 MHz wide channels with the video carriers 1.25 MHz above the low ends of the channels. The channel endpoints are at multiples of one MHz; for example, channel 2 is 54 to 60 MHz. The newer Harmonically Related Carrier (HRC) systems also use 6 MHz-wide channels, but place the video carriers at exact multiples of 6 MHz. The Channel 2 video carrier is at 54 MHz in the HRC system.

The table gives frequencies and channel indications for both methods of numbering and allocations. In all cases, the sound carrier is 4.5 MHz above the video carrier. The video signals are AM and the audio signals are FM. The audio carrier is about 15 dB lower in rf level

than the video carrier.

CATV operators must be capable of quantitatively measuring leaks at all frequencies. The FCC CATV radiation limits and required measurement methods are shown in Fig. 22-2. These limits are so low that it takes a very sensitive receiver to make actual measurements per Part 76 of the rules when the system is in compliance. Low noise preamplifiers will be needed on the field strength meters. Some CATV companies are not well equipped to make these measurements.

In addition to the radiation limits, paragraph 76.613 of FCC rules states that CATV operators must take prompt measures to eliminate any harmful interference. This includes any obstruction or repeated interruption to any radio communication service. Although some cable operators have claimed that so long as they comply with paragraph 76.609 they have no further obligation, FCC has so far held that any harmful interference from CATV to amateurs or others must be stopped no matter what the measurements per 76.609 show.

Experience has shown that leaks of sufficient magnitude to cause harmful QRM to amateur and aircraft vhf communications are usually well in excess of the requirements of 76.609. Exceptions to this might be weak signal work such as moonbounce (EME), where very high sensitivity receivers and high gain antennas are used.

Table 22-1. CATV Channels Coinciding with Amateur and Aircraft ELT Frequencies.

LETTER CHANNEL	NUMBER CHANNEL	STANDARD VIDEO CARRIER	H. R. C. VIDEO CARRIER	POSSIBLE QRM (VIDEO/AUDIO)
A	14	121.25	120.0	Aircraft, ELT (V)
D	17	139.25	138.0	CAP (A)
E	18	145.25	144.0	Amateur 2 meter, CAP (V, A)
J	23	217.25	216.0	Amateur 1-1/4 meter (A)
K	24	223.25	222.0	Amateur 1-1/4 meter (V)
N	27	241.25	240.0	ELT (V)
UU		421.25	420.0	Amateur 70 cm (V, A)
VV		427.25	426.0	Amateur 70 cm (V, A)
WW		433.25	432.0	Amateur 70 cm (V, A)
XX		439.25	438.0	Amateur 70 cm (V, A)
YY		445.25	444.0	Amateur 70 cm (V, A)
ZZ		451.25	450.0	Amateur 70 cm (V)
T-7		7.0		Amateur 40 meter (V)
T-8		13.0		Amateur 20 meter (V)
T-10		25.0		Amateur 10 meter, CB (V, A)

T channels are "upstream" channels, not directly viewed in homes. Interference from leakage tends to be worse on the lower channels.

76.605(a)(12) - ...radiation from a cable television system shall be measured in accordance with procedures outlined in 76.609(h), and shall be limited as follows:

FREQUENCY	RADIATION LIMIT (MICROVOLTS/METER)	DISTANCE (FEET)
Up to and including 54 MHz	15	100
Over 54 up to and including 216 MHz	20	10
Over 216 MHz	15	100

76.609(h) - Measurements to determine the field strength of radio freuency energy radiated by cable television systems shall be made in accordance with standard engineerig procedures. Measurements made on frequencies above 25 MHz shall include the following:

(1) A field strength meter of adequate accuracy using a horizontal dipole antenna shall be employed.

(2) Field strength shall be expressed in terms of the rms value of synchonizing peak for each cable television channel for which radiation can be measured.

(3) The dipole antenna shall be placed 10 feet above the ground and positioned directly below the system components. Where such placement results in separation of less than 10 feet between the center of the dipole antenna and the system components, the dipole shall be repositioned to provide a separation of 10 feet.

(4) The horizontal dipole antenna shall be rotated about a vertical axis and the maximum meter reading shall be used.

(5) Measurements shall be made where other conductors are 10 or more feet away from the measuring antenna.

Fig. 22-2. Radiation limits from the FCC rules, Part 76.

FCC regulations require cable systems carrying signals in the aeronautical bands (108 to 136 and 225 to 400 MHz) to be monitored at least once every three months for rf leakage, over the entire length of the system. Many operators meet this requirement by putting a special signal on the cable in the FM broadcast band. Using sensitive FM receivers in all the vehicles of their installation fleet, the company monitors for leakage of this signal. Ordinary FM automobile receivers can be used, but special fix-tuned receivers with squelch circuits are preferred. This keeps the installers from tuning off to their favorite FM stations!

In addition to the FM tone modulation, the special signal is stepped in amplitude. There are four levels covering a 20 dB range. The signal steps through each level every one-third second. This stepped signal gives a good indication of the distance to the leak, because only when the leak is very close will all four steps be detected.

Hunting Cable System Leaks

If you observe strong TV carriers on your receiver at the frequencies of Table 1, or if your amateur operations cause television interference to any of the listed channels, a main line cable leakage problem is likely, particularly if you QRM your nearby cable-equipped neighbors also. Your transmitter hunting skills can help locate the leak and speed the solution, even though the problem may not be your fault.

First determine which cable system is doing the radiating. Each municipality is usually served by its own separate system, and it is possible for across-the-alley neighbors to be on different systems when towns border one another.

Find someone on the offending system with a cable connection to his FM receiver, and tune the FM band looking for the cuckoo, a double-tone signal typically put on the cable about 5 dB above the other FM band signals. It will probably be near 108 megahertz, since the vhf Omnirange system frequencies are a prime area of concern for leaks. The special signal may have only a single tone instead of the cuckoo sound. If the FM receiver has an S-meter, it will fluctuate with the levels of the amplitude steps. Determine the exact frequency of the signal, which should not correspond to that of any on-the-air FM station in the area.

The auto or portable FM receiver used for hunting the cuckoo should be as sensitive as possible, and should have good selectivity to minimize problems with nearby on-the-air stations. If there are strong on-the-air stations on adjacent frequencies, it is necessary to turn off the receiver automatic frequency control (AFC) circuit to keep the receiver from being pulled off frequency. If the receiver has AFC but not a defeat switch, it may be necessary to bypass the AFC circuit within the radio to use it for this purpose.

An FM band preamplifier ahead of the receiver may aid in detection of the cuckoo at greater range from the leak. Such preamplifiers for automotive use are available from some audio dealers. To be effective, the preamp must have a low noise figure and good cross modulation performance, so as not be affected by strong local signals.

An ordinary FM yagi can be hooked to the receiver antenna terminals to provide gain and give a directional bearing. Horizontal polarization will work best. Directivity is sharpest if 75 ohm coax and appropriate baluns are used instead of 300 ohm twin lead, as shown in Fig. 22-3.

If you don't have the FM yagi, you can still hunt CATV QRM quite effectively so long as you can detect the cuckoo. Since the signal must be coming from the cable system, you can just follow the overhead lines and observe the signal level rise or fall. In underground systems check the above-ground amplifiers and distribution component housings as the most likely sources.

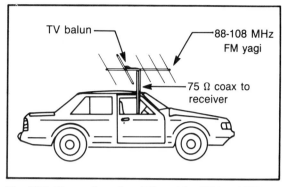

Fig. 22-3. Connection of an FM yagi for FM band DFing.

Because of the cuckoo's amplitude stepping, a receiver S-meter is not required. When furthest away, only the highest amplitude step will be detected, and it may not quiet the receiver. More steps will become audible as the leak is approached. When all steps are heard full-quieting, you'll know you're very close.

It may be difficult to determine the exact point of leakage without pole climbing, because a broken shield can act like a many wavelength antenna. Other wires such as guys and telephone lines can re-radiate the signal. Once you have found the strong signal point, note the pole number and let the CATV company take over.

Do not climb the pole or attempt to do any work on the cable company equipment. This constitutes trespassing. You might be accused of being a cable pirate. Whether the signal comes from the main cable or from a subscriber's home, notify the engineering department of the CATV franchise. The company has the obligation to satisfy the FCC radiation rules.

Other Leak-Finding Methods

If there is no cuckoo on your CATV system, the FM auto receiver can still be used to find strong leaks by tuning it to a cable FM station frequency that is not used on the air locally. Of course there are no stepped power levels to aid in the hunt.

There are three other ways to successfully track down CATVI. One is to use a battery-operated television receiver and yagi TV antenna to search out radiation at a vhf channel which is used on the cable but not on the air. Because of the wide bandwidth of TV signals, the signal-to-noise ratio of the detected radiation is poorest using this method, but in areas where there are no FM or mid-band signals on the cable, this may be the only way to hunt the radiation.

Multichannel (vhf/uhf or channel 2-13) antennas are not recommended for this purpose because they do not provide much gain on any one channel, and their directional characteristics are very suspect. Use a cut-to-channel yagi for the particular channel you are using for the hunt. These antennas are made by the major manufacturers, such as Sitco, Winegard, and Jerrold, but won't be easy to find in your area if there is no over-the-air station on that channel.

For hunting purposes, you can make your own yagi out of wire and wood or PVC pipe, using the same standard formulas as would be used for a ham band yagi. Use a gamma match or balun with coax feed. Horizontal polarization will probably provide best results, but try vertical polarization also to be sure. Designed-to-channel yagis have narrow bandwidth and will have degraded directional performance on channels for which they are not made.

When first detected, the weak TV signal appears as a rolling horizontal or diagonal bar on the screen. As the radiation source is approached, a snowy picture develops, and snow diminishes as you get closer to the leak.

Hunting with a TV picture has one interesting advantage over hunting an audio signal. Multipath due to reflections is visible as "ghosts" on the screen. Since each line of the picture is scanned left to right, the leftmost image is more nearly direct than the ghosts to its right. When turning the antenna, remember that when the intensity of the leftmost image is increasing, this indicates an increase in the direct signal. Increase in the ghosts indicates that more multipath is being received from that direction.

Signal strength of TV signals can be measured by putting an S-meter on the TV set. Use a sensitive voltmeter on the i-f automatic gain control (AGC) line. This point is not difficult to find using schematics in the service literature. Observe safety precautions in bringing out leads from a TV set, keeping in mind that some sets have one side of the ac line connected to the chassis. All add-ons should be well insulated.

CATV radiation tends to decrease with increasing frequency. On the other hand, hunting is easier on higher frequencies due to better directivity and smaller antennas. If you have a choice of cable channels on which to hunt, choose the highest channel on which the leakage can be detected.

The second method is the most sensitive and should be used to find weak radiation at amateur frequencies. Use a high gain quad or yagi with an amateur receiver tuned to the video carrier frequency per the table. An SSB receiver with a low-noise preamp gives the greatest sensitivity. Use whichever polarization provides the greatest signal strength.

Switched antenna DF units are not recommended for CATVI hunting. Most are not set up for hunting on FM or TV bands. A long line radiating along its length fills in the pattern nulls and causes major inaccuracies in switched pattern units. Some companies use Doppler units such as the Doppler Systems models with their FM receiver for directional indications, but this works for strong leakage signals only. They do not have the high sensitivity necessary to hunt weak radiation.

Phil Karn, KA9Q, has found that the Doppler effect can be used to help locate CATVI without having to use a special Doppler DF. All you need is a receiver with BFO for the frequency range where you want to search. A 2 meter SSB receiver works for hunting at midband channel E or 18. He hunts by driving along the cable and observing the small amount of apparent frequency shift that occurs as the receiving antenna passes the point of leakage. This coincides with maximum S-meter reading.

There isn't much shift at 2 meters, only about 9 hertz at 40 miles per hour. The faster you drive, the more shift will be seen. (Do you think the patrolman will understand your need to exceed the speed limit in a residential zone?) As the leak is passed, the shift goes from +9 to −9 for a total of 18 hertz shift. This small tone change is audible if the tone is low in pitch to begin with. Use the receiver incremental tuning control to get as low pitched a heterodyne as possible on the leakage signal and still be within the i-f filter passband. It's a lot easier to discern the difference between 250 and 268 hertz than the difference between 2500 and 2518 Hz.

The Doppler shift is less sharp if the leak is behind a house than if it's right out on the street next to the car, because the range to the leak does not change as abruptly when the leak is further from the road. Multiple leaks give confusing results.

The Upstream Path

In many CATV systems, the frequencies 5 to 30 MHz are used to carry signals from subscribers or program sources to the main distribution point. A few systems use frequencies as high as 120 MHz for this purpose. Uses include program origination, meter reading, and data services. Amateur and CB stations can cause interference to the entire cable system instead of just a neighborhood if their hf signals get into this upstream path. Minor cable discontinuities which may not cause noticeable radiation at vhf can allow severe ingress at 10 or 11 meters.

There is little that the individual amateur or CBer can do about such a problem, provided that there are no breaks or improper connections within his own house. CATV companies are often hard pressed to find the point of ingress because once it appears at the head end, it could come from anywhere in the system. Some systems now have remotely controlled switches in the upstream branches, which divide the system to isolate the source for troubleshooting and system protection.

HUNTING POWER LINE NOISE

Interference from power lines and devices connected to them can be one of the most frustrating problems an amateur radio or CB operator can face. Sometimes it seems that packing up and moving is the only way to get rid of the QRN. Don't do it yet. Use your T-hunting skills to find the problem. If you can locate a QRN source that causes TV or radio interference to your neighbors, you may become a local hero, particularly if you were blamed for the TVI!

Power companies have an interest in solving interference problems. After all, the energy going into making rf noise is usually being taken from the lines on their side of the meter. While most companies do not have the manpower resources to patrol their lines for noise or to find minor noise sources, they usually take steps to repair noisy line hardware when the exact source is pointed out to them.

Just as it was for CATVI, it is important here to have an understanding of the causes and characteristics of power line QRN before grabbing the equipment and beginning the search. We'll break down the problem into two major areas: interference from the lines themselves and interference from customer equipment.

Overhead Line QRN

Much of the frustration of power line noise results from its intermittent nature. Some noise is present only in dry or windy weather. Sometimes it's there only in wet weather. The exact causes are not usually fully explained in amateur radio texts. It turns out that there are two different types of this QRN. The good news is that they can be readily distinguished from one another, while the bad news is that one of them is very difficult to cure.

Characteristics of Spark Gap Noise

The most common (and fortunately most easy to cure) QRN source from overhead lines in residential areas is spark gap interference. Electrical stresses between hardware at different potentials can exceed the critical voltage breakdown value, and sparks occur. This breakdown value is a function of the separation, the geometry of the parts, and the humidity. Spark gap interference is worst in dry weather, as we'll show later.

Arcs between HV conductors or to ground are very rare as the separation is too great unless damage has occurred. Such arcs result in immediate breaker trips and interrupted service. Instead, the arcs are usually between hardware items which have become separated slightly from each other due to corrosion or flexing. These items include crossarm bolts, ground wire staples, tie wires, and suspension insulator socket joints.

The insulated item is in a high energy field, so it charges up and arcs across the air or corrosion path. Since it is small and thus has low capacitance, the potentials equalize in a fraction of a microsecond. The arc extinguishes, the charge builds up again, and the process repeats indefinitely, so long as the field gradient is high enough.

Figure 22-4 shows how spark pulses occur at the positive and negative peaks of the power line frequency, when the gradients are highest. The actual spark repetition rate may be four per millisecond or so during that time. The noise sounds like a 120 Hz buzz in AM and SSB receivers because the pulse groups are occurring at that rate. Many receiver noise blankers are ineffective against such bursts of noise which last for milliseconds.

In wet or humid weather, partial conduction paths are present across the spark gap paths due to moisture in the air and pole wood, and due to the solubility of the corrosion products. The separated hardware is no longer fully insulated, and does not charge up and arc. So this type of power line interference is characteristically present only in dry weather.

Rf radiation from a spark gap source is worst at low frequencies, falling off steadily up to 200 kHz. The long power lines serve as good antennas at low frequencies, and sometimes the QRN can be heard for miles. The radiation level is then constant from 200 kHz up to about 30 MHz, and again falls off at a rate of about 6 dB per octave through the vhf/uhf spectrum. This means that the broadcast band and all hf ham bands are adversely affected, with line resonances determining which is worst. The higher vhf bands and TV channels are least affected.

Once the source of spark gap interference is found, it's easy for the power company to correct. In the majority of cases, the problem will be traced to loose hardware on wooden poles. Wood absorbs moisture, causing it to

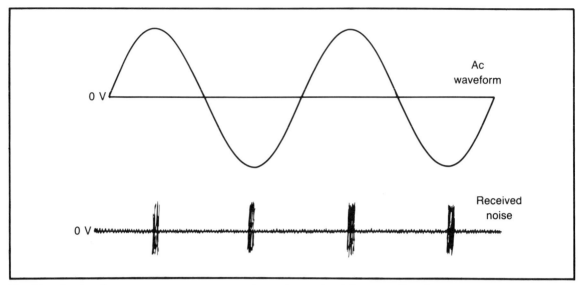

Fig. 22-4. Spark pulse groups occur at positive and negative peaks of the power line sine wave.

swell. It then shrinks slightly as it dries, resulting in stresses to any hardware on wooden poles. Noise from loose hardware may also be initiated by windy weather.

Corrosion anywhere near the line can also be the culprit. On occasion QRN has been caused by arcing between parts of a fence near a power line. The parts were at different potentials due to corrosion. Also check out ungrounded guy wires, usually broken up with insulators, for inadvertent contact with other wires.

Characteristics of Corona

Corona, the other form of power line interference, is more insidious and difficult to solve, but fortunately is more rare. It results from ionization of the air around a HV conductor, even in the absence of another conductor close enough for arcs. When very severe, corona can be seen at night as a blue glow around the source. The corona occurs on both the positive and negative half cycles of the sine wave, but it turns out that there are differences between them. The positive polarity corona is of greatest concern. It can cause severe interference in the AM band, but falls off rapidly with frequency and is usually of little concern at vhf.

Corona generation is a function of the conductors themselves and of the voltages present. The problem is worst with lines carrying 230 kV or more. Corona can be generated by any sharp point or line surface irregularity, including those made by corrosion, trash, or bird residue. But single points are not nearly as much of a problem as multiple irregularities made by water, snow, or ice all along the line. This is why corona QRN is worst in bad weather, and may disappear on the warm sunny day when you decide to hunt it down.

Corona production is very much a function of the line design, and there is usually little that can be done to quickly cure it once it is located. Many new lines are now being built with multiple (bundled) conductors for each phase. This increases the effective diameter of the wire and thus reduces the gradients that cause corona from moisture. Still it's not a good idea to choose a house near a new 230 kV line as a DX chasing location.

Electrical Appliance Noise

Certain home electrical items such as electric blankets and the brush type motors used in sewing machines, electric razors, and hair dryers are notorious for causing wideband noise. It is easily identified and can be cured by standard techniques, although the intermittent nature of their use usually results in a "grin and bear it" attitude.

Other items can cause QRN which is more nearly continuous and harder to trace. Aquarium heaters, even the so-called interference-free ones, have thermostats which love to arc. Everyone has a doorbell transformer hidden in an out of the way spot. A secondary or load short can cause the thermal protector to open and arc repetitively, or the primary can develop an intermittent open that arcs across. In K0OV's neighborhood, everyone had the same brand of gas range, with two oven thermostats that would arc and wipe out 10 meters for many seconds each minute. Of course as Murphy's Law would have it, neighborhood meal preparation time always coincided with evening OSCAR Mode A passes, when 10 meter weak signal reception was important.

Hunting Down the Noise

The preceding information can help localize the power line interference problem even before the search is begun. Noise that is present only in dry and/or windy weather is probably spark gap noise in distribution lines from 5 to 75 kV. Wet weather noise is more likely to be corona related, coming from higher voltage transmission lines. Noise that is present all the time or only for certain hours of every day is probably in a secondary circuit or someone's home.

The HV line noise sources also have characteristic waveforms which may be useful in sorting out the problem. The waveforms may be displayed on an oscilloscope connected to the output of an AM receiver, or they can be observed on a TV screen if the interference is severe enough to cause TVI. They are shown in Fig. 22-5.

Spark gap noise and most appliance thermostat noise occurs at both the positive and negative peaks of the wave, producing two noise bursts per cycle, 180 degrees apart. Corona noise is most prevalent only at the positive half of the waveform, but since it occurs mostly on three phase lines, there may be three bursts 120 degrees apart. Because of the geometry of the lines, the QRN may appear to be almost continuous. There are also some differences in the appearance of the bands, as shown. Spark gap noise is dash-like, and corona noise is more dot-like.

For successful power line QRN hunting, an AM receiver covering a wide frequency range is needed. A battery-operated 500 kHz to 30 MHz general coverage unit will do, and if it covers up into the vhf region, it's even better. Hunting should always be done on the highest frequency possible to minimize the waveguide effect. The power wires and the ground form a sort of waveguide which propagates the noise wave along the wires, radiating it as it goes. This can make locating the point of noise generation very difficult at low frequencies such as the AM broadcast band. There are peaks and nulls along the line at wavelength intervals. At the higher frequencies, the wave does not propagate as well, making it easier to find the source.

Directional antennas are not mandatory for power line QRN trackdown, and may be confusing at mf and hf due to line radiation. It's often easier to just follow the line, listening on the highest frequency where the noise is heard and using the S-meter. A car broadcast antenna or 8-foot mobile antenna should be a sufficient antenna for this. For a starting bearing, a 10 meter beam may be useful, but do not rely on the bearing too heavily, as re-radiation from the line may cause inaccuracy.

A yagi or quad antenna at vhf may help to pinpoint the offending pole when you get close. Switched antenna RDF units will give erroneous indications because there is no rf carrier to hunt, only noise pulses. Your 2 meter amplitude sniffer may work well, as will an aircraft band receiver with a small beam or loop. A vhf-FM amateur handheld rig or police band receiver will probably not be useful because of limiting effects.

It's quite possible that the noise is audible from atop the pole. An arc will probably sound like a buzz, while corona makes a hissing or frying sound. If the problem cannot be pinpointed at vhf, it may be necessary to use an even higher frequency for the receiver and beam, such as 450 MHz.

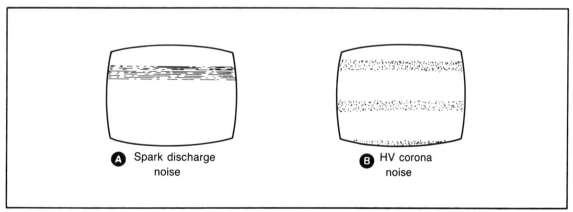

Fig. 22-5. Comparison of spark gap and corona QRM waveforms on TV screen images.

Once the HV line causing the QRN is pinpointed, it's time to call in the power company. Ask for the engineering supervisor, and give him as much information as possible, including the pole number and details of the interference. This information is particularly important if the QRN is intermittent. Offer to let them hear the problem on your equipment. **NEVER, UNDER ANY CIRCUMSTANCES, MAKE DIRECT OR INDIRECT CONTACT WITH THE POWER LINE!**

What if the T-hunt trail takes you to a neighbor's house (or your own)? Enlist the neighbor's cooperation if you can. He probably knows about the noise and will want to help. Rather than trying to trace the noise along the wires in the walls, use the process of elimination. Turn off the individual circuit breakers, or remove branch circuit fuses one at a time to identify the branch supplying the noisemaker. Then check everything on the offending branch by disconnecting items one at a time.

When you find the offending item, there are several ways to filter it to stop the electrical noise pollution. They are covered adequately in the standard amateur radio texts and are not repeated here. Be sure to apply the filtering right at the appliance and not at the power plug. An eight foot cord makes a very good antenna at 10 and 11 meters!

Appliance manufacturers may be able to help with a stubborn QRN problem, but don't rely too heavily on technical advice from installers and distributors. In the case of the oven thermostats mentioned earlier, K0OV learned how technically naive the manufacturer's representatives were. He told them of the problem, and the conversation then went like this:

REP: "Oh, you need a resistor."

K0OV: "A resistor? Are you sure? Where do I install it?"

REP: "Yeah, a resistor. Right across the two terminals on the thermostat."

K0OV: "You're sure it's a resistor? What value resistor?"

REP: "Well let me see . . . I have one here . . . Here it is . . . It says 'point-oh-one, one kV' "

Fortunately, a .01 microfarad *capacitor* did cure the oven thermostat's radio interference.

Chapter 23

Looking Ahead

Since the early days described in Chapter 1, amateurs and professionals have taken advantage of the latest technology to improve the science of radio direction finding. Digital ICs have made possible very inexpensive Doppler DFs. You can now have computerized triangulation in your own car. This book has shown how to do both. As rf and digital technology continues to move forward, more high-tech DF concepts heretofore prohibitively expensive will be within the reach of the amateur. Now that computerized LORAN-C navigation systems for small boats are available and affordable, how soon will the miniature mobile three channel TDOA system be here?

In this chapter we'll look at T-hunting in your future from two standpoints. First, a forward look at how satellites will do more to assist search and rescue volunteers and professionals in locating aircraft and watercraft in distress. Satellite DF systems, pioneered in part by amateur radio enthusiasts, are making global T-hunting a reality. Finally, to cap it all off, glance at hunters and their psyches. This whimsical look may give you a clue as to what kind of hunter you will become.

T-HUNTING FROM ORBIT

In an earlier chapter we extolled the advantages of hunting from an airplane. A logical extension is to use an orbiting satellite for an RDF platform. A number of satellites could provide detection and tracking of ELTs or other signals of interest anywhere on the surface of the earth.

Such a system, the SARSAT/COSPAS network, is now being put to work for search and rescue under an international agreement. Amateur radio enthusiasts will be proud to learn that experiments using a ham-built OSCAR (Orbital Satellite Carrying Amateur Radio) satellite paved the way for this lifesaving innovation.

How It's Done

Low altitude satellites, such as Phase II OSCARs and some NOAA weather spacecraft, are suitable for the task. The signal strength at only 900 miles up is 28 dB greater than a geostationary bird (which always stays above one point on the earth). In the proper orbit, one satellite can cover every point on the earth's surface at least once a day. Best of all, the Doppler shift associated with the satellite's movement helps to DF the signal.

Some simple orbital mechanics show how this works. We will assume a perfectly circular orbit to simplify the calculations.

The period of a satellite, which is the time it takes to circle the earth once, is a function of its height described by Kepler's third law of planetary motion. If a satellite is 22,300 miles over the equator, it circles the earth at the same rate as the earth turns beneath it, and thus appears to remain stationary over one point. If it is below about 80 miles, it cannot sustain orbit due to air friction. Figure 23-1 gives period versus altitude above earth for satellites in low circular orbit. At 900 nautical miles, the height of OSCAR 6, the period is about an hour and 55 minutes.

If OSCAR just went around the equator, as in Fig. 23-2, it would never cover all the earth's surface, because at its low height, it only "sees" a section of the surface 4990 miles in diameter (see Fig. 23-3). To allow coverage of the entire earth, OSCAR is given an inclination, which means that on each revolution it crosses the equator at an angle i, as shown in Fig. 23-4. With its inclination of 101.6 degrees and height of 900 miles, OSCAR 6 covered all the earth's surface, including the North and South Poles.

For low orbit OSCARs, the satellite appears over a particular part of the earth at very close to once every 24 hours. Such a sun-synchronous orbit is very useful be-

cause the satellite appears at about the same time each day. There are two periods of availability 12 hours apart. For OSCAR 6, the northbound orbits were in the evenings and southbound orbits were in the mornings everywhere on earth except near the poles. At latitudes higher than 67 degrees north or south, the satellite was available on every orbit.

So the satellite sweeps over all of the earth's surface in a predictable way. How then does it DF an ELT on the ground? It uses our old friend, the Doppler effect.

As mentioned in an earlier chapter, the Doppler effect causes the apparent change of pitch of the horn of a car or train as it speeds by the observer. The pitch is higher as the vehicle approaches, and abruptly changes to lower as the sound source goes by. Similarly, as a satellite passes over an ELT on the ground, the decreasing radial distance causes the ELT signal to appear higher in frequency to the receiver in the satellite. At the point of closest approach (PCA), the apparent frequency shifts lower as the satellite begins to move away from the ELT.

This effect is most pronounced when the satellite passes directly over the ELT, as in Fig. 23-5A. With passes to the side (B), the positive and negative shifts are less because the relative radial velocity is less. If the sat-

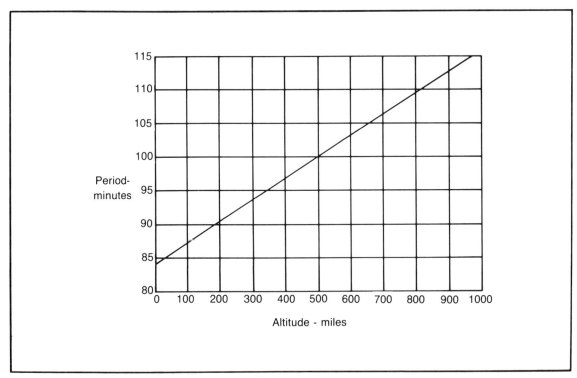

Fig. 23-1. Graph of circular satellite period versus altitude.

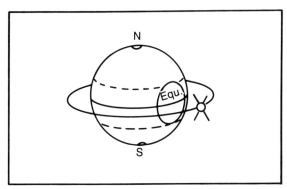

Fig. 23-2. Earth satellites which revolve only around the equator do not cover all the earth's surface.

miles or better accuracy could be gotten almost immediately, and in a few minutes the computer could develop a fix to within one to five miles, with the speed depending on the capacity of the computer.

From this beginning, the SARSAT/COSPAS program was born. This cooperative venture between the USA, Canada, France, and the USSR is moving toward a complete international network of compatible ELT relay stations to provide rapid DFing of ELT transmissions anywhere on earth. The free world nations' units are called SARSATs, standing for Search and Rescue Satellite. The COSPAS units are being launched under an agreement with the Soviet Ministry of Merchant Marine.

ellite knows exactly where it is in space, it can use the shape and timing of these curves to pinpoint the ELT on earth.

Hams Prove The Principle

In 1975 the Communications Research Center of the Federal Communications Department in Canada first tried to demonstrate the principle of ELT location via satellite. Experiments were conducted using the OSCAR 6 Amateur Radio satellite. After about 60 simulations, it was concluded that an initial fix with about 70 ground

SARSAT/COSPAS Proves Its Worth

For the Canadians, who pioneered the development of this principle, the payoff was almost immediate. The first COSPAS transponder was launched aboard the Soviet Union's COSMOS 1383 in late July, 1982. On September 10 of that year it was called upon to find a Cessna 172 lost somewhere in northern British Columbia the day before.

The ELT was heard and DF'ed, and soon the wreckage was located in 50-foot trees. All three passengers survived, and probably owe their lives to satellite DFing. It is interesting to note that when the Cessna crashed, it

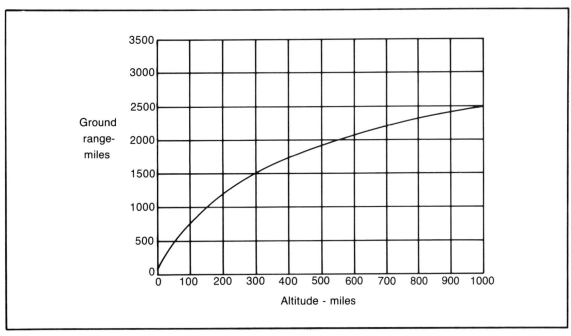

Fig. 23-3. Radius of ground range circle as seen by a satellite at low altitude.

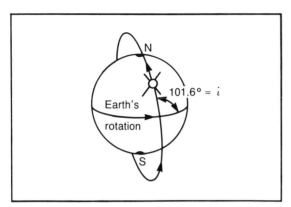

Fig. 23-4. OSCAR 6's inclination of 101.6 degrees allowed coverage of the entire earth.

was continuing informally a search for another light plane which had gone down two months earlier. That earlier search had been officially called off after Canadian rescue forces had reportedly spent two million dollars in the unsuccessful attempt.

SARSAT/COSPAS has continued to be an excellent example of international cooperation in space. Program officials estimate that 200 lives were saved in the first 18 months of the program due to location of downed aircraft or distressed boats.

It's Not a Panacea

So satellites are already making the need for ground monitoring of 121.5 by flyers and ham repeaters obsolete, right? Not at all. Though their accuracy and ability to detect signals from wilderness areas is impressive, the day of continuous surveillance of all points on earth by many SARSATs is a long way off.

Consider the polar orbit of a satellite like OSCAR 6. As mentioned earlier, any point on the continental USA is passed by only two or three northbound and two or three southbound passes a day. If a plane goes down five minutes after a westerly pass, it may be ten hours or so before that satellite is in range again. Also, the store-and-forward method of the advanced uhf system introduces a significant time delay. So there's still a need for more ground monitoring in desolate areas, to confirm orbital detection and to possibly save valuable time.

THE PSYCHOLOGY OF SPORT HUNTING

Sports psychology is now very popular, as athletic competitors figure out how their heads rule their bodies. Well, mental performance is of utmost importance in T-hunting, too. It not only affects your scores, but how much you enjoy hunting, ultimately determining how long you remain a regular hunter.

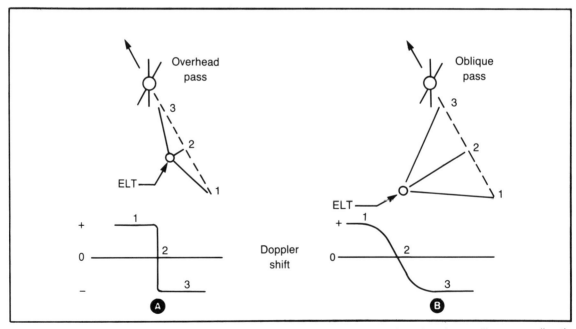

Fig. 23-5. An abrupt change of frequency occurs at Point of Closest Approach (PCA) when the satellite passes directly overhead at (A). The change is less for a non-overhead pass at (B).

Think of your own personality, and what kind of hunter it makes you. Match yourself with some of these T-hunting stereotypes:

☐ Gus, the Glory Seeker. Gus is the epitome of high expectations. He heard about T-hunting and began to brag about how he'd show everyone how it's done. He was far too impatient to build his own gear, so he ran out to buy the best commercial DF unit he could find. He came in far behind the pack in the first few hunts, and was never seen on a hunt again. (Note: Gus is the same fellow who told everyone that he wasn't going to mess with the Novice license, but go straight for the Extra in one month. He finally got his Technician in a year and a half and on the fourth try. He blamed his poor progress on the class instructor.)

☐ Steve, the Scientist. Steve thoroughly analyzes everything. He planned his installation and built it step-by-step. Undaunted by his initial failures, he reviewed each hunt and figured out what to do better. He is not afraid to try something new even if it means losing the hunt, so his setup is a bit different (and better) each time. Occasionally Steve gets angry about his own or his equipment performance, but never at other hunters or the hider. He hunts regularly and often, and wins his share. He may eventually lose interest if he runs out of new ideas to try.

☐ Lucky Lou. Lou built a simple quad, with no attenuator, and won his first hunt. Very excited and proud, he told all his friends. He lost the second hunt because he had no attenuator for the strong signal. After grudgingly building an attenuator, he lost the third hunt because he didn't have a sniffer. Rather than build a sniffer, he took out his frustration on the hider. Lou has finished back in the field on the last few hunts, but he's had an excuse every time, such as, "I don't know the roads like you natives," or, "The signal wasn't strong enough," or, "I had a bad partner." Occasionally he pouts or shows his temper. He is hunting less and less often now, and will probably soon drop out entirely.

☐ Steady Sam. Sam put together a simple setup and worked on it as required to get it working well. He hunts regularly, often bringing a guest. He didn't win the first few times, but steadily improved to be a frequent winner. Seldom has he added to his setup, but his scores are consistently good. Sam enjoys the camaraderie and teasing of the other hunters. He is always pleasant and cooperative, and is a good jammer hunter, too.

Recognize any of these people? Since they are exaggerated caricatures, you most likely aren't like any particular one, but a combination of two or more. In any case, the better you know your own characteristics, the better you can control the course of your hunting career.

Notice that there is no "Seth, the Super Hunter." No one wins on every outing—at least not over long periods of time. The competition is always improving, and a certain amount of luck enters into the picture, too. If you are satisfied with nothing less than 100 percent victories, you are doomed to be a disappointed DFer.

These suggestions will add to your enjoyment of sport hunting:

☐ Learn something from every hunt. Pay attention to bearings that led you astray. Find out why they did, to avoid it happening next time. Figure out ways to improve the gear or its installation. Learn the terrain and roads of unfamiliar territory.

☐ Hunt often. The more you participate, the better you'll get. If you can't go out with your own gear on a particular hunt, try to ride along and help other hunters. You may learn a few tricks from them.

☐ Keep improving your setup. If weak signal or sniffer hunts are likely, get ready for them. Upgrade your direction indicators, maps, lights, and so forth to keep up with the skills of the hiders.

☐ Don't give up. You may do poorly at first, but you *will* improve!

☐ Encourage new hunters. New blood makes hunting more enjoyable for everyone. Invite another ham, a family member, or a neighborhood youngster to go along with you on a hunt. You'll enjoy their reactions.

☐ Enjoy yourself, even when it doesn't go well. You don't have to win every time. If you can't lose gracefully, you aren't going to have much fun at T-hunting.

Bet you thought this section was going to give you ideas about how to "psych out" the other hunters. Sorry, you'll have to figure out those for yourself. It's more important to understand what's going on in your own head than to mess around in the heads of others.

Happy hunting!

Appendices

Appendix A

Manufacturers and Organizations

A & A Engineering
2521 W. La Palma, Unit K
Anaheim, CA 92801
(714) 952-2114

Advanced Receiver Research
Box 1242
Burlington, CT 06013
(203) 582-9409

BMG Engineering
9935 Garibaldi Ave.
Temple City, CA 91780
(213) 285-6963

Circuit Board Specialists _No longer in existance_
P.O. Box 969
Pueblo, CO 81002
(303) 542-5083
(719)

Civil Air Patrol, National HQ
Maxwell Air Force Base
Alabama 36112

Condor Systems, Inc.
3500-B Thomas Road
Santa Clara, CA 95050
(408) 988-2400

Datong Electronics, Ltd.
Spence Mills, Mill Lane
Bramley, Leeds LS13 3HE England
(0532) 552461

Digi-Key Corporation
P.O. Box 677
Thief River Falls, MN 56701-9988
(218) 681-6674

Doppler Systems
111 E. Moon Valley Drive
Phoenix, AZ 85022
(602) 998-1151

Dow Key Division of Kilovac Corp.
P.O. Box 4422
Santa Barbara, CA 93103
(805) 684-4560

EM Systems, Inc.
290 Santa Ana Ct.
Sunnyvale, CA 94086
(408) 733-0611

Gates Energy Products
1050 So. Broadway
P.O. Box 5887
Denver, CO 80217
(303) 744-4806

Globe-Union, Inc.
Battery Division
5757 No. Green Bay Ave.
Milwaukee, WI 53201
(414) 228-2393

Gould, Inc.
Portable Battery Division
931 N. Vandalia St.,
St. Paul, MN 55114
(612) 645-8531 (800) 328-9146

Hamtronics, Inc.
65-X Moul Rd.
Hilton, NY 14468
(716) 392-9430

Happy Flyers
1811 Hillman Ave.
Belmont, CA 94002

International Crystal Mfg. Co.
10 N. Lee
Oklahoma City, OK 73102
(405) 236-3741

L-Tronics
5546 Cathedral Oaks, Road
Santa Barbara, CA 93111
(805) 967-4859

Lunar Electronics
2775 Kurtz St., Suite 11
San Diego, CA 92110
(619) 299-9740

National Association of Search and Rescue
P.O. Box 2123
La Jolla, CA 92038

Ocean Applied Research
10447 Roselle St.
San Diego, CA 92121
(714) 453-4013

Palomar Engineers
P.O. Box 456
Escondido, CA
(714) 747-3343

Panasonic Industrial Company
Battery Sales Division
P.O. Box 1511
Secaucus, NJ 07094
(201) 348-5266

Piezo Technology, Inc.
P.O. Box 7859
Orlando, FL 32854
(305) 298-2000

RCA Solid State Division
P.O. Box 3200
Somerville, NJ 08876

Regency Electronics, Inc.
7707 Records St.
Indianapolis, IN 46226
(317) 545-4281

Tech-Comm, Inc.
5001 Hiatus Road
Sunrise, FL 33321
(305) 749-1776

Watkins-Johnson Co.
CEI Division
700 Quince Orchard Road
Gaithersburg, MD 20878
(301) 948-7550

Yuasa Battery (America), Inc.
9728 Alburtus Ave., P.O. Box 3748
Santa Fe Springs, CA 90670
(213) 698-2275

Appendix B

References

The books and articles listed below are good sources of detailed information about the subjects discussed in this book. This is by no means a complete bibliography of RDF. You will find other highly technical books on the general topic of RDF in technical libraries. The references here were chosen because they provide expanded coverage of specific topics treated in this book, such as Doppler DFs and interferometers.

Boyd, et al, *Electronic Countermeasures*, Peninsula Publishing, Los Altos Hills, CA, 1978

Cunningham, "DF Breakthrough!," *73 Magazine*, June 1981, p. 32

Fisk, "Helical Resonator Design Techniques," *QST*, June 1976, p. 11

Geiser, "Double-Ducky Direction Finder," *QST*, July 1981, p. 11

Gething, "High Frequency Direction Finding," *Proceedings of the IEE*, Vol. 113, No. 1, January 1966

Goodman & Chaffin, "Strategy and Technique for Location of ELT Transmitters," New Mexico ELT Location Team, Albuquerque, 1982

Johnson & Jasik, *Antenna Engineering Handbook*, Second Edition, McGraw-Hill, 1984

Jones & Reynolds, "Ionospheric Perturbations and their Effect on the Accuracy of Hf Direction Finding," *The Radio and Electronics Engineer*, January/February 1975, p. 63

Postlethwaite, *Airborne Radio Direction Finding*, Happy Flyers, Belmont, CA, 1978

Rice, "The ZL/DF Special," *73 Magazine*, March 1981, p. 40

Rogers, "A DoppleScAnt," *QST*, May 1978, p. 24

Sparks, "Build this C-T Quad Beam for Reduced Size," *QST*, April 1977, p. 29

Steinman (K1ET), editor, *Handbook for Local Interference Committees*, ARRL, Newington, CT, 1982

Index

Index

160-meter radiosport, 237

A

80-meter DF set, block diagram for, 236
Adcock antennas, 1, 249
  state-of-the-art systems for, 261
add-on sideband detector
  construction of, 159
  receiver connection of, 161
  schematic of, 160, 162-163
  use of, 161
alkaline batteries, 214
all-day hunt, 187
aluminum yagis, 29
amplitude variations, 18
antenna mounts, 71-86
antenna patterns, 40
antennas
  Adcock, 249, 250
  close-in hunting, 173
  customized for hiding, 203
  directional, 1
  directive gain, 8
  distributed, 206
  ferrite rod use in, 248
  hidden, 200
  lens, 261
  lossy coax, 207
  non-vehicular, 234-238

null steering, 292
patterns for, 40
rod, 237
rotating, 260
sniffing, 173
switching of, 207
two-element driven array, 236
two-meter rhombic, 204
vehicle mounts for, 71-86
ARRL external attenuator, 58
audible signal strength indicators, 50
automatic antenna switching, 64
automatic attenuation control, 67-70

B

baiting, 282, 285
balanced H Adcock direction finder, 4
bar graph driver direction indicator, 96
base stations, 249
baseline averaging, 114
batteries
  alkaline, 214
  lead-acid, 212
  nickel-cadmium, 209
  primary, 213
battery charger, schematic for, 211
battery minder, 178

beamwidth, 39
bearings, 14-17
  changes in, 22
  determining inaccuracies in, 113
  forty-five/ninety technique for, 19
  stairstep method for, 19
beat frequency oscillator, 6
Bellini, 1
bias, 66
BMGs, 18
body fade, 165
bow and beam bearing technique, 19
bridge circuits, 52
Budenbom, H.T., 121
buffer board, 164
bunny box, 219-233
  audio and timing for, 219
  crystal control for, 229
  keying for, 228
  other bands and variations for, 229
  physical layout of, 219
  power for, 228
  power regulators and intermittent mode switching in, 232
  RF synthesizer for, 220
  schematic of output stages in, 230
  schematic of triplers in, 230
  triplers and final amplifier for, 226

C

cable television leakage, 296
  coinciding frequencies causing, 297
  company's hunting method for, 297
  hunting for, 298
  other methods for finding, 299
  upstream path of, 300
chaining, 201
Civil Air Patrol, 144
close-in hunting, 165-181
  antennas for, 173
  sealing up receivers for, 166
  systematic sniffing for, 176
clothing requirements, 12
coast effect, 257
coax cable
  characteristics of commercial, 205
  mil-spec characteristics chart for, 205
commercial direction finding systems, 260-268
compass, 12
compass bearings, 15
computer hunting, 271
continuous sub-audible tone access, 292
coordinates, 269, 270
cordless microphones, 217
corona, 302, 303
counter/timer circuit, 292

D

Datong Electronics, 140
deflection generation, 249, 255
dipswitches, 227
direction finders
  double-ducky, 100
  L-per, 102
  tactical mobile, 264
  time-difference-of-arrival, 265
direction indicators, 86, 96
  accuracy of, 89
  digital, 97
directional access lockout, 292
directional antennas, 1
directive gain antennas, 8
distance, estimation of, 21
distributed antenna, 206
door mounts, 74
doppler DF units, 8, 18, 120-141
  16-LED display for, 134
  commercial, 137
  control unit for, 139
  Datong Electronics mobile, 140
  hunting with, 136
  low signal level lockout for, 135
  modifications and improvements to, 134
  operation of, 120
  other applications for, 141

radio amateurs use of, 122
RDF systems using, 137
Roanoke Vhf, 123
Doppler shift, 121
double-ducky direction finder, 100
Drake rigs, 198
driver circuit, 68
dual gate metal oxide semiconductors (MOSFET), 66
DX signals, 239

E

electrical appliance noise, 302
emergency locator transmitters (ELT), 142, 151
  amateur detection of, 146
emergency position indicating radio beacon (EPIRB), 143
external attenuators, 55
  antenna switcher use with, 63
  ARRL, 58
  high-performance, 57
  "indestructible", 62
  slide switch, 56
  switched resistive, 55
  table of PI resistor values for, 58
  use of toggle switches with, 59
  waveguide, 61

F

ferrite rod, 248
field strength meters, 19
final amplifiers, 226
fixed site direction finding, 256-259
  calibrating station for, 259
  HF problems in, 257
  setting up station site for, 258
  Vhf vagaries in, 256
flashlight, 12
FM gain block ICs, 53
FM hunting, 6
FM quieting, 154
FM receivers, noise meter for, 156
forty-five/ninety technique, 19
fox, 190-218
  antennas for hiding when, 200
  checklist for, 218
  finding perfect spot for, 190
  hints on hiding, 217
  preparing the Tee for, 191
  remote control for, 213
  simple transmitter cycler for, 192
  tape recorder for, 191
  tone boxes for, 193
  transmitter power for, 207
frequency selection, 227, 228
front-to-back ratio, 39, 40

G

GaAs-Fet preamplifiers, 153
garbage can antenna, 28

H

Happy Flyers direction finder, 106-113
  circuit diagram for, 107
  controlled reflection performance of, 106
  deluxe edition of, 108
  vhf/DF system diagram for, 109
Happy Flyers organization, 146
helical resonator, 179
Hertz, 1
HF, 257
HF AM/SSB receivers, 51
homing DF units, 8, 99-119 165
  determining bearing inaccuracies in, 113
  double ducky direction finder as, 100
  Happy Flyers DF for, 106-113
  K6BMG SuperDF as, 115
  little L-per direction finder for, 102
  side-step and baseline averaging techniques for, 114
  switched cardioid pattern, 100
  testing the, 118
hunting below 50 MHz, 239-255
  2 to 15 MHz loops for, 246
  Adcocks for base stations while, 249
  balanced and unbalanced loops for, 241
  locating DX signals while, 239
  loop technique for, 240
  oscilloscope display for, 249
  preamps for, 244
  setting up for loop, 246
  unidirectional loop systems for, 245
hunting without a vehicle, 234-238
  Asiatic competition in, 235
  championship rules for, 235
  equipment for, 236
  European competition in, 234
hunts, 182-189
  avoiding arguments during, 185
  evaluating performance in, 188
  increasing attendance of, 189
  maps for, 272
  miscellaneous rules for, 185
  organizing the team for, 281
  other ideas for, 186
  sample rules for, 183-184
  scoring chart for, 188
  use of odometers in, 185
  when and where of, 182
  winning criteria for, 184
  writing rules for, 182

I

interferometer
  measuring distance with, 150
  theory of, 147

triple channel, 264
wide aperture, 148
intermittent mode switching, 232
internal attenuators, 63
   altering bias of, 66
   supply voltage control of gain in, 65
International Amateur Radio Union (IARU), 234
inverter, 94, 95

J

jammers
   countermeasures against, 292
   dealing with, 284
   filing charges against, 290
   ignoring, 293
jamming, 282, 283
   submitting reports on, 287
   volunteers to control, 287
JFET preamplifier, 153
   schematic of, 154
junction field effect transistor (JFET), 52
Jutland, battle of, 2

K

K6BMG Super DF kit, 115-118
   assembly of, 116
   hunting with, 117
keying, 228
knocking down signals, 55-70
Kolster, Dr. F. A., 1

L

L-per direction finder, 18, 102
   configurations for, 104
   controlled reflection performance of, 102
   improvements for, 106
   models and frequency ranges for, 103
   switch box for, 105
lateral tilts of the ionosphere (LAT), 239
laws and regulations, 286
lead-acid batteries, 212
LED meters, 48
   adding a, 48
   bar graph, 49, 50
   use of, 49
LED ring, 95
lens antennas, 261
linear display, 253
long baseline signal location, 267
long-persistence phosphor, 253
looking ahead, 305-309
loops
   2 to 15 MHz, 246
   300-ohm TV twin lead, 242
   balanced, 241
   setting up for hunting with, 246

shielded, 243
shielded 75-meter, 247
shielded coax, 243
unbalanced, 241
unidirectional systems of, 245
unshielded, 241
LORAN-C, 305
lossy coax antenna, 207

M

magnetic declination, 14
   United States map of, 16
manufacturers, 313-314
map lights, 12
maps, 11, 270, 272
Marconi, 1
marine DF units, 9, 10
meter driver, 47
military direction finding systems, 260-268
mirror mounts, 73
mischief and malice, 281-293
   handling jammers and, 284
   prosecution of, 285
   using psychology to control, 284
mobile computerized triangulation system, 269-280
   correcting errors in, 279
   improving the system of, 279
   mobile set-up for, 275
   off-map references and, 273
   other uses of, 280
   program #1 for, 270
   program #2 for, 271
   program example for, 276-278
   program explanation for, 274
   using a computer to hunt with, 271
mobile DF unit, 140
multipath, 253
multiple receiver sites, 293

N

navigating by wire, 295
network analyzer response, 59
nickel-cadmium batteries, 209
   capacities of, 210
night hunting, 177
no-holds-barred hunt, 187
NOAA satellites, 305
noise meter, 156
noisemaker, 221
novelty hunts, 187
null, 40
null steering antenna, 292
null-hunting antennas, 43

O

obscenity, 283
off-map references, 273
op-amp, 47
organizations, 313-314

OSCAR 6, 308
oscilloscope display, 249
   all-electronic, 255
   circuitry of, 251
   operation of, 251
   schematic for, 254
overhead line QRN, 301

P

Phase II OSCARs, 305
phased arrays, 41
PIN diode attenuator, 70
pirate patch, 291
point of closest approach (PCA), 308
polarization, 41
pot attenuators, 63, 64
power line noise, 301
   hunting out, 303
power regulator, 232
preamplifiers, 244
   build-it-yourself JFET, 153
   loop combined with, 244
primary batteries, 213
   capacities of, 214

Q

quad antennas
   building details for, 31
   configurations for, 39
   creative configurations for, 38
   element lengths and spacing for, 32
   helpful hints for, 36
   shrunken, 174
   stiff-wire, 35
   strung-wire, 32, 33
   strung-wire, spreader detail for, 34

R

radiation limits, 298
radiosport, 236
   160-meter, 237
   American, 237
   two-meter, 237
RDF, 1-6
   countermeasures against, 268
   early history of, 1
   future of, 305-309
   military use of, 4-6
   other uses for, 294-304
   problems of, 4
   satellite, 305
   sport hunting of, 6
   World War II ham operators of, 4
   World War II use of, 3
receivers, sealing up, 166
references, 315
relay keying, 198
remote control, 213
   radio, 214
repeater shut-off, 283
requested transmission hunt, 187

reverse jamming, 282
RF synthesizer, 220
  block diagram for, 222
  parts layout for, 223
  schematic and parts list for, 224-225
Roanoke Vhf Doppler DF unit, 123
  antenna construction details for, 129
  antenna unit for, 128
  block diagram for, 123
  calibration of, 133
  checkout of, 131
  construction of, 126-131
  detail circuit description of, 123
  parts list for, 126
  schematic of, 124-125
  waveforms and timing diagram for, 132
rod antenna, 237
rotary switches, 228
rotating antenna, 121, 260
Rounds, H.J., 1

S

S-meters, 44-54
  amplified external, 46
  audible signal strength indicator as, 50
  bridge circuit as, 52
  external, 45
  FM gain block ICs, 53
  HF AM/SSB receivers use of, 51
  LED meters in, 48
  linearity evaluation of, 47
  performance curve for, 20
  schematic of, 45
  single IC, 54
  Vhf circuits in, 44
SARSAT/COSPAS network, 305, 307
satellites, 305
  circular period vs. altitude, 306
  ground range circle of, 307
  point of closest approach, 308
Schmitt trigger oscillator circuit, 198
search and rescue hunting, 142-150
  advanced interferometer techniques for, 147
  airborne, 146
  amateur detection of ELT alarms in, 146
  Civil Air Patrol and, 144
  Happy Flyers organization and, 146
  US Coast Guard Auxiliary and, 145
selsyn, 90, 92, 93
shrunken quad antenna, 174
  pattern for, 175
side-step averaging, 114
sideband detectors, 158

add-on, 159
sidelobe level, 39-41
signature analysis, 161
signature recognition lockout, 293
simple tone notcher, 291
slide switch external attenuators, 56
sniff-amp, 168
  construction of, 170
  schematic of, 170-171
sniffers, 165-181
  alternate uses for, 181
  antennas for, 173
  battery minder for, 178
  better selectivity for, 178
  deluxing of, 177
  listening to signal using, 180
  night hunting, 177
  primitive types of, 167
  sniff-amp type, 168
  systematic, 176
spark gap noise, 301, 303
sport hunting, 308
stairstep method, 19
stalking the wild fox, 294
stiff-wire quad antennas, 35
  center mounting Tee on, 38
  construction detail for, 37
Stone, J., 1
strung-wire quad antennas, 32
switched cardioid pattern homing unit, 100
switched resistive external attenuators, 55
switching, 207
  relay driver schematic for, 208
synchro, 90
systematic ionospheric tilts (SIT), 239

T

T-hunting from orbit, 305
through-the-window mounts, 71
time-difference-of-arrival direction finder, 265
  schematic for, 266
timed hunt, 19
toggle switch attenuator, 60
tone boxes
  check-out and programming of, 199
  circuit description of, 194
  construction of, 195
  improvements to, 200
  un-music, 193
topographic maps, 11
Tosi, 1
transmitter cycler, 192
transmitter hunting, getting started in, 7-12
transmitters
  chargers for, 210
  powering of, 207

traveling ionospheric disturbances (TID), 239
triangulation, 89, 177
  coordinate system of, 269
  DX signal, 240
triple channel interferometers, 264
triplers, 226
two-element driven array antenna, 236
two-meter loops, 26, 27
two-meter radiosport, 237
two-meter rhombic antenna, 204

U

un-music tone box, 193
  CW programming map for, 194
  schematic and parts list for, 196-197
unidirectional loop systems, 245
United States Geological Survey maps, 11
upstream path of cable television leakage, 300
US Coast Guard Auxiliary (USCGA), 145

V

vehicles, 12
  direction indicators for, 86
  door mounts for, 74
  equipping, 71-98
  hole in the roof antenna mounts for, 79-82
  legal considerations for mounts on, 97
  mirror mounts for, 73
  sun roofs and convertibles as, 79
  through the window mounts for, 71
  window brackets for, 82
  window coverings for, 82
  window inserts for, 86
vhf, anomalies of, 256
vhf choices, 8
vhf directional antenna hunting, 6, 26-43
  aluminum yagis used in, 29
  evaluating antenna performance in, 40
  instant hunting ideas for, 28
  measuring beam and quad performance in, 38
  phased arrays in, 41
  quad antennas and, 31
  simple antennas for, 26
  ZL special antenna in, 42
vhf mobile hunting
  bearings without a compass when, 15
  closing in on, 18, 21
  guessing distance while, 18
  importance of high ground in, 14

initial bearing for, 14
magnetic declination, 14
running bearings when, .18
start bearing for, 17
team techniques for, 23
techniques for, 13-25
voltage controlled oscillator, 258

W

Watson-Watt two-channel direction
  finder, 261
wave interference (WI), 239
waveguide attenuator, 61

weak signal hunting, 151-164
  build-it-yourself preamp for, 153
  FM quieting for, 154
  grabbing signal when, 152
  noise meter for FM receivers
    when, 156
  reasons for, 151
  sideband detectors for AM and
    FM, 158
  signature analysis when, 161
wide aperture wullenweber system,
  262
window box, 85

window brackets, 82
window coverings, 82
window inserts, 86
window mounts, 83
Wullenweber system, 4
  wide aperture, 262

Y

yagis, 29
  choosing polarization for, 30
  three vs. four elements on, 30

Z

ZL special antennas, 42

Other Bestsellers of Related Interest

Old Time Radios!: Restoration and Repair
Joseph J. Carr
Restore classic vacuum-tube and transistor radios easily and inexpensively! This book gives you the transistor theory and practice you can use on radios produced in the '50s and early '60s. You'll review the history, theory, and practical operation behind these old-time home radio sets, and find the instructions and schematics you need to repair or rebuild them.
0-07-155735-0   **$17.95 Paper**

Build Your Own Shortwave Antennas, 2nd Edition
Andrew Yoder
This revised, bestselling beginner's guide tells how to achieve optimal performance from shortwave radios and save money by building simple but powerful antennas and matching them to receivers.
0-07-076534-0   **$16.95 Paper**

Shortwave Listener's Handbook (The), 4th Edition
Hank Bennett/David T. Hardy/Andrew Yoder
Includes updated frequency listings, broadcast schedules, and club data, plus the latest information on short-wave receivers and antennas.
0-8306-4347-8, #0-07-076534-0   **$19.95**

Shortwave Listener's Q & A Book (The)
Anita Louise McCormick
Everything you always wanted to know about getting started in shortwave listening. Question-and-answer format makes it easy to choose a radio and tune into broadcasts from around the world.
0-07-044774-8   **$12.95 Paper**

Shortwave Radio Listening for Beginners
Anita Louise McCormick
A fast and easy introduction to the fascinating world of international shortwave radio listening. Covers history, equipment, terminology, station profiles, broadcast schedules, resources, and more.
0-8306-4135-1   **$11.95 Paper**
0-8306-4136-X   **$19.95 Hard**

Tuning In to RF Scanning: From Police to Satellite Bands
Bob Kay
From the leading U.S. authority on radio frequency scanning, this friendly guide provides accessible advice, with keys to unlocking often-overlooked frequencies and information on monitoring cellular and cordless telephones.
0-07-033964-3   **$14.95 Paper**
0-07-033963-5   **$24.95 Hard**

Shortwave Listening on the Road: The World Traveler's Guide
Andrew Yoder
This book is written for the large number of shortwave listeners who travel frequently.
0-07-076509-X   **$14.95 Paper**

How to Order

 Call 1-800-822-8158
24 hours a day,
7 days a week
in U.S. and Canada

 Mail this coupon to:
McGraw-Hill, Inc.
P.O. Box 182067,
Columbus, OH 43218-2607

 Fax your order to:
614-759-3644

 EMAIL
70007.1531@COMPUSERVE.COM
COMPUSERVE: GO MH

Shipping and Handling Charges

Order Amount	Within U.S.	Outside U.S.
Less than $15	$3.50	$5.50
$15.00 - $24.99	$4.00	$6.00
$25.00 - $49.99	$5.00	$7.00
$50.00 - $74.49	$6.00	$8.00
$75.00 - and up	$7.00	$9.00

EASY ORDER FORM—
SATISFACTION GUARANTEED

Ship to:

Name _____

Address _____

City/State/Zip _____

Daytime Telephone No. _____

Thank you for your order!

ITEM NO.	QUANTITY	AMT.

Method of Payment:

☐ Check or money order enclosed (payable to McGraw-Hill)

☐ **VISA** ☐ **DISCOVER**

☐ **AMERICAN EXPRESS Cards** ☐ **MasterCard**

Shipping & Handling charge from chart below	
Subtotal	
Please add applicable state & local sales tax	
TOTAL	

Account No. ☐☐☐☐☐☐☐☐☐☐☐☐☐☐☐☐☐☐

Signature _____ Exp. Date _____
Order invalid without signature

**In a hurry? Call 1-800-822-8158 anytime,
day or night, or visit your local bookstore.**

Code = BC15ZZA